T0300993

"Professor Bluffstone has crafted a book full of plain-language explanations of economic concepts and well-chosen examples from around the world. The book's global orientation is vital given our species' vast impacts on the planet's ecosystem services and the importance of understanding how these impacts result from decisions the billions of us make every day in our own local contexts. Students with little prior exposure to economics will especially welcome the book's emphasis on using economics to develop practical solutions to a wide range of tough environmental problems."

Jeffrey R. Vincent, *Ph.D. Clarence F. Korstian Professor of Forest Economics & Management, Duke University, USA*

"[This book] is a valuable resource for researchers, students, and instructors interested in an accessible source of information on the economics of conservation, sustainability and climate change. The book contains insights, tools, and policy options to address key issues of our time – especially climate change and pollution. A unique aspect of the book is the inclusion of perspectives from the Global South – an important contribution that facilitates a global view of challenging resource and environmental issues. I recommend this volume to those interested in acquiring or enhancing their knowledge of global environmental challenges and learning about economic methods to support policy development."

Vic Adamowicz, *Distinguished University Professor and Vice Dean, University of Alberta, Canada*

"It is a well-known fact that the world is faced with a series of local and global environmental challenges. To deal with these, we need to apply our best research, more often than not with multiple disciplines involved. Finally, there is an up-to-date textbook that provides the tools of environmental economics to address local and global environmental issues, including an important focus on applications in the Global South, and that is encouraging interdisciplinary approaches. I will actively encourage its use in teaching within the whole global Environment for Development network, and beyond."

Gunnar Köhlin, *Director Environment for Development, University of Gothenburg, Sweden*

ENVIRONMENTAL ECONOMICS AND ECOSYSTEM SERVICES

Environmental Economics and Ecosystem Services provides a rigorous yet accessible introduction to environmental economics, using ecosystem services as the underlying framework. Assuming no prior knowledge of economics, and using a conversational writing style, the focus is on exploring society's linkages with the environment and how economics can help solve key environmental problems.

Structured in three parts, the book first introduces readers to the key theories in environmental economics and ecosystem services, and then explores the challenges of conservation. The final section examines environmental policy options, such as cap-and-trade, behavioral nudges, community-based natural resource management and carbon taxes. There is a strong international focus throughout the book, with real-life examples taken from North America, Europe, Asia, Africa and other regions. Students are supported by a range of pedagogical features, including chapter objectives, chapter summaries, discussion questions and further reading suggestions. In addition, the book offers worked examples, analytical problems and "Challenge Yourself" boxes to develop critical thinking skills. Lecture slides and answers to questions for discussion and practice problems are available for instructors.

This is the ideal text for introductory courses in environmental economics, ecological economics, economics of sustainability, environmental management, environmental policy and ecosystem services.

Randall Bluffstone is an environmental and natural resource economist, Professor of Economics and Director of the Institute for Economics and the Environment at Portland State University, USA.

ROUTLEDGE TEXTBOOKS IN ENVIRONMENTAL AND AGRICULTURAL ECONOMICS

US Agricultural and Food Policies
Economic Choices and Consequences
Gerald D. Toland, Jr., William Nganje, and Raphael Onyeaghala

Global Food Security
What Matters?
Zhang-Yue Zhou

Economics of Agricultural Development
World Food Systems and Resource Use
George W. Norton, Jeffrey Alwang and William A. Masters

The Economics of Farm Management
A Global Perspective
Kent Olson and John Westra

Agricultural Policy in the United States
Evolution and Economics
James L. Novak, Larry D. Sanders and Amy D. Hagerman

Energy Economics, Second Edition
Peter M. Schwarz

Forestry Economics, Second Edition
A Managerial Approach
John E. Wagner

Environmental Economics and Ecosystem Services
Randall Bluffstone

For more information about this series, please visit: www.routledge.com/Routledge-Textbooks-in-Environmental-and-Agricultural-Economics/book-series/TEAE

Designed cover image: Getty Images

First published 2025
by Routledge
4 Park Square, Milton Park, Abingdon, Oxon OX14 4RN

and by Routledge
605 Third Avenue, New York, NY 10158

Routledge is an imprint of the Taylor & Francis Group, an informa business

British Library Cataloguing-in-Publication Data
A catalogue record for this book is available from the British Library

Library of Congress Cataloging-in-Publication Data
Names: Bluffstone, Randall, 1960- author.
Title: Environmental economics and ecosystem services / Randall Bluffstone.
Description: First edition. | Abingdon, Oxon ; New York, NY : Routledge, 2025. |
Series: Textbooks in environmental agriculture and economics |
Includes bibliographical references and index.
Identifiers: LCCN 2024037031 | ISBN 9781032311364 (hbk) | ISBN 9781032311302 (pbk) |
ISBN 9781003308225 (ebk)
Subjects: LCSH: Environmental economics. | Ecosystem services.
Classification: LCC HC79.E5 B595 2025 | DDC 333.72-dc23/eng/20241205
LC record available at https://lccn.loc.gov/2024037031

ISBN: 978-1-032-31136-4 (hbk)
ISBN: 978-1-032-31130-2 (pbk)
ISBN: 978-1-003-30822-5 (ebk)

DOI: 10.4324/9781003308225

Typeset in Interstate
by Newgen Publishing UK

Access the Support Material: www.routledge.com/9781032311302

ENVIRONMENTAL ECONOMICS AND ECOSYSTEM SERVICES

Randall Bluffstone

Routledge
Taylor & Francis Group

LONDON AND NEW YORK

CONTENTS

FIGURES

PHOTOS

TABLES

ABOUT THE AUTHOR

Randy Bluffstone is an environmental and natural resource economist, Professor of Economics and Director of the Institute for Economics and the Environment at Portland State University. He has taught and conducted research at the intersection of economics and the environment for the past 25 years, including on climate change, energy, pollution control and deforestation in low-income countries. Bluffstone is the author of over 50 published peer-reviewed journal articles and has co-edited three books on economics and the environment. Randy received his Ph.D. in economics from Boston University in 1993 and from 1983 to 1985 was a US Peace Corps volunteer in Nepal.

ACKNOWLEDGEMENTS

Looking back on the computer files associated with what has become *Environmental Economics and Ecosystem Services*, I see that I have been working on this book with various emphases and degrees of dedication for about 15 years. With such a long history, it is therefore not surprising that I have a number of people to thank.

I would first like to thank the many talented students at Portland State University who helped me formulate, refine and debug the ideas presented in this book. They kindly put up with my mistakes and took the time to tell me what worked for them and what was not effective. Their input and patience undoubtably increased the quality of *Environmental Economics and Ecosystem Services*. I also gratefully acknowledge participants in two South Asian Network for Development and Environmental Economics (SANDEE) summer schools who allowed me to road test many of the topics presented in this book, as well as the support and important ideas of SANDEE Coordinator Mani Nepal and co-founder Sir Partha Dasgupta.

Colleagues in the environmental economics community have been extremely helpful as I developed this book. Though errors and omissions are all mine, I would like to especially highlight the important role of researchers affiliated with the Environment for Development (EfD) Initiative. Many EfD family members have been enormous influences on the trajectory of this book. They are truly too numerous to mention individually, but they know who they are. Number one on the list is EfD Director Gunnar Köhlin, who was a very early and staunch supporter of this effort. I would also like to mention the many research-related conversations with Thomas Sterner, who helped me coalesce a number of key ideas. Perhaps most importantly, EfD has not only been critical to my research agenda. It has also helped me understand the critical importance of including and carefully considering the experiences, challenges and aspirations of people in the Global South.

Several colleagues in the economics community kindly reviewed and provided comments on draft chapters of this book. These comments lifted up what seemed to be working and helped me avoid some critical errors. These reviewers included Patricia Atkinson, Edward Barbier, Aditya Bhattacharjea, Carlos Chavez, Sahan Dissanayake, Ruth Dittrich, David Fuente, Grant Jacobsen, William Jaeger, Marc Jeuland, Daniel LaFave, Anthony Leiman, Alejandro Lopez-Feldman, Joel Magnuson, Arnab Mitra, Noelwah Netusil and E. Somanathan. Their help and support are greatly appreciated.

I would like to acknowledge several research assistants who significantly contributed to this book. First among them is Ma Chan, who made most of the graphs, helped me obtain necessary copyright permissions, commented on all chapters and dedicated considerable effort to helping me finalize *Environmental Economics and Ecosystem Services*. Similarly, former students Colin Gibson, Ming He, Jake Kennedy and Miriam Silverman, as well as others, contributed in important ways. I am very grateful to Portland State University for funding their work on the book.

Michelle Gallagher, who is my editor at Routledge in the Taylor & Francis Group, gave important guidance and commented on several chapters. Routledge Senior Editorial Assistant Chloe Herbert helped me stay on top of this project, made sure I had the resources I needed to be successful and prompted me as needed. I thank both for their support and patience as finalizing this book took much longer than expected. Copy editor Drew Stanley dramatically improved the quality of each chapter in the book.

Numerous family members have supported me throughout this process, including expressing interest and asking me how the project was going. Zoe and Ari were willing to listen to me talk about the book and always encouraged me to make sure it became a reality. Without the truly unflagging support of my wife Marla, *Environmental Economics and Ecosystem Services* would never have been completed. She provided encouragement, helped with design and took on additional work in many, many ways, giving me space so I could work on and ultimately complete this book. I love her even more for so freely making these sacrifices.

Randy Bluffstone

SECTION I
Economic Fundamentals, Climate Change and Sustainability

1 Introduction to the Book, Ecosystem Services and the Sustainable Development Goals

What You Will Learn in this Chapter

- The difference between environmental assets/natural resources and ecosystem services
- The importance of the Millennium Ecosystem Assessment in highlighting what nature does for people
- About planetary boundaries
- The role of the Sustainable Development Goals in international development
- Goals of the book and what to expect

1.1 Recap and Introduction to the Chapter

This first chapter of *Environmental Economics and Ecosystem Services* is about natural resources, ecosystem services, the environmental challenges we humans who find ourselves in charge of the planet face, and the key international agreements related to sustainable development. It also discusses how this book hopes to contribute to understanding human-driven environmental problems and the solutions that have worked around the world.

Every chapter of the book begins with a short recap of previous material, so before we consider what economics might offer in terms of understanding, I will ask you to please think about the priors you bring to this discussion. How would you "recap" your previous engagement with the topic of environmental protection? Would you say your thinking has had a human-centered or perhaps even an "economic" component? If you are mainly interested in the environment, why do you want to understand the economic approach to environmental protection? If you primarily come to *Environmental Economics and Ecosystem Services* from an economic perspective, why would you like to apply that lens to understanding environmental issues?

Please take a moment to consider these questions.
Environmental Economics and Ecosystem Services is about humans and their relationship with the natural world, the environmental problems they cause and the policy solutions that

DOI: 10.4324/9781003308225-2

humans have developed that seem to work. Economics is one of many disciplines that can potentially contribute to understanding environmental problems and solutions, but the most important problems are neither easy to understand nor easy to solve; it will take many disciplines and perspectives to really comprehend the challenges and think through solutions.

The economics of the environment is actually a relatively new field, having only started in the late 1960s, which is much later than other disciplines that also analyze environmental issues. But in those approximately 55 years, the contributions of economics and the subdisciplines of environmental economics, resource economics and ecological economics to understanding environmental problems have been significant. It is to highlight these contributions and insights in a rigorous and forthright way and make them available in a user-friendly form that this book was developed.

We first turn our attention to the topic of ecosystem services. In Section 1.3 we consider some of the key human-caused environmental issues in the world today, including climate change, followed by an overview of the main international agreements that guide international sustainable development. Finally, in the last section of the chapter we discuss the *raison d'être* for this book and overview its sections.

1.2 Approach of this Book, Ecosystem Services and the Millennium Ecosystem Assessment

This book approaches the relationship between humans and the natural world from the perspective of *ecosystem services*, which are the services that nature provides to humans. The idea and importance of ecosystem services were popularized in 2005 with the publication of the Millennium Ecosystem Assessment (MEA, 2005). The MEA was conducted between 2001 and 2005 under the auspices of the United Nations (UN) by over 1300 experts who attempted to answer the following questions:

1. How have ecosystems and their services changed? What has caused these changes?
2. How have these changes affected human well-being?
3. How might ecosystems change in the future and what are the implications for human well-being? What options exist to enhance the conservation of ecosystems and their contributions to human well-being?

Box 1.1 The Key Points on Natural Resources/Environmental Assets and Ecosystem Services

- Natural resources/environmental assets are *stocks*, which are relatively fixed for significant periods of time.
- Ecosystem services are *flows*, which occur and can change over time.
- Ecosystem services are produced by natural resources.
- Humans manage natural resources.

Throughout this book, we use words such as "nature," "environment," "natural resources," "environmental assets" and "natural capital" pretty much interchangeably. Examples of environmental assets include clean air, a cool free-flowing river, a natural forest and agricultural soil fertility. All of these natural resources are *stocks*, meaning they are fixed over reasonable periods of time. These environmental assets produce, among other things, ecosystem services for humans. Natural capital therefore produces *flows*[1] of ecosystem services that accrue over time.

The focus of the MEA was solidly on what nature does for people and how human decisions feed back onto nature. The MEA identified four classes of ecosystem services. *Provisioning services* are living or non-living materials people extract from nature. Examples include stone, animals, minerals, oil, captured fish, timber and mushrooms. These products are directly used by people for building materials, fuels and food. Notice that when we take provisioning ecosystem services we extract them from the natural world, changing and typically degrading natural resources in the process.

Box 1.2 Classes of Ecosystem Services in the Millennium Ecosystem Assessment

- **Provisioning services** - directly used by humans (e.g., oil, food, fiber).
- **Regulating services** - ecological integrity that indirectly supports human well-being (e.g., air quality, disease regulation, erosion control).
- **Cultural services** - directly or indirectly used for spiritual or aesthetic purposes (e.g., eco-tourism).
- **Supporting services** - fundamental ecosystem functions (e.g., nitrogen cycle, primary productivity, soil formation).

Pollution by people into the air, water and land also alters and degrades natural resources. The service of nature to process these pollutants too can be thought of as a provisioning service. Indeed, when we think about it, the capacity of the natural world to process our waste is actually pretty handy! For example, where would we be if trees did not absorb carbon dioxide (CO_2) and transform it into biomass, such as wood or stems, helping us deal with some of our carbon pollution? Absolute magic! They also knock down particles floating in the air onto the soil so they cannot enter our lungs, potentially causing cancer. Microbes in the soil biodegrade organic materials, such as food wastes, paper and yard wastes, turning them into compost that can improve soil fertility - again, 100% magic and very helpful for humans. These so-called *sink functions* of the environment are therefore very, very important provisioning services.

Regulating services are the ecological integrity that indirectly supports human well-being. I emphasize the word *indirectly*, because regulating services arise because of indirect effects of ecosystem quality changes on humans. For example, humans enjoy better respiratory health when air quality improves. Better-quality forests reduce soil erosion and can indirectly lead to more food production. Increased biological diversity (i.e., biodiversity)[2] resulting from better habitat may help reduce the prevalence of infectious diseases, such as malaria. Notice that in all these examples people are indirectly "using" natural resources, but there is no need to extract from or otherwise change the environment to get benefits.

Challenge Yourself

Can you name two provisioning and two cultural ecosystem services provided by nature near your home?

Cultural services provide aesthetic or spiritual value to people. Examples might include enjoyment of a beautiful sunset, a swim in the ocean or a hike through a forest. These are all cases where humans directly use natural resources to get ecosystem services, but do not significantly degrade them in the process. Cultural services can also be much less direct and not based on usage. For example, many cultures worship parts of nature - mountains, lakes, rivers, species, the sun - as sacred. The spiritual values these natural resources offer people are not due to them being used at all, but simply due to them existing.

Similarly, many of us get benefits from the existence and thriving of certain species or even whole ecosystems, without ever using or even seeing them. For example, some of us may feel sad that, because of global warming, the icy habitat of polar bears is declining and the species is therefore under serious stress. Conversely, those of us who feel this way would materially benefit if steps were taken to improve polar bear habitat.

Challenge Yourself

Can you name two institutions that rely heavily on cultural eco-system services? What are those critical ecosystem services?

Finally, *supporting services* are the basic ecosystem functions that keep the planet running. These are the operating systems of the Earth, such as the carbon cycle, which processes carbon from the air and soil, or soil formation, which makes sure plants can grow and facilitates the magic of primary productivity in which plants take carbon, water and sunlight and produce biomass.

The MEA evaluated changes in the quality of natural resources and the ecosystem services they generate. They concluded that many services - especially if they are not subject to market forces (i.e., they are not bought and sold) - were in decline. More on this topic in Chapters 3 and 8 of this book. They also attempted to map ecosystem service declines onto changes in human welfare across five dimensions: 1) security; 2) basic material for a good life; 3) health; 4) good social relations; and 5) freedom of choice and action. They found strong linkages between many natural resources, the ecosystem services they generate and these five aspects of human well-being. In the next section of this chapter, we discuss some of the contemporary evidence on human-caused environmental problems that threaten natural resources, ecosystem services and human well-being.

You may be struck that all this discussion is extremely human-focused and I think you would be correct in that assessment. Let us stick with this approach just for a minute, but also think of important human institutions and indeed the economy much more broadly than just money transactions and markets. For our purposes, let us even suppose the "economy" includes not only market institutions with buyers and sellers but also the many other institutions that in one way or another improve human welfare using resources.[3]

Please take a moment to think of all the human-created institutions you regularly encounter that try to improve the human condition.

Perhaps schools, coffee shops, hiking associations, Apple, governments, sports leagues and teams, auto repair businesses, political parties, clubs, environmental advocacy organizations, religious institutions, natural resource user groups, running and writers' groups would come to mind. All these groups – market-focused or otherwise – have as their goals to make the world a bit better, mainly for humans, and use resources to make it happen. All such groups to a greater or lesser degree need inputs from the environment. These inputs might be trails, forests and good weather for hiking associations. Sports leagues need clean air to avoid breathing pollution. Travel companies help travelers visit beautiful scenery around the world, which can be part of the environment. Militaries need open spaces to conduct training exercises.

Households, businesses, clubs, schools and political parties need energy, which can come from renewable (e.g., water, solar or wind) or fossil fuel resources, to have the heat and electricity they need, for example, to comfortably have their meetings and classes. Beer companies need fresh water, hops, energy and the raw materials to make glass so they can bottle their beer. Jewelry companies require gold, which comes from mines. Please notice that people use a variety of ecosystem services, only some of which are provisioning services.

These groups and the people who compose them also use the environment to dispose of their solid, waterborne and gaseous wastes. That is, all these human institutions pollute, which will have implications for the environment and reduce the quality of inputs for humans.

In sum, as we think about the interaction between people and natural resources, from the human standpoint we have two main ecosystem service dimensions: 1) inputs and 2) sink services.

A critical first step in analyzing these issues is to develop a general framework for analysis – a way of thinking or a "model" – that represents the key interactions. Models are depictions of contexts, situations or people, in our case human-environment ones, that are stripped of non-essential elements, so we can see the fundamental relationships and potentially draw inferences.

Challenge Yourself

What are some models with which you are familiar?

As we think about such a model, let us pull back from a completely human-focused approach for just a second. Figure 1.1 presents a nature-focused model of economics and the environment. In our nature-focused model, the economy is set within the natural world and as these human institutions, individuals and families interact with the environment, they are subject to the standard laws and limits of the natural world; the economy therefore faces the same physical, biological and ecological laws that govern other species. This assumption seems reasonable. After all, why shouldn't humans be subject to the Second Law of Thermodynamics, which says that energy tends to move from ordered to disordered states when it is transformed? Don't humans have ecological carrying capacities in the same way that other mammal species experience population limits?

This logic is reasonable and no doubt ultimately correct. But when exactly is "ultimately?" Maybe soon, given the climate change situation, which we will discuss in the next chapter, but perhaps not. This question of ecological constraints, limits to the growth of human consumption, population and progress, and when increasing scarcity will kick in has been the subject of speculation, investigation and modeling for at least 50 years (e.g., see Simon, 1980; Nordhaus et al., 1992[4]; Turner, 2008; Meadows et al., 1972). We leave this discussion here, but in the next subsection of this chapter will discuss some of the most important environmental limits we seem to have transgressed.

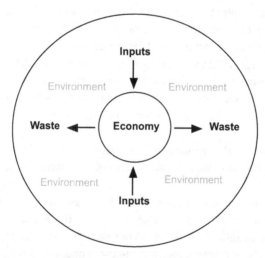

Figure 1.1 Nature-Centric Model of Economics and the Environment

But beyond the question of to what degree humans are or will be subject to ecological and other limits, there are complications with modeling the interface between humans and the environment from a nature-centric perspective. For example, and of critical importance, the units of the natural world do not line up well with those of the human world.

Nature runs on energy flows that, among other things, produce biomass - leaves, tree trunks and animal fat, to name a few.[5] Economies run on human-focused metrics, such as hours worked and volunteered, kilograms produced, miles travelled and, of course, money. Let's focus on money, which is a human construct and one might argue an artificial one. Why not just convert everything now denominated in money to energy? Certainly, a number of the metrics just mentioned could at least in principle be framed as energy, but is that a manageable modeling approach? More fundamentally, is there anything especially important at the human/environment interface that would not be amenable to scrapping money?

Actually, probably yes. For example, framing human skills, insight, creativity, social skills, emotional intelligence and innovation in terms of energy expended would at minimum be missing the point. These features, which on Earth are basically unique to the human species, often have very limited connections with energy units. Not that it does not take effort to operationalize creativity - we've all had great ideas that came to nothing because we did not work at them - but sweat is not the heart of the matter and we could therefore miss the point if we place energy at the center of our human-environment model. Furthermore, human creativity and innovation have been absolutely critical for the amazing progress of our species during the last 10,000 years. It is therefore perhaps not surprising that these assets are also some of the most rewarded, and the way innovative people get paid is often with money.

Aspiring to fully model humans' economic activity within the context of nature is a noble endeavor, but it turns out that even putting human economies front-and-center offers plenty of challenge.[6] In this book, we therefore sidestep the very significant modeling issues associated with placing economies fully within the natural context. The implications of this simplification - and a simplification it is! - are illustrated in Fig. 1.2, which presents an economy-centered model of economics and the environment. As is also true in the nature-centric model, people use a broad range of natural inputs, as well as the waste-absorbing features of the environment. The wastes and other damage to natural resources by people then feed back into the ability of the environment to provide ecosystem services. The pieces

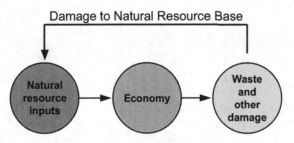

Figure 1.2 Econo-Centric Model of Economics and the Environment

missing from this model are other species and an explicit modeling of natural processes, which are both in the background and outside the model.

This model is neither complete nor "correct," but as I will try to demonstrate throughout this book, such a human-focused approach has a lot to offer. In the coming sections we will explore why humans sometimes (often?) trash the natural world and how to frame – from a human perspective – the challenge of protecting natural resources. We will also investigate some of the most forward-thinking ways humans have been trying to solve environmental problems around the world. All this said, there are implications of a strictly economy-focused approach. For example, in the econo-centric model the only reason natural resource damages, such as species extinctions, matter is because they negatively affect humans.

As of 2024, the human species is, for better or worse, in charge of the planet. Indeed, human actions are so consequential that many scientists even propose that our current geologic era should be called the Anthropocene rather than the Holocene, which has been our era for the past 11,000 years, because human domination of the Earth has been going on for over two centuries (Crutzen, 2002). Given this reality, perhaps it is worth looking at environmental issues a bit from the human perspective. But before diving into that large topic, let us overview some of the most important natural resource challenges facing humans today.

1.3 Key Aspects of the Environmental Policy Challenge

The world certainly has lots of human-caused environmental problems, that is for sure. In arid areas there is often serious overuse of water resources by humans, particularly for agriculture, which can leave people, hydropower systems and ecosystems high and dry. In many urban areas, especially but hardly exclusively in lower-income countries, air quality can be terrible, with concentrations of pollutants such as dust, sulfur dioxide and ground-level ozone many, many times recommended levels.

In some cities, such as Kathmandu, Nepal, it is sometimes difficult for many people to breathe. As shown in Box 1.3, on February 24, 2018, the air pollution in Kathmandu reached highly dangerous levels. Those who live in cities like Accra, Addis Ababa, Beijing, Delhi, Kathmandu, Mexico City and Nairobi probably have shorter and less healthful lives because of breathing heavily polluted air.

Box 1.3 Air Pollution in Kathmandu, Nepal

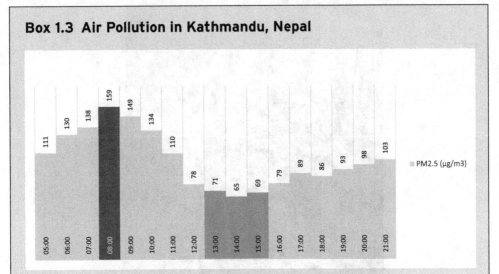

Figure 1.3 PM$_{2.5}$ Concentrations in Kathmandu Valley (February 24, 2018)

Nepal Government standard 40 ↔g/m³. World Health Organization standard 25 ↔g/m³.

>150 ↔g/m³ dangerous. 75-149 ↔g/m³ highly polluted. Face mask advised. 41-75 ↔g/m³ harmful for vulnerable populations.

Source: *República*, Kathmandu, February 25, 2018, p. 4. Based on government data from 17 air quality monitoring stations.

Rivers, streams and lakes can also be amazingly contaminated, particularly in low- and middle-income countries. People may use water from these sources for drinking, washing and other uses, sometimes with disastrous health consequences, particularly for children.

Tropical and sub-tropical regions bounding the equator are some of the most biodiverse regions in the world, but it is exactly in these areas that forest loss and other habitat destruction have been most important. Even experts across a variety of disciplines may be unable to identify, much less quantify, all the ecological and economic consequences of species and habitat losses with anything resembling certainty.

Past industrial pollution can yield holdover issues of toxic contaminants in land and water that may create serious ecological and health problems, perhaps requiring that areas be closed to use. Legacy pollution can also be extremely expensive or impossible to clean up and responsible parties may be very difficult to identify and hold accountable.

And then there is anthropogenic[7] (i.e., human-caused) climate change, which has the potential to alter the basic functioning of the global environment. This challenge is one of our biggest and potentially most difficult issues to address, because of its worldwide scope and deep roots in the energy structure of economies. From a scientific perspective, it is generally agreed that as of 2023 the Earth's climate, typically measured at various places on the Earth's surface, has warmed by an estimated 1.18 degrees Celsius (2.12 degrees Fahrenheit)[8] compared with the 20th-century average.[9] The warmest year on record was, as

Photo 1.1 A Trash-Filled Canal in Nepal
Source: Author.

of this writing... last year, 2023, and the period 2014–2023 included the 10 warmest years since recordkeeping started in 1850! This warming cannot be attributed to idiosyncratic events that are likely to reverse themselves within the foreseeable future (NOAA, 2023) and it is beyond doubt that humans are driving this planetary warming (IPCC, 2023). The Earth's average surface temperature has been climbing steadily since 1970 and essentially since World War II, but unfortunately, because of our past activities the temperature will continue to increase.

Stabilizing the temperature of the planet is one of the most important challenges – full stop – that humans face. Even 40 years ago it would have been inconceivable to most people that something as big as the atmosphere and the surface temperature of the Earth could be appreciably affected by human activities. But it turned out that we were wrong and in reality we *are* warming the planet through emissions of greenhouse gases into the atmosphere.

And of course, as we are facing all these human-caused environmental challenges, the 8 billion+ people of the world have aspirations, which include meaningful, productive and prosperous lives. Particularly when we look broadly across the world and sincerely include the needs and desires of low-income people in lower-income countries, it is hard to imagine saving the world without taking account of human livelihoods, particularly at the lower end of the global income distribution. Of course, human aspirations and values typically also include a clean and healthy environment, a topic we will take up in Chapters 3 and 4.

If we look across the last 50 years, in a variety of contexts and places humans have dramatically improved environmental quality, creating the need to understand the lessons from those success stories. For example, in many areas around the world, air and water quality has dramatically improved over the last half-century. There are numerous examples, but one that comes immediately to mind, because it is near where I grew up, is water quality in the US and Canada Great Lakes Region. Fifty years ago, there were widespread water pollution, ecosystem decline and lack of human access due to the risks associated with contact with Great Lakes water. Circa 2024, water quality has dramatically improved and many previously hazardous areas are fully fishable, swimming is possible and ecosystems function much better than in the past.

Air quality in many cities around the world was horrible 70 years ago and while there are still numerous and serious problems, mainly in higher-income countries air quality by a variety of measures has dramatically improved. These changes have occurred even as production, driving, consumption, etc. substantially increased.

Furthermore, while the amount of disposed solid waste has increased over time, again particularly in the higher-income world, disposal into land is much more organized than in the past. In most areas of the higher-income[10] world "landfills," often with sophisticated environmental controls, have replaced "dumps," with all the esthetic, human health and ecological improvements those changes imply. Trash randomly dumped on and in the land and water in the form of littering in many places is significantly less than a half-century ago and human norms to recycle and reduce solid waste have emerged as significant trends.

So, among all the environmental problems and potential success stories, on net what are the most important challenges? In the last 15 years a robust literature has emerged focusing on estimating the environmental boundaries that humans must stay within if the Earth is to retain a sufficiently healthy character that it can provide critical ecosystem services for humans and other species. This literature has also evaluated which boundaries are most under threat.

Perhaps first proposed by Rockström et al. (2009), the planetary boundaries approach draws on Earth system science to identify what is typically referred to as a "safe operating space" for humanity. These spaces are defined with respect to environmental levels that were in place before humans began industrial activities in the second half of the 18th century, although safe spaces are generally at least somewhat above pre-industrial levels.

Rockström et al. (2009), which seems to be the seminal[11] paper, evaluated ten Earth system processes, but at that time were only able to define eight of them. These processes, their proposed boundaries and their current statuses are given in Table 1.1. The first Earth process is climate change, which I have already mentioned and will be discussed in detail in Chapter 2. Climate change is measured as atmospheric CO_2 concentration in parts per million and also in terms of radiative forcing in watts per m^2. CO_2 and other greenhouse gases in the air are mainly due to burning fossil fuels for energy.

The second process is biodiversity loss, which is measured as the rate of species extinctions. The third and fourth Earth processes focus mainly on chemical fertilizers used in agriculture. Nitrogen fertilizer is derived from atmospheric nitrogen, which is then converted to non-gas forms. The process is sort of like emitting CO_2 into the atmosphere, except in reverse. Instead of taking something from the planet and sending it up into the air, materials

Table 1.1 Planetary Boundaries from Rockström et al. (2009)

Earth System Process	Units	Proposed Boundary	Current Status	Pre-Industrial Value
Climate change: metric 1	CO_2 concentration in parts per million	350	387	280
Climate change: metric 2	Radiative forcing in watts per m^2	1	1.5	0
Biodiversity loss	Extinctions (species/million species/year)	10	>100	0.1–1.0
Nitrogen cycle	N_2 removed from the atmosphere (millions of tons/year)	35	121	0
Phosphorus cycle	Flows into ocean (millions of tons/year)	11	8.5–9.0	–1.0
Stratospheric ozone	Ozone concentration	276	283	290
Ocean acidification	Saturation state	2.75	2.90	3.44
Freshwater use	km^3/year	4000	2600	415
Change in land use	Percentage of global area converted to agriculture	15	11.7	Negligible
Atmospheric aerosol loading	Particle concentration	To be determined		
Chemical pollution	To be determined			

from the atmosphere are chemically manipulated using some variant of the Haber–Bosch process. This process, which was developed in the early 1900s and has since revolutionized agriculture, captures atmospheric nitrogen and creates ammonia, which can be used for industry and to promote plant growth.[12] Unfortunately, during the year and century-level time scales that are most relevant to humans, nitrogen captured from the air will remain on the planet and much of it ends up in water where it can dramatically disrupt aquatic ecosystems. Phosphorus is also an element needed for plant growth. It is manufactured from phosphate in rocks and also largely ends up in the water.

Process five focuses on stratospheric ozone (O_3). The stratosphere is one of the layers of the Earth's atmosphere and is anywhere between 4 and 31 miles above the planet's surface (NOAA, undated). It is important, because it is home to an ozone layer, which protects the Earth and its inhabitants from excessive ultraviolet radiation that can cause cancer. Process six is ocean acidification, which is related to climate change, because CO_2 in the atmosphere gets absorbed by the ocean. As anyone who likes carbonated drinks knows, CO_2 is acidic, which is why it burns your tongue when you take a sip. Ocean acidification is important, because it disrupts marine[13] ecosystem processes, including the formation of shells by crustaceans, such as clams, oysters and mussels.

The authors also evaluate changes in land area from natural states to agriculture and withdrawal of freshwater from underground and surface sources (processes 7 and 8). Finally, Rockström et al. (2009) lay out the need to set boundaries for aerosol loading (i.e., particulate pollution that lingers in the air) and chemical pollution, but were not able to define those boundaries.

As of 2009, the authors estimated that humans had exceeded planetary boundaries involving climate change, biodiversity loss and nitrogen cycle processes. At that time, CO_2

concentration was about 10% over the proposed limit,[14] which was itself 25% above the pre-industrial baseline. Biodiversity loss, measured by the extinction rate in 2009, was about 1000 times the baseline rate and 10 times the proposed boundary. Similarly, nitrogen cycle disruption was estimated to be very significant. Compared with the estimated pre-industrial level of 0, the 2009 emissions were 121 million tons per year, which was about 4 times the proposed boundary. The authors concluded that these three areas where planetary boundaries were exceeded needed to be urgently addressed to avoid disrupting Earth's processes.

Since 2009, a lot of work has been done on planetary boundaries. Processes and metrics have in some cases shifted, but over time the original conclusions have only intensified and expanded. Steffen et al. (2015) reformulated biodiversity loss into biosphere[15] integrity, which is made up of genetic diversity and functional integrity. They also added human-made compounds and new life forms, which they called "novel entities." They found that the gen- etic diversity and nitrogen cycle planetary boundaries continued to be grossly exceeded, as well as climate change to a lesser extent. They also estimated that phosphorus cycle and land system change limits were also exceeded, but they were not yet able to compare observed levels of functional biosphere integrity, aerosols and novel entities with proposed planetary boundaries.

By 2023 more was known. Richardson et al. (2023) figured out ways to appropriately measure novel entities, aerosol loadings and functional biosphere integrity. They find that of nine Earth processes, only three (stratospheric ozone depletion, atmospheric aerosol loading and ocean acidification) are within the planet's safe operating space; six processes are being disrupted by humans beyond what they estimate as safe levels.

In sum, there are a lot of environmental problems created by people and therefore a lot of work to do at the intersection of people and the planet. Though there have been some important successes, which will be discussed throughout this book and will be analyzed from a policy standpoint in Section 3, there are many aspects of the human-nature interface that are out of balance. Why these imbalances exist from an economic standpoint and what are some ways to think about human-focused solutions are the business of the rest of this book.

But any solutions to environmental problems have to take into account that so many people have not yet really had their shots at prosperity. Standard, market-focused economic "development" of the countries where people in the Global South live will need to be ser- iously considered as we address areas where we are exceeding planetary safe operating spaces. In other words, *sustainable economic development* is of critical importance. We therefore now discuss the Sustainable Development Goals, which guide global development policies around the world through 2030.

1.4 The Sustainable Development Goals

The Sustainable Development Goals (SDGs) are important because they have been the road map for the world's program of economic development since 2015 and will continue to guide this important work until 2030. As shown in Fig. 1.4, there are 17 SDGs, all of which are proposed to be achieved by 2030.

Figure 1.4 The Sustainable Development Goals

Source: https://www.un.org/sustainabledevelopment/news/communications-material/

For our purposes, the SDGs are of keen interest, because they not only focus on standard, market-focused economic development, but also seek to meaningfully include environmental protection. As noted on the UN SDG website, what is formally called the *2030 Agenda for Sustainable Development* "...provides a blueprint for peace and prosperity for people and the planet, now and into the future" (United Nations, undated).

These goals were adopted by the UN General Assembly after the expiration of the Millennium Development Goals (MDGs), which guided development policy between 2000 and 2015. There were only eight MDGs, in some cases explicitly addressing issues that are not included in the SDGs and sometimes more aggregated than the SDGs. They also did not explicitly call out environmental goals as do the SDGs. The MDGs are listed in Box 1.4.[16]

Box 1.4 The Millennium Development Goals

1. Eradicate Extreme Poverty and Hunger
2. Achieve Universal Primary Education
3. Promote Gender Equality and Empower Women
4. Reduce Child Mortality
5. Improve Maternal Health
6. Combat HIV/AIDS, Malaria and Other Diseases
7. Ensure Environmental Sustainability
8. Global Partnership for Development

Ten of the SDGs are basically human-focused (no poverty, zero hunger, reduce inequality, gender equity, etc.), but the remaining seven have important environmental aspects. These SDGs include life on land, life below water, climate action, sustainable cities and communities, clean water and sanitation and affordable and clean energy. Goal #12 focuses on responsible consumption and production, a topic we will return to within the context of consumption in Chapter 5.

At a number of points in this book we will refer to the SDGs. **Please take a moment to familiarize yourself with these goals.** The SDGs are very important for us, because they are focused on improving the welfare of the majority of the world's population and are positioned squarely at the intersection of human well-being and the environment.

Each of the SDGs has targets and indicators associated with them, and the UN Secretary General is required to annually report on progress. You can read these annual progress reports at United Nations (undated). For example, SDG #1 is to "End Poverty in All its Forms Everywhere." It has 7 targets against which to gauge progress and indicators associated with each target (total of 13). Target 1 associated with SDG 1 is "by 2030, eradicate extreme poverty for all people everywhere, currently measured as people living on less than $1.25 per day." The indicator for this target is the proportion of people living below this income cutoff, evaluated across a number of sub-populations, including age, gender, employment status and rural/urban.[17] The World Bank projects that this target will not be achieved, with about 8% of all people (600 million) still living in extreme poverty in 2030 (World Bank, undated).[18]

1.5 The What, Why and How of this Book

As I mentioned at the start of this chapter, *Environmental Economics and Ecosystem Services* is about humans and the natural world, the environmental problems humans cause and the policy solutions that people have developed. No economics background is presumed, so any theoretical material or analytical techniques that are necessary to understand environmental issues are fully presented between the covers of this text. Whether you are a student of economics who wants to apply economics to environmental problems or a student of environmental issues who wants to add economics to your toolkit (good call!), you should find material that helps you achieve your goals.

Because all these environmental challenges and successes are human caused, this book focuses on the interplay between humans and the environment. At least viewed through an economic lens, which is admittedly not the only one, this book suggests that it is only by thinking through our relationship with natural resources and the associated incentives that we can craft workable and effective policy instruments that reduce environmental problems.

To address these issues, we will cover both ideas and analytical techniques. This book reviews the critical fundamentals and some of the key cutting-edge ideas in environmental and natural resource economics. It also tries to provide you with an understanding of the most important economic modeling tools and empirical techniques for analyzing environmental issues.

The world may sometimes seem very small, but there is also enormous diversity. Around the globe there are many, many different ecological zones and social contexts that need to be included and considered. Though global environmental problems, such as climate change, are probably paramount as of 2024, it is also true that terrestrial (i.e., land-based) and water-based ecosystems can differ significantly across the world. The problems of tropical forests are very different from those of savanna ecosystems and coastal ecosystems differ dramatically from those located away from saltwater. These realizations imply that environmental problems we might be trying to understand and solve can differ dramatically across the globe.

Furthermore, how people interact with natural resources varies. People live in different circumstances, have vastly different income levels, material standards and livelihood strategies. In the lowest-income countries people may be getting along on an average of *$1000 or less of buying power per year per person*. For those in the higher-income world who may not believe this statistic, I would emphasize – yes *really*.

To be in a World Bank-defined high-income country, the average income must be over $13,000 per person per year, but the highest-income countries have three to six times that level of income per person[19] and in any case substantially less than half of all countries are high-income. With such huge dissimilarities in average incomes across countries, plus many other important differences in economic opportunities across the globe,[20] cultural norms, people's material circumstances, preferences and perceived tradeoffs vis-à-vis the environment are bound to differ. These differences in country-level circumstances yield diverse demands on nature, environmental problems and solutions across the globe, which we need to seriously consider.

The world has become much more urbanized over the last 50 years, with now over half the world's population living in cities. These concentrated populations typically earn their livelihoods in trade, manufacturing and government, among other economic sectors, and air and water pollution may be especially important local environmental problems. There is also the approximately 44% of humanity who live in rural areas where farming, mining, fishing and animal husbandry, among other jobs, are the main ways to make a living. The environmental issues of rural areas may include pollution, but often focus most heavily on natural resource issues, such as deforestation, agricultural expansion, over-grazing, land degradation from mining and other forms of extraction and over-fishing. The differences in issues across the urban-rural divide, of course, also need to be reflected in our analysis of humans' management of natural resources.

As we will see throughout this book, a lot of the solutions to environmental problems involve governments at the local, regional, province/state and national levels. Government effectiveness differs dramatically around the world and, perhaps not surprisingly, higher per person incomes fund – and are facilitated by – more effective governments. On average, the World Bank reports that governments in countries classified as high income are four to six times more effective overall, in terms of regulatory quality, adherence to the rule of law (including protection of property rights) and control of corruption than low-income countries.[21]

As we search for institutions to implement solutions to environmental problems, we need to recognize that the degree to which governments have the capacity and have been shown to deliver for their populations differs across the globe and perhaps systematically by income level. In situations where governments are less effective, other human institutions often are called upon to step up to at least partially fill gaps.

With such a diversity of ecological, human and institutional circumstances, *Environmental Economics and Ecosystem Services* tries to present a variety of economic interactions with the environment. This approach is intended to be inclusive and allow readers from around the world the opportunity to see their own circumstances and have *their* problems discussed within the pages of the book. It also, though, allows all of us to become more conversant with others' situations. As a result, we may find approaches from elsewhere that transfer well to our circumstances and we get the chance to appreciate others' circumstances. We might also start to understand that the issues, perspectives and approaches that are most salient and resonant for us may not apply equally well in other settings.

Environmental Economics and Ecosystem Services is organized into three sections. Section I focuses on **Economic Fundamentals, Climate Change and Sustainability** and consists of seven chapters. After discussing climate science and policy in Chapter 2, the next two chapters focus on the tools needed to analyze environmental problems from the economic perspective. These tools include notions of economic value, price concepts and the supply-demand market model. We will also discuss how to estimate the benefits of environmental assets and ecosystem services in monetary terms.

The remaining three chapters in Section I examine consumption and the human population as potential contributors to environmental problems. In Chapter 5 we consider consumption and evaluate the economic meaning of consumption and some perspectives on "over" consumption. Chapter 6 discusses human population and assesses where "over" population might be an issue and what might be the key drivers. Chapter 7 revisits consumption and to some degree also population, catapulting those topics into a dynamic framework around defining and measuring sustainability. In this chapter, we learn the economic perspective on sustainability and discuss contemporary measures of environmental, social and economic sustainability.

Section II of *Environmental Economics and Ecosystem Services* is titled **The Challenge of Protecting Natural Resources**. It asks and attempts to answer the critical questions 1) "Why are natural resources so difficult to manage?" and 2) "What are the human incentives that make natural resource management challenging?" Chapter 8 looks deeply at natural resources and the ecosystem services they provide and evaluates which services are most at risk and why. It also introduces environmental cost-benefit analysis, which is one framework within which to compare alternative outcomes.

Chapters 9 and 10 present alternative models – really analytical frameworks – of the incentive problems associated with natural resource management. These chapters look at environmental protection incentives from externality, property rights and collective action perspectives, which are the main analytical frameworks for analyzing environmental "moral hazard."

Chapters 11 and 12 move a bit more into political explanations of incentive problems by focusing on economic subsidies. Subsides occur when governments use their budgets or regulatory powers to benefit private actors, often producers, resulting in distortions, such as especially low prices. Chapter 11 presents the key situations and ways governments involve themselves in markets and tweak market prices. It also examines agricultural subsidies, including for irrigation water in the arid American west.

Chapter 12 dials in on energy production and consumption, which is an especially important source of air pollution, and defines the meaning of energy subsidies. The chapter catalogues the existence and levels of energy subsidies around the world. It also examines the case of the European planned economies that were part of or allied with the Soviet Union, which is a particularly egregious historical example of energy subsidies that led to serious environmental consequences.

Section III brings us to **Solving Environmental Problems around the World**. The seven chapters in this section present a variety of environmental policy instruments, including some of the most innovative in use today. In Chapter 13, we begin with critical social innovations to improve natural resource collective action.

Pollution control is one of the most important types of environmental protection policy. We therefore devote Chapters 14-18 of the book to that topic. In Chapter 14 we consider the economic theory of pollution control, including cost-effective abatement and key pollution policy instruments. Green pricing – using taxes and fees to steer economies in greener directions – is a critical class of pollution control policies. Chapter 15 examines green pricing to reduce pollution within the context of the theory and practice of taxes on measured pollution. Chapter 16 continues our exploration of green pricing by applying our tools to carbon taxes and environmental product taxes, then considers possibilities for green tax reforms, which try to shift tax systems toward taxing pollution and away from standard types of taxes.

We next consider the theory and practice of green markets: government-developed pollution markets, such as cap-and-trade, that try to cost-effectively achieve pollution reduction goals. Chapter 17 presents the theory of green markets and discusses the extensive US experience with pollution markets. Green markets are important tools for mitigating climate change and from the very beginning have been part of international climate policy. Chapter 18 therefore discusses "pollution trading" within the context of international climate agreements and some examples of projects that were completed under these programs. The chapter discusses so-called "voluntary" carbon markets, but also presents the international experience with "compliance" markets around the world, including the European Union Emissions Trading Scheme, which has been running since 2005. Finally, Chapter 19 attempts to weave together the key intellectual threads developed throughout the book into a set of general conclusions and outstanding issues.

1.6 Chapter Conclusions

Environmental Economics and Ecosystem Services is about humans' economic relationships with the environment. Purely for the sake of feasibility, the book puts humans front-and-center and largely takes natural processes and the interests of other species as given. This "econo-centric" emphasis naturally leads us to an approach that focuses on ecosystem services, which are the benefits nature provides to people.

The Millennium Ecosystem Assessment, which was published in 2005, highlighted this way of thinking about the human–environment interface and emphasized how nature promotes human well-being. It also focused on the decline of many services, particularly those that are not bought and sold in markets.

The chapter discussed some the critical environmental problems in the world today, including many that will continue to be explored in the remainder of the book. The planetary boundaries literature highlights the most important ways people are interfering with and endangering critical Earth processes. As measurement and other analytical techniques have progressed during the past 15 years, this body of work suggests that humans may be transgressing 2/3 of Earth process boundaries.

Improving human interactions with the natural world is critical, but it is also true that people have aspirations they would like to achieve, hundreds of millions of people are surviving on less than $1000 per year and many face other barriers to fulfillment. Under UN auspices, countries have agreed to actively work to improve the human condition, while also recognizing the important role nature plays in human welfare. The Sustainable Development Goals guide this process of human and natural development through the year 2030. Given their importance, we will rely heavily on this and other sustainability frameworks throughout this book.

Issues for Discussion

1.   Models are abstracted versions of reality, which are created to portray key aspects of that reality. Please name two models with which you are familiar, discuss the reality they are meant to represent and talk about the reasons those models are helpful.
2.   Can you name two provisioning and two cultural services that are especially important to you? Please discuss why these ecosystem services fit into these MEA categories.
3.   Can you name three regulating services not mentioned in the chapter? In what ways are your three examples of "indirect" benefit to people?
4.   Choose one of the 17 SDGs. What are the targets and indicators to judge progress toward meeting that goal? Do they seem appropriate to you? Why or why not?
5.   The UN Secretary General is required to report on progress toward achieving the SDGs on an annual basis. These progress reports are available at https://sdgs.un.org/goals. Choose one SDG and look at the most recent progress report. Is the world making suffi-cient progress? What, if anything, is missing?

Notes

1 Throughout this book, I will use italics to denote technical terminology that I would like you to note.
2 It is common in the scientific literature to use Latin abbreviations: "i.e." means "that is" and is used to rephrase something that was previously said. The abbreviation "e.g." just stands for "for example." "Etc." is short for et cetera, which means "and so forth" and avoids the need to provide a list of similar things. "Circa" means approximately and is especially used for dates. These and perhaps one or two other such common abbreviations derived from Latin will be utilized throughout this book.
3 We will discuss how markets and trade have the potential to improve human welfare starting in Chapter 3. Let us not presume that everyone agrees on the meaning of "improved" human welfare, i.e., politics exist!
4 Chapter citations indicate that other authors' information or evidence are used at those points in the chapter. Citations are very important for scientific writing because they help us acknowledge others' work. The full references related to citations are then listed at the end of chapters. Et al. is a Latin phrase that is commonly used with citations. This abbreviation for the Latin phrase "et alia" means "and others." It indicates that there were at least three authors of the publication, but in the citation only the last name of the first author is listed. If there are only one or two authors, the last name(s) will be followed by the year of publication. Unless there are many authors, in general all authors are included in the references.
5 We will discuss the importance of nature-focused measures of sustainability in Chapter 7 and at a number of points in this book we will analyze energy issues.
6 Ecological economics at least partly focuses on nature-centered approaches and there is a well-respected academic journal titled *Ecological Economics* (https://www.sciencedirect.com/journal/eco logical-economics). Natural resource economics, which focuses mainly on terrestrial environmental assets, and environmental economics, which addresses issues related to pollution, generally take human-centered approaches.
7 "Anthropo-" is a prefix derived from the Greek language, which means relating to humans. For example, anthropology is one of the social sciences that studies humans. The suffix "-genic" is also derived from Greek and means producing. Anthropogenic therefore means "human produced."
8 Degree Celsius is abbreviated as °C and degree Fahrenheit is abbreviated as °F. °F = °C*(9/5) + 32.
9 An average (sometimes called the *mean*) is just the sum of all the values divided by the number of values. Your grade point average (sum of all the numerical grade values divided by the number of classes you've taken), goals per game by a football team (total goals/number of games) and average height in a class in cm or inches (sum of all heights divided by number of students) are all examples of averages.

10 The country-level terminology related to people and places with systematically different income levels can be confusing. High-income countries are sometimes referred to as "developed" or in the past "first world." Lower-income countries have sometimes been referred to as "developing" or in the past "third world," with the second world being countries, such as Cuba, North Korea and the former Soviet Union, with planned economies. The contemporary terminology seems to be "Global South" and "Global North" to refer to lower and higher-income countries. I will therefore use these labels or the neutral terms "lower-income" and "higher-income." Of course, within all countries there are very high income and low income people.

11 A seminal contribution is the original research that inspired subsequent literature.

12 See https://en.wikipedia.org/wiki/Haber_process for more details.

13 i.e., saltwater.

14 Throughout this book, we will be using percentages and percentage changes. A percentage is just a way to present a portion of something and can also be expressed as a decimal. For example, 0.20 is the same as 20%. It is also a way to evaluate if amounts are significant. For example, we might want to know to what degree the CO_2 concentration is over the proposed limit value. The 10% would therefore be calculated as the following: (actual CO_2 concentration – CO_2 concentration limit)/(CO_2 concentration limit).

15 The biosphere is made up of the parts of the planet where life exists.

16 The Millennium Development Goals 2015 final report, which indicates that in some important respects progress was made, is available at https://www.un.org/millenniumgoals/

17 See https://unstats.un.org/sdgs/indicators/indicators-list/ for a full list of the 169 targets and 248 indicators, 13 of which repeat.

18 The World Bank is an international development bank that is part of the UN system. It is supported by member countries who contribute to its budget, which is used to give loans and grants to lower-income countries and provide them with technical assistance. It is also a major and authoritative source of information and we will rely heavily on World Bank data throughout this book. Among other information that it makes available for free on its website, it classifies countries by income per person. As of 2023, almost 700 million people across the world were classified as extremely poor by the World Bank, which now is defined as living on less than $2.15 per day or $785/person/year. About half the world's population lives on less than $2500/person/year (World Bank, undated).

19 For example, the 2022 average income in France was about $45,000, $53,000 in Belgium, $80,000 in Ireland and $95,000 in Norway.

20 For example, the money in many countries is difficult to convert to other currencies, making travel and purchases from abroad difficult. Banking in some countries, particularly in rural areas, may also be very limited. This means having safe places to save and low-cost ways to borrow money may also be limited. Most people around the world do not have the opportunity to use credit cards.

21 See https://datahelpdesk.worldbank.org/knowledgebase/articles/906519 for the official World Bank income cutoffs and classifications of all countries in the world.

Further Reading and References

Crutzen, P. 2002. "The Geology of Mankind." *Nature* 415 (6867): 23.

IPCC. 2023. Summary for Policymakers. In: *Climate Change 2023: Synthesis Report. Contribution of Working Groups I, II and III to the Sixth Assessment Report of the Intergovernmental Panel on Climate Change* (Core Writing Team, H. Lee and J. Romero, eds.). IPCC: Geneva, Switzerland.

Meadows, D.H., D.L. Meadows, J. Randers and W.W. Behrens III. 1972. *The Limits to Growth: A Report for the Club of Rome's Project on the Predicament of Mankind*. Universe Books: New York. Retrieved from https://policycommons.net/artifacts/1529440/the-limits-to-growth/2219251/ February 14, 2024.

Millennium Ecosystem Assessment (MEA). 2005. *Ecosystems and Human Well-Being: Synthesis*. Island Press, Washington, DC.

National Oceanic and Atmospheric Administration (NOAA). 2023. *Global Climate Report*. Retrieved from https://www.ncei.noaa.gov/access/monitoring/monthly-report/global/202313 September 2024.

National Oceanic and Atmospheric Administration (NOAA). Undated. *Layers of the Atmosphere*. Retrieved from https://www.noaa.gov/jetstream/atmosphere/layers-of-atmosphere#:~:text=called%20 the%20stratopause.-,Stratosphere,but%20very%20little%20water%20vapor February 2024.

Nordhaus, W.D., R.N. Stavins and M. Weitzman. 1992. "Lethal Model 2: The Limits to Growth Revisited." *Brookings Papers on Economic Activity* 1992 (2): 1-59.

O'Neill, D.W., A.L. Fanning, W.F. Lamb and J.K. Steinberger. 2018. "A Good Life for all within Planetary Boundaries." *Nature Sustainability* 1: 88-95.

Richardson, D., W. Steffen, W. Lucht, J. Bendsten, S.E. Cornell, J.F. Donges, M. Drüke, I. Fetzer, G. Bala, W. von Bloh et al. 2023. "Earth beyond Six of Nine Planetary Boundaries." *Scientific Advances* 9: eadh2458.

Rockström, J., W. Steffen, K. Noone, Å. Persson, F.S. Chapin III, E.F. Lambin, T.M. Lenton, M. Scheffer, C. Folke, H.J. Schellnhuber et al. 2009. "A Safe Operating Space for Humanity." *Nature* 461 (24): 472-475.

Simon, J. 1980. "Resources, Population, Environment: An Oversupply of False Bad News." *Science* 208 (4451): 1431-1437.

Steffen, W., K. Richardson, J. Rockström, S.E. Cornell, I. Fetzer, E.M. Bennett, R. Biggs, S.R. Carpenter, W. de Vries, C.A. de Wit et al. 2015. "Planetary Boundaries: Guiding Human Development on a Changing Planet." *Science* 347 (6223): 1259855.

Turner, G. 2008. "A Comparison of *The Limits to Growth* with 30 Years of Reality." *Global Environmental Change* 18 (3): 397-411.

United Nations. Undated. *The 17 Goals*. Department of Economic and Social Affairs: Sustainable Development. Retrieved from https://sdgs.un.org/goals February 2024.

World Bank. Undated. *Poverty Overview*. Retrieved from https://www.worldbank.org/en/topic/poverty/ overview February 2024.

2 Climate Science and International Climate Agreements

What You Will Learn in this Chapter

- The fundamentals of climate science and global warming
- Which gases are greenhouse gases
- How much the average surface temperature of the Earth has increased
- Answer to the question "Is the current period of warming unusual?"
- How CO_2 concentrations in the atmosphere have changed over time
- The countries that are responsible for climate change
- What a safe level of greenhouse gas emissions is
- About the main international climate change agreements

2.1 Recap and Introduction to the Chapter

Chapter 1 offered an orientation to the ecosystem services view of natural resource management and discussed the key human-caused threats to the environment. The chapter tried to make clear that, in harmony with the 2005 Millennium Ecosystem Assessment and the ecosystem services approach more generally, the book mainly utilizes a human-focused perspective on the human–environment interface. The Sustainable Development Goals, which are guiding the United Nations sustainable economic development agenda through 2030, also adopt a people-centered slant. A key reason for adopting such an approach is practicality. A more general analytical framework that puts the natural world at the center can be very challenging to work with and concentrating on people offers the opportunity to focus in on human incentives for effective natural resource management.

There are a host of environmental problems created by people and the literature on planetary boundaries calls out some of the most important ones. These issues include conversion of natural land to other uses, overloading the nitrogen capacity of the planet, increasing extinctions by at least ten orders of magnitude compared with historical levels and changing the climate of the planet. Most notably, human actions have meant that on average the planet has become significantly warmer. As of this writing, the last ten years were the warmest recorded since 1850.

DOI: 10.4324/9781003308225-3

Though the other Earth process threats discussed by planetary boundaries authors are no doubt equally worthy of in-depth discussion, we use this chapter to dial in on the singular challenge of global climate change. We start this investigation with some important climate change fundamentals and discuss the best contemporary estimates of planetary warming, the greenhouse gases causing human-caused climate change and which countries are most contributing/have contributed to this global problem. The international community has made some important agreements to curb climate change dating back to the early 1990s. The chapter therefore concludes by discussing the Framework Convention on Climate Change and its subsidiary agreements, the Kyoto Protocol of 1997 and the Paris Agreement of 2015.

2.2 Brief Overview of Climate Science and Global Warming

Though I am not by any means a climate scientist, because of the importance of climate change – perhaps more evocatively known as global warming – we all need to know a bit about climate science and climate epochs (i.e., climate periods). Weather and climate are often confused, because they are related, but differ in timeframe. Climates, which vary across the globe, are just the average of weather, including temperature, precipitation and other weather variables, over a long timeframe.

The US Government National Oceanic and Atmospheric Administration (NOAA), which is the US agency that is responsible for, among other things, climate monitoring, uses a 30-year climate timeframe. "Normal" weather for NOAA therefore typically means the 30-year average, but it also uses longer timeframes to assess whether climate is changing. Of critical importance is that climates are highly predictable, but weather is not. Climate change is therefore not best thought of as having too much or too little hot weather or rain at any point in time. It is about the development of a new normal weather, because something has fundamentally changed for the planet.

A critical climate metric is the temperature measured at the Earth's surface. Of course, surface temperature varies across the planet (e.g., far north and south are colder than near the equator) and depends on whether we measure on land or the ocean, with land typically having higher temperatures. The average surface temperature is the statistic that is most often quoted.

NOAA, which is one of the most reputable sources of climate information in the world, annually publishes an online global climate report that presents the global average surface temperature, which as of 2023 was 15.08 degrees Celsius (59.12 degrees Fahrenheit) (NOAA, 2023). Perhaps not surprisingly, in discussions of climate *change*, what typically gets discussed is not the temperature level, but the *change* in temperature compared with a baseline estimate.

NOAA uses the average of the 20th century as that baseline and calls any difference from the 20th-century average the temperature "anomaly," signaling that this difference is unusual and not as it should be. As of 2023, that temperature anomaly was 1.18°C (2.12°F), making it the warmest year on record. September 2023 saw the biggest global temperature anomaly ever recorded; compared with the average of all the 20th-century Septembers, the 2023 version was 1.44°C (2.59°F) warmer (NOAA, 2023).

Box 2.1 Earth's Oceans Much Warmer than Normal and Scientists Don't Know Why

Since early 2023, the world's oceans have been substantially warmer than usual and scientists at least for now do not know the reason. This situation has a number of consequences.

1. The sea ice near Antarctica is not growing as usual.
2. Warmer temperatures in the Atlantic Ocean could mean a strong and long hurricane season.
3. Arctic sea ice may not form as expected, potentially slowing down ocean circulation, which affects the climate.

As Brian McNoldy of the University of Miami put it: "The North Atlantic has been record-breakingly warm for almost a year. It's just astonishing. Like, it doesn't seem real."

Source: Gelles (2024).

The oceans tend to stabilize the surface temperature, because they absorb heat and warm relatively slowly, so NOAA distinguishes land and ocean temperatures. Though the land was 1.79°C (3.22°F) above the 20th-century average, the ocean was only 0.91°C (1.64°F) warmer. Global warming varies not only across land versus ocean and by month, but also by hemisphere. The northern hemisphere has warmed basically double that of the southern hemisphere and the Arctic region in the far north of the planet has a 2023 temperature anomaly more than twice that of the globe as a whole (2.55°C and 4.59°F). The Antarctic region at the far south of the Earth has seen very little warming, with a temperature anomaly of only 0.15°C (0.27°F). This information and much more granular detail are available from NOAA (2023).

A second important source of climate-related information is the *Inter-Governmental Panel on Climate Change* (IPCC), which is the scientific arm of the 1992 United Nations Framework Convention on Climate Change (UN FCCC), which we will discuss very soon. The IPCC publishes periodic assessment reports, including a summary for policymakers, the sixth of which was released in 2023. The IPCC brings together international climate experts to assess the climate literature and synthesize that huge body of information.

Unlike NOAA, the IPCC uses the 1850-1900 period as a baseline.[1] Using this comparator, IPCC (2023) finds that global average surface temperature between 2011 and 2020 was 1.1°C higher than in 1850-1900 and basically all of this increase was due to human activities. The authors further note that the Earth's average temperature increased "faster since 1970 than in any other 50-year period over at least the last 2000 years."

What has been the Earth's long-run historical climate, which is often called the paleoclimate? Let us start with the geologically recent period going back about 20,000 years. This was right at the end of the last glacial maximum, when ice sheet and glacier coverage on the Earth last peaked. There then began a period of climate instability and over the next roughly 8500 years the temperature rose and fell significantly.

The northern island of Greenland is of keen interest for tracking the temperature of the paleoclimate, because ice has been present in Greenland for many thousands of years and that same ice remains today. Scientists take ice cores, analyze the gases from those cores and draw inferences about, among other things, temperatures.

Scientists know from this ice core evidence that around 14,000 years ago the average temperature rose by about 30°F and then fell by an even larger amount, reaching a trough about 11,500 years ago. At the end of a cold period called the Younger Dryas, there was a quite abrupt shift and average temperature rose dramatically, including an increase of approximately 10°C in just one decade (Finchon, 2013; Alley, 2014; Alley, 2000)! Since then, the average temperature has been remarkably constant.

The climatic shift that started almost 12,000 years ago began a period called the *Holocene*, which has been extraordinarily stable and extends until today. Compared with the previous millennia humans have experienced, the Holocene has been warmer, wetter and, in much of the world, especially suitable for a variety of human activities, including agriculture.

Most humans roamed and foraged for their livings in small bands until roughly the start of the Holocene, when some people began settling down. They first foraged in one place and in some regions later developed sedentary agriculture, which started around 10,000 years ago (Harari, 2015, pp. 77-83). Though the stable climate may not fully explain the transition from mobile lifestyles to permanent settlements, it was likely an important contributor (Dow and Reed, 2015). Though Harari (2015) and perhaps others argue that sedentary agriculture was a raw deal for people, the transition to permanent human settlement resulted in technical change and investments in more complex social structures that eventually led to the enormous human progress of the last ten millennia (Dow and Reed, 2015).

Over the last 14,000 years, the climate has varied, but the Earth has been habitable. Figure 2.1 suggests that over the last 500 million years the planet has often been much, much hotter than it is today and likely not at all habitable by humans. As shown in Fig. 2.1, when the average temperature hits 18.33-21.1°C (65-70°F), the Earth can be expected to eventually completely lose its polar ice caps (Michon and Lindsey, 2023 based on Wing, undated).

The 2023 average temperature was 15.08°C (59.12°F), meaning that if the average surface temperature were to rise by 3.5°C, the world could be on track to be ice-free. This paleoclimatic situation would not be unusual, as there have been many periods over the long history of the Earth when the average surface temperature was above 21.1°C (70°F) or even 27°C (80°F).

2.3 Greenhouse Gas Emissions and Concentrations in the Atmosphere

Greenhouse gases (GHGs) emitted by humans have caused basically all the contemporary global warming observed since the late 1800s/early 1900s and this conclusion is beyond doubt (Stips et al., 2016; IPCC, 2023). There are three main GHGs – carbon dioxide (CO_2), nitrous oxide (N_2O) and methane (CH_4), plus certain synthetic chemicals that make up a minority of global GHGs, and water vapor, which is a naturally occurring gas. These gases throw off the balance between the solar radiation coming into the atmosphere from the sun

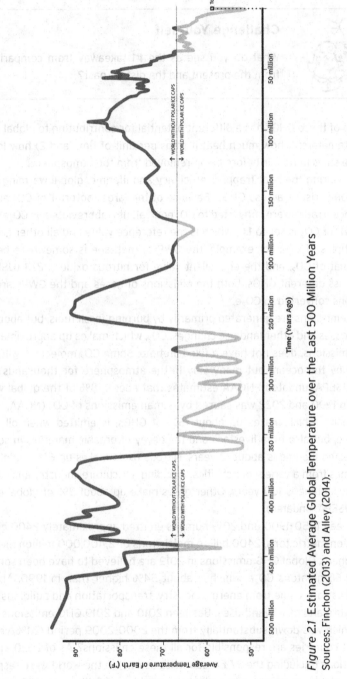

Figure 2.1 Estimated Average Global Temperature over the Last 500 Million Years

Sources: Finchon (2013) and Alley (2014).

and re-radiation back out into space, because they absorb the radiation, trapping heat in the air.

Challenge Yourself

What do you see as the #1 takeaway from comparing information from the present and the distant past?

Each of these GHGs has a different potential for contributing to global warming based on two parameters: 1) how much heat it traps per unit of time and 2) how long a molecule on average stays in the air before being removed from the atmosphere.

Considering the heat trapping efficiency and lifetime, global warming potentials (GWPs) allow comparisons across GHGs. Because of the large footprint of CO_2 among GHGs, other gas concentrations are converted to CO_2 equivalents (abbreviated as CO_2e) using their GWPs. The GWP of CO_2 is set to 1.0, which is the reference value and all other gases are expressed as multiples of CO_2. For example, the GWP of methane is somewhere between 27 and 30 times that of CO_2, and the equivalent value for nitrous oxide is 273 (USEPA, 2024). When we discuss different GHGs, both the emissions of gases and the GWPs are considered, with emissions converted to CO_2e.

CO_2 emissions are generated primarily by burning fossil fuels, but about 11% comes from forest losses and other land use changes. CO_2, which makes up approximately 76% of global CO_2e emissions, does not have a fixed lifetime. Some CO_2 molecules will be absorbed very quickly by the ocean, but many stay in the atmosphere for thousands of years (USEPA, 2024; USEPA, undated). NOAA estimates that about 78% of the global warming observed between 1990 and 2022 was caused by human emissions of CO_2 (NOAA, undated).

Methane, which makes up about 16% of GHGs, is emitted when oil, gas and coal are produced, but also from livestock and the decay of organic materials in solid waste landfills. Its expected lifetime is about 12 years. Nitrous oxide makes up 6% of global GHG emissions and comes from a variety of activities, including agriculture, industry and combustion of fossil fuels. Its lifetime is 109 years. Other gases make up about 2% of global emissions (USEPA, 2024; USEPA, undated).

Between 1850–1900 and 2019 humans emitted approximately 2400 gigatons CO_2e (i.e., 2.4 trillion metric tons, 2400 billion metric tons or 2,400,000 million metric tons) into the atmosphere. Global GHG emissions in 2019 are believed to have been somewhere between 54 and 65 gigatons CO_2e, which is about 54% higher than in 1990. About 80% of total 2019 emissions came from energy, industry, transportation and buildings and the rest from agriculture and other land uses. Between 2010 and 2019 GHG emissions grew by 1.3% per year, which was down substantially from the 2000–2009 period (2.1%/year) (IPCC, 2023).

What countries are responsible for all these emissions? As of 2020, the top 10 emitting jurisdictions (including the 27 EU countries as one) in the world were responsible for about 65% of global emissions; the rest of the world (or ROW) - over 150 countries - made up

Table 2.1 2020 Greenhouse Gas Emissions, Population, GDP and Emissions per Person of the Top Ten Emitters and the Rest of the World

Country/Region	Emissions (Millions of Metric Tons CO_2e)[a,b]	% Global GHG Emissions	Population (Millions)	Emissions per Person (CO_2e Tons/Person)	Total Value of Production (GDP in Millions of $US)	$GDP/ Ton GHG Emissions
World	47,513[a]	100%	7820	6.08	$85,257,737	$1801
China	12,296	25.88%	1411	8.71	$14,687,744	$1195
US	5289	11.13%	332	15.95	$21,322,950	$4032
India	3167	6.67%	1396	2.27	$2,674,852	$845
European Union (27)	2958	6.23%	448	6.61	$15,381,419	$5200
Russia	1800	3.79%	144	12.49	$1,493,076	$829
Indonesia	1476	3.11%	272	5.43	$1,059,055	$718
Brazil	1470	3.09%	213	6.90	$1,476,107	$1004
Japan	1063	2.24%	126	8.42	$5,055,587	$4756
Iran	845	1.78%	87	9.68	$239,735	$284
Canada	732	1.54%	38	19.26	$1,655,685	$2262
ROW[c]	16,417	34.56%	3353	4.90	$20,531,508	$1251

Sources: ClimateWatch and World Resources Institute (2021) for GHG information. World Bank (undated) for GDP and population information.

a The abbreviation for million metric tons is Mt, which is the abbreviation used in the remainder of this book.
b This estimate is a bit less than the IPCC (2023) estimate.
c ROW = Rest of the World

only 35%. The emissions of the highest-emitting jurisdictions plus the rest of the world are presented in Table 2.1.

We see from Table 2.1, which circa 2024 appear to be among the most authoritative published data available, that China was responsible for 12.3 billion tons or about 26% of global emissions, followed by the US at 11% and India at 7%. The European Union, which as of early in 2020 had 27 countries, emitted 6% of the global total.

In addition to total emissions by country, Table 2.1 also presents emissions per person in the world and by country, as well as emissions per $US of *Gross Domestic Product* (GDP). We will discuss GDP in detail in Chapter 5, which focuses on consumption as an explanation for humans' effects on the environment. For now, let us therefore just mention that GDP is a measure of production within the borders of a country.

For the atmosphere, average surface temperature and the climate, only global GHG emissions matter, because carbon pollution is *uniformly mixed*, which means the location of GHG emissions does not affect the damages they cause. But location certainly matters for assigning *responsibility* for global warming.

"Blame" for destabilizing the climate is closely related to measures of appropriate shares of past and future global carbon budgets. A complication is that the economies of countries differ and have different populations. One possibility for assigning responsibility and estimating GHG headroom is to adjust for the population of each country by calculating the emissions per person, with low emissions (e.g., emissions below the global average) implying room to increase. Being above the global average could be interpreted as over-using the atmosphere for carbon pollution.

By the measure of emissions/person, Canada, the US and Russia look terrible, with emissions per person two to four times the 2020 global average of about six tons per person per year. This per person approach is used for some of the sustainability measures discussed in Chapter 7. The argument could also be made that, because most CO_2e comes from burning fossil fuels that are used to produce goods and services for people, the goal should be to be *very efficient* in our production methods. If a country produces a lot of value per ton of GHG emissions, perhaps that lets them off the hook for high total emissions or high emissions per person? After all, these countries are getting more of something good (production) for every ton of something bad (CO_2e) they emit into the air. By this production efficiency measure, the European Union is at the top of the list, producing over $5,000 in value for every ton of CO_2e emitted, followed by Japan, the US, Canada and China, which are all above the world average.

Please take a moment to think about the stress different countries put on the climate. Do you think *adding* to GHG concentrations in a given year can be justified on any measure? What is an appropriate method of assigning responsibility when GHG emissions increase?

Table 2.2 presents information on total CO_2e emissions in Mt and as a percentage of global emissions in 1990, which is 30 years before the information in Table 2.1. Let us compare these data with those from 2020.

What differences do you see? How have total emissions and the distribution of emissions across countries evolved over time? Please think about these questions before reading on.

First, comparing Tables 2.1 and 2.2 we note that global emissions increased dramatically over the 30 years. In 1990, emissions were 33 Gt, but by 2020 the total was 48 Gt, an increase of 46% compared with the 1990 level.[2] The shares of China and India both increased dramatically, while the shares of most other top emitters declined.

Table 2.2 1990 Greenhouse Gas Emissions of the 2020 Top 10 Emitters and the Rest of the World

Country/Region	Emissions (Millions of Metric Tons CO_2e)	% Global Emissions
World	32,661	100%
China	2892	8.85%
US	5447	16.68%
India	1020	3.12%
European Union (27)	4270	13.07%
Russia	2624	8.03%
Indonesia	1141	3.49%
Brazil	1643	5.03%
Japan	1106	3.39%
Iran	300	0.92%
Canada	732	2.24%
ROW	11,486	35.18%

Source: ClimateWatch and World Resources Institute (2021).

Challenge Yourself

In your view, based on the data which countries should be responsible in international agreements for addressing global climate change?

The declines in the *total emissions* of the European Union and its world share are especially notable in the data, but this observation likely has something to do with the dissolution of the Soviet Union, which collapsed in 1991. Post-Soviet countries all had major GHG declines since 1990, and most of these countries have since joined the EU. We will discuss this topic in more detail in Chapter 12 when we analyze energy subsidies.

Most of the other 2020 top 10 emitters (except Iran and Brazil) also reduced emissions at least somewhat over the 30-year period. These declines happened even as emissions from China and, to a lesser extent, India, absolutely exploded. The 150+ countries outside the top 10 emitters also increased their emissions very significantly, but their percentage of world emissions stayed basically constant while China and India increased their total emissions three- to fourfold over 30 years.

The data we have discussed so far focus on emissions in 2020 and 1990 (i.e., single-year data), but GHG emissions started in the late 1700s/early 1800s and the global warming situation we find ourselves in has built up over time. From a *cumulative* GHG emissions standpoint, who is responsible?

Figure 2.2 presents estimates of cumulative CO_2-only emissions, excluding emissions from land use changes, from 1750 to 2021. You will recall that CO_2 is responsible for almost 80% of global warming and almost 90% of CO_2 emissions are from burning fossil fuels. These data suggest that the responsibility picture is not all that different from those from the single-year emissions data, albeit with a bit of reordering of the countries. Based on cumulative emissions, the US is responsible for about 1/4 of the warming due to CO_2 emissions that the Earth has seen to date, followed by the European Union (16%), China (14%), Russia (7%) and Japan (4%). The US, the EU and Japan together make up about 45% of historical emissions.

As Popovich and Plumer (2021) point out, though, another way to cut these data is in terms of richer versus poorer countries; they include India, Russia and China in the Global South. The authors note that 23 countries in the Global North are responsible for half of all the CO_2 that has been emitted since industrialization started and the other countries account for the other half.

Since 1958, direct CO_2 (i.e., not CO_2e) atmospheric concentration measurements have been made at a NOAA observatory located near the Mauna Loa Volcano in the US state of Hawaii. As shown in Fig. 2.3, in 1958, which was the first year CO_2 was directly measured, CO_2 averaged about 315 parts per million (ppm). By December 2023, the average was 422 ppm, which is a change of about 34%. Note how small are the concentrations – out of every *1 million* cubic meters in the atmosphere, currently 422 of them are CO_2. No wonder people never imagined we could affect the climate! It turns out, though, that small concentrations have large warming impacts. Who knew?

Cumulative CO₂ emissions (tons)

Cumulative emissions are the running sum of CO₂ emissions produced from fossil fuels and industry¹ since 1750. Land use change is not included.

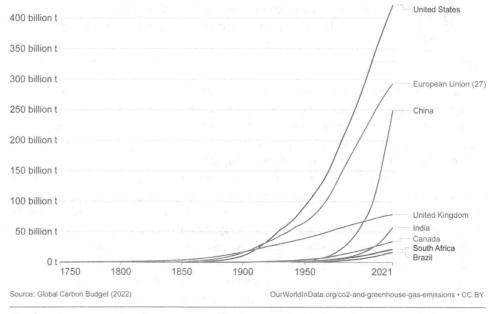

Source: Global Carbon Budget (2022)   OurWorldInData.org/co2-and-greenhouse-gas-emissions • CC BY

1. Fossil emissions: Fossil emissions measure the quantity of carbon dioxide (CO₂) emitted from the burning of fossil fuels, and directly from industrial processes such as cement and steel production. Fossil CO₂ includes emissions from coal, oil, gas, flaring, cement, steel, and other industrial processes. Fossil emissions do not include land use change, deforestation, soils, or vegetation.

Figure 2.2 Estimated Cumulative Emissions of CO₂ 1750-2021 for Major Historical Emitters
Source: Michon (2023) (from Our World in Data).

Figure 2.4 focuses on the last four years. As we see from the figure, in the recent past CO$_2$ concentrations have continued to increase and based on Fig. 2.3, have, if anything, increased at an increasing rate during the past decade. This observation is confirmed by Lindsey (2023) and it turns out that only in the last 11 years have average CO$_2$ concentrations increased by more than 2 ppm per year. It is also interesting to note that CO$_2$ concentrations bounce around a bit during the year and are not just one number. Indeed, in the Hawaiian summer carbon concentrations are highest and in the winter they are lowest.

How unusual are such CO$_2$ concentrations? The answer, as you may have already guessed, is *VERY* unusual indeed. IPCC (2023) concludes with high confidence that the 2019 atmospheric concentrations of CO$_2$, which at that time averaged 410 ppm (now higher), "**were higher than at any time in at least 2 million years**" (emphasis added). With very high confidence the authors also conclude that the 2019 methane and nitrous oxide concentrations had not been seen in at least 800,000 years.

The estimated CO$_2$ concentrations in the atmosphere as read from the paleoclimate record over the past 800,000 years are shown in Fig. 2.5. Based on ice core evidence, current CO$_2$ concentrations outstrip the past maximum (which occurred over 300,000 years ago) by about 41% – a huge difference.

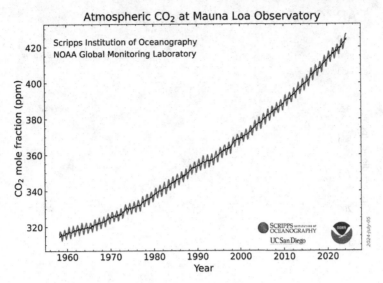

Figure 2.3 Atmospheric Concentrations of CO_2 1958-2023
Source: NOAA Global Monitoring Laboratory (2023).

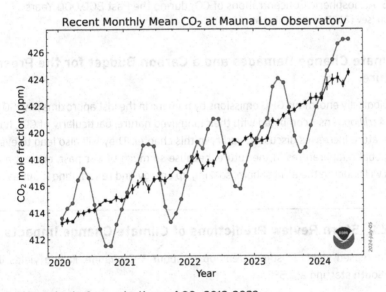

Figure 2.4 Atmospheric Concentrations of CO_2 2019-2023
Source: NOAA Global Monitoring Laboratory (2023).

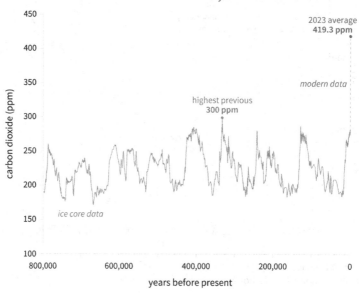

Figure 2.5 Atmospheric Concentrations of CO_2 during the Past 800,000 Years
Sources: Lindsey (2023) and Lüthi et al. (2008).

2.4 Climate Change Damages and a Carbon Budget for the Present and Future

These geologically enormous GHG emissions by humans in the last approximately 150 years – a total of 2.4 *trillion* tons – combined with their long-lived nature, particularly of CO_2, have led to the temperature increases discussed earlier in this chapter. They will also lead to even higher average global temperatures in the future, because so much of our past, current and future emissions will linger in the atmosphere, soaking up energy and re-radiating it out as heat.

Box 2.2 Stern Review Predictions of Climate Change Impacts

Food: Increased yields in some areas up to about 3°C change. Falling yields in the Global South starting at 1.5°C.

Water: Small glaciers disappearing now. Significant decrease in water availability after 2°C increase. Coastal cities threatened by sea level rise at 4°C.

Ecosystems: Coral reef damage now. Rising number of species face extinction after 2°C increase.

Extreme Weather: Rising frequency and intensity of droughts, fires, storms, flooding and heat waves at 2°C and particularly after 3°C of warming.

Risk of Abrupt, Large Climate Shifts: Especially after 4°C of warming.

Source: Stern (2007).

But how bad will the future climate be? Not surprisingly, it all depends on by how much and how quickly we reduce GHG emissions.[3] Lots of research, some of which was included in IPCC (2023), has tried to predict future temperature increases based on our GHG emissions, but most estimates only project out until the year 2100.

There are a variety of scenarios that are mainly based on how much we reduce emissions, particularly in the near term. Some scenarios presented in IPCC (2023) in which we quickly reduce global emissions predict that the global average temperature increase compared with the 1850–1900 average will exceed 1.5°C (an important target from an international policy perspective, as we will soon see) and then drop back below that threshold. IPCC (2023) estimates that during the 2081–2100 period, depending on global emissions, the temperature increase could be anywhere from 1.4°C (2.52°F) (very low GHG scenario) to 4.4°C (7.92°F) (very high GHG scenario).

The emissions that will keep global average surface temperature within limits can actually be expressed as a "carbon budget." As Michon (2023) notes, "The world can only emit a certain amount of carbon dioxide if we want a reasonable chance of limiting global warming to no more than [for example] 2°C above pre-industrial temperatures."

Figure 2.6 presents carbon budgets for keeping the increase to 1.5°C, 1.7°C and 2°C – considering emissions through 2022. To have a 50:50 shot at keeping the average surface temperature increase to 1.5°C, humans need to limit CO_2e emissions to about 380 (giga) billion tons. This amount is about 8 years' worth of emissions at the 2020 level (see Table 2.1) of about 47 gigatons. Humans seem to have burned through too many fossil fuels to make that a feasible goal.

The 1.7°C and 2°C goals may be more feasible, because we have maybe 12–25 years to stop all CO_2e emissions and achieve what is typically called *carbon neutrality*. Unfortunately, as discussed below and in Box 2.2, an increase of 2°C is not a safe increase for the world and the stated international goal is the seemingly infeasible 1.5°C increase.

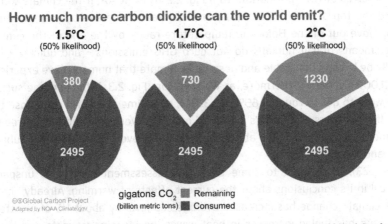

Figure 2.6 Estimated Global Carbon Budgets with Various Increases in Global Surface Temperature

Source: Michon (2023).

How to allocate these CO_2e emissions across countries? This topic is quite contentious, because of differing historical emissions, populations and economies. One possibility that is sometimes mentioned is to allocate the remaining carbon budget based on an equal share by population. Using this metric, the 1.7°C temperature increase limit would mean, ignoring population increases, the approximately 8 billion people on Earth would each get about 91 tons of CO_2.

Challenge Yourself

The carbon budget to keep the climate within safe limits is very tight. Taking into account everything you know about the situation, how would you suggest this carbon be allocated?

As shown in Table 2.1, that allocation would last the US about six years at its 2020 emissions rate, but at the world average emissions rate of six tons per person per year (roughly the European Union per person emissions), it would last two to three times longer. Even at the world average rate of emissions the carbon budget is very tight.

If instead the carbon budget were distributed based on per person *historical* emissions, even with a goal of staying below 2°C, the US has already emitted about four times its allocation based on population. Russia has "over-spent" its per person carbon budget by a factor of three, the European Union is about 40% over budget and even Brazil is modestly in the red. China, by contrast, has lots of headroom and can more than double its emissions and stay within its CO_2e budget limit. By this historical emissions metric, India, the most populous country in the world, could increase its emissions by almost seven-fold and still do its part to limit climate change (Michon, 2023).

What are the implications of the different future warming possibilities? Let us first remind ourselves of the information from Fig. 2.2 from NOAA. If the climate were to end up toward the top of the IPCC (2023) possibility range, the average temperature could very well move out of the Holocene temperature range by the end of the century. This possibility means that if humans do not curb GHG emissions – and quickly – we could potentially be leaving the stable and hospitable climate that humans have experienced for the last 11,000+ years. Furthermore, as we saw from Fig. 2.3, if the average surface temperature were to get to roughly 65°F (18.33°C), over time the world could lose the polar ice caps. If this occurs, the Earth's ice would melt, the world will end up ice-free until the planet processes the GHGs in the atmosphere, and sea levels would rise very substantially, flooding land areas.

IPCC (2023), in contrast to some past IPCC assessment reports, is unsparing and unequivocal in its conclusions about the current effects of warming. Already, the authors conclude, climate change has increased average sea level by about 1/5 of a meter (about 8 inches) and has fueled increases in heat waves, heavy rain, droughts and hurricanes/cyclones/typhoons.[4] As many as 3.5 billion people are highly vulnerable to the effects of climate change due to extreme weather events and people in lower-income settings have the least adaptive capacity.

Climate change has slowed growth in agricultural productivity and affected water availability in many regions. Global warming has also increased human mortality and illness, including mental health, due to increasing temperatures and trauma from extreme weather events. With high confidence, IPCC (2023) notes that climate change is right now driving displacement of people in Africa, Asia and North America and causing economic damages in especially environment-dependent sectors, such as fishing, forestry and tourism (IPCC, 2023).

If and when the planet ends up on average 3°C (4.65°F) above the 1850-1900 baseline, the world will look very different. According to IPCC (2023), in many areas of the world the hottest day of the year will be 4-5°C (6-8°F) more than in 2020.

Before reading on, imagine for a moment what it would feel like in your part of the world to add 4-5°C to your current maximum temperature.

In Portland, Oregon, where I live, maximum summer temperatures can reach 105°F (41°C). The typical maximum will therefore move to perhaps 113°F (46°C) if the Earth plateaus at 3°C (4.65°F) above the 1850-1900 baseline. Section B. 2.4 of IPCC (2023) perhaps says it all:

> With further warming, climate change risks will become increasingly complex and more difficult to manage. Multiple climatic and non-climatic risk drivers will interact, resulting in compounding overall risk and risks cascading across sectors and regions. Climate-driven food insecurity and supply instability, for example, are projected to increase with increasing global warming, interacting with non-climatic risk drivers such as competition for land between urban expansion and food production, pandemics and conflict.
>
> (Section B. 2.4)

There is much more to digest in IPCC (2023) and I would like to invite you to read the document, which is quite accessible, as it is a summary for policymakers. With all this bad news, record GHG emissions, high temperatures and environmental effects, what has the international community been doing to try to stave off the worst of these potential outcomes? We now turn our attention to that topic.

2.5 International Climate Policy

The United Nations Framework Convention on Climate Change (UN FCCC) is the foundational international agreement that brings countries together to address climate change. The goal of the UN FCCC, as stated in Article 2, is to stabilize GHG concentrations "... at a level that would prevent dangerous anthropogenic (human induced) interference with the climate system." It furthermore states that "such a level should be achieved within a timeframe sufficient to allow ecosystems to adapt naturally to climate change, to ensure that food production is not threatened, and to enable economic development to proceed in a sustainable manner."[5]

The UN FCCC was agreed in Rio de Janeiro, Brazil as part of the so-called Earth Summit held in 1992 and countries could also sign the Convention at that time. Other environmental conventions agreed in Rio include the UN Convention on Biological Diversity and the Convention to Combat Desertification. After the Earth Summit, countries could sign the Convention at the United Nations headquarters in New York and by June 1993 the Convention had been signed by 166 countries. There are currently 198 members, which are called "parties," to the UN FCCC, basically all the countries in the world.

The Convention, which entered into force in March 1994 after 50 parties ratified or otherwise approved it, is a slim 25 pages long and therefore only sketches out how the international community might achieve its main goal. A few items are, however, very clear. First, the UN FCCC places responsibility squarely on 25 high-income countries that at the time were all members of the Organisation for Economic Co-operation and Development (OECD) and these countries were listed in Annex 1 to the UN FCCC. The countries were allocated primary responsibility for addressing global warming, because they had emitted the bulk of the historical emissions as of 1992.[6]

Over time, other countries chose to join Annex 1, taking on the associated responsibilities. Currently, a total of 43 parties are listed in Annex 1. The earliest Annex 1 opt-ins were the Eastern European countries who were in the process of looking to the West after the collapse of the Soviet Union in August 1991.

Challenge Yourself

With 20-20 hindsight, was it a good idea for the UN FCCC to make only Annex 1 countries responsible for addressing global warming?

The UN FCCC has a number of elements that remain highly relevant today. These include a requirement that the parties to the convention must develop and publish periodic national GHG inventories and Annex 1 countries must adopt national policies to mitigate climate change. There is an Annex 2 that includes a list of higher-income countries that are responsible for providing finance and technical assistance to lower-income countries – what the Convention calls "developing" countries. Financial support for the Global South has therefore been a feature of international climate policy from the very beginning.

The UN FCCC established what is called a *secretariat* to implement the Convention and a subsidiary scientific body, which became the IPCC. The IPCC existed before the Convention, but was soon brought under the UN FCCC to conduct needed scientific studies. In the previous section, we reviewed some of the results from IPCC (2023).

The Convention also established the Conference of the Parties (COP), which is the group representing all the parties, as the supreme authority over the UN FCCC. The COP can make any agreements necessary to fulfill the objective of the Convention, including subsidiary agreements, such as protocols. Since 1992, the COP has met periodically and in recent years met annually, usually in December. The 2023 COP (COP 28) was held in Dubai in the United Arab Emirates.

Particularly important COPs took place in 1997 in Kyoto, Japan, where the Kyoto Protocol was signed, and in 2015 in Paris, when the Paris Agreement was made. The Kyoto Protocol was adopted on December 11, 1997, but went into force only in 2005. As of 2024, there were 192 parties to the Protocol. It was an important step toward making the UN FCCC operational in the sense that countries with obligations knew what they needed to do and those without requirements understood the opportunities to opt in.[7]

As the Kyoto Protocol is an agreement that is subsidiary to the UN FCCC, it is not surprising that the burden to "adopt policies and measures on mitigation and to report periodically" is on higher-income countries and mainly on those listed in UN FCCC Annex 1. Annex B to the Protocol lists the 37 countries that accepted quantitative emissions reduction targets, which collectively added up to an approximately 5% reduction in CO_2e (over six GHGs) during the 2008-2012 period. As is the case for the UN FCCC, the baseline for comparison is 1990.

For example, the European Union, which in 1997 was made up of 15 countries (now 27 after the 2020 withdrawal of the UK), plus a number of countries that have since become members, committed to an 8% reduction. Hungary, Japan and Poland agreed to 6% reductions compared with 1990 and several countries either agreed to cap emissions at 1990 levels (e.g., New Zealand and Russia) or proposed increases (e.g., Norway, Australia and Iceland).

Box 2.3 Kyoto Protocol Flexible Market Mechanisms

The Kyoto Protocol allows three market-based mechanisms:

1. **Joint Implementation (JI)** – agreements between UN FCCC Annex 1 countries where entities from one country reduce GHG emissions and another country uses those reductions to meet its commitments.
2. **Clean Development Mechanism (CDM)** – similar to Joint Implementation, except between Annex 1 and non-Annex 1 countries.
3. **International Emissions Trading** – Kyoto Protocol Annex B countries that overachieve their emissions targets can sell the excess capacity to other Annex B countries.

Source: UN FCCC (undated).

Following directly from the UN FCCC, the Kyoto Protocol offered countries substantial flexibility in achieving its commitments. The motivation for giving flexibility was to allow countries to be more ambitious in their commitments and provide ways to reduce GHG emissions at lower cost. This goal of so-called *cost-effectiveness* is a major theme in the economics of pollution control and a topic we will investigate in detail in Section III of this book, which is titled **Solving Environmental Problems around the World**.

The rationale for allowing the three flexible *market mechanisms* listed in Box 2.3 is the following:

> These mechanisms ideally encourage GHG abatement to start where it is most cost-effective, for example, in the developing world. *It does not matter where emissions are reduced, as long as they are removed from the atmosphere.* This has the parallel benefits of stimulating green investment in developing countries and including the private sector in this endeavor to cut and hold steady GHG emissions at a safe level. It also makes leap-frogging – that is, the possibility of skipping the use of older, dirtier technology for newer, cleaner infrastructure and systems, with obvious longer-term benefits – more economical.
>
> (UN FCCC, undated a; emphasis added)

The US and Canada were both parties to the Kyoto Protocol, with the US agreeing to a 7% reduction and Canada taking on a 6% target. The US requires that any protocols be ratified by the Senate, which is the upper house of the legislature. The Senate did not ratify the Kyoto Protocol, so the US never formally took on its 7% Kyoto target. Canada exited the agreement in December 2012.

Challenge Yourself

In 1997, the US was the biggest GHG emitter in the world. It never ratified the Kyoto Protocol and therefore never fully took on its target. Why do you think the US did not ratify the Kyoto Protocol?

Though the Kyoto Protocol is recognized for establishing flexible mechanisms to address climate change (e.g., see Böhringer, 2003), there are many who view the Protocol as a failure (e.g., see Helm, 2012; Rosen, 2015). I will leave it to you to look at these and other documents and we will circle back to the Kyoto Protocol in Chapter 18, when we discuss flexible market mechanisms. At this stage, I would note that recent rigorous evidence suggests that the agreement reduced GHGs by approximately 7% compared with what would have otherwise been the case (Maamoun, 2019).

The Kyoto Protocol was amended in Doha, Qatar in 2012, adding a commitment period from 2013 to 2020, but by that point work had already started on an agreement to succeed Kyoto. The Paris Agreement was adopted at COP 21 in December 2015 by 196 Parties to the UN FCCC and went into force in November of the following year. It is the agreement that guides international cooperation on climate change through 2030.

The goals of both the Kyoto Protocol and the Paris Agreement were to operationalize the UN FCCC and limit global warming. The Paris Agreement's goal is to hold "'the increase in the global average temperature to well below 2°C above pre-industrial levels' and pursue efforts 'to limit the temperature increase to 1.5°C above pre-industrial levels'" (UN FCCC, undated b). COP 26 and 27 held in 2022 and 2023 encouraged countries to increase their levels of ambition to try to keep warming below 1.5°C.

Though the goals are similar, the two subsidiary agreements work very differently. Whereas Kyoto limited obligations essentially to UN FCCC Annex 1 countries, Paris includes virtually all countries. Whereas Kyoto asked countries to sign up for quantitative, binding emissions limits, Paris Agreement commitments are voluntary and depend on country-level circumstances. These commitments are expressed in terms of what are called *Nationally Determined Contributions* (NDCs), which can be very different from each other. Some are long, some are brief, some focus mainly on emissions reductions and others talk about what changes they plan to make in their economies that they anticipate will lead to emissions reductions. Some countries say they will take X, Y and Z steps only if financing is provided by other countries. These are referred to as "contingent" commitments.

Please review the NDC registry, which has all countries' most up-to-date commitments and discusses what each country has promised to do to reduce climate change.[8]

Where is the enforcement in such an approach? As noted in Aldy and Stavins (2008),[9] sovereign countries cannot be compelled to act. It is therefore necessary to *encourage* participation in what they call the "post-Kyoto climate regime," while also creating incentives for participants to fulfill their commitments. To help encourage and incentivize, Article 13 of the Paris Agreement specifies what is called the "enhanced transparency framework."

This part of the Agreement utilizes public information to the extent countries are willing and able to provide it, to monitor emissions, track progress toward achieving NDCs, report on support countries give to and receive from other countries and create internal incentives for compliance. The idea here is that all parties know what other parties are doing and will do their best to fulfill their obligations and increase their ambition over time.

In terms of the public-facing reporting that is the basis of the transparency framework, all countries must submit national communications to the UN FCCC every four years. These communications contain a variety of information, including on GHG emissions, climate vulnerability and financing.[10] They must also submit NDC progress reports every two years. GHG inventory requirements are more stringent for Annex 1 countries, which are required to submit separate annual GHG inventory reports, using consistent up-to-date methods that cover all GHGs from 1990 until two years before the submission date.[11]

Challenge Yourself

What do you think of the Paris Agreement? Does the architecture seem up to the challenge?

Non-Annex 1 countries are required to submit inventories every other year and are "encouraged" to use methodologies from 1996 for three gases (CO_2, methane and nitrogen dioxide) "as appropriate, to the extent possible." They are not required to include all years and they report with a four-year lag rather than two years as for Annex 1 countries (Goodwin and Kizzier, 2018). All countries are required to update their NDCs every five years with increased ambition, but in 2021, as it became clear that existing NDCs were insufficient to keep warming below 1.5°C, all countries were asked to "revisit and strengthen the targets in their NDCs in 2022" (United Nations, undated).[12]

Critical parts of the Paris Agreement focus on finance, technology development, technology transfer and capacity building for the Global South. You might recall from Chapter 1 that these are also aspects of the Sustainable Development Goals (SDGs). The Paris Agreement does not specifically mention the SDGs, because the SDGs only went into force in January 2016, but throughout the document there are numerous mentions of sustainable development. Many places in the Paris Agreement also discuss support to lower-income countries and it "...reaffirms that developed countries should take the lead in providing financial assistance to countries that are less endowed and more vulnerable..." (UN FCCC, undated b).

The Paris Agreement is notable, because it brought the human world together to address global warming and it is not clear this would have happened with another climate architecture. According to UNEP (2023), which is the 2023 version of the annual United Nations Environment Programme GHG emissions gap report, the agreement has also brought down CO_2e emissions compared with what would have otherwise been the case.

Unfortunately, though, global CO_2e emissions have continued to increase over time and in 2022 were at a new record high of 57.4 Gt (1.2% above 2021). Even if the NDCs were fully implemented, by 2030 emissions would be 3% *higher* than in 2015. This is even as emissions need to *fall* by 28% between 2015 and 2030 to put the world on track to stay below 2°C of warming by the end of the century (UNEP, 2023).

Based on the current NDCs, the Earth is on track for about 2.9°C of warming by the end of the century, which will put the world's climate well outside the realm of past human experience. UNEP (2023) estimates with 50% confidence that emissions need to be 41Gt CO_2e per year for the world to limit warming to 2°C, but it forecasts that even if the current NDCs were achieved, we would emit about 14 Gt CO_2e over that amount. It is especially disturbing to read that even after achieving the current NDCs, to be on track for 2°C, 24 Gt CO_2e will need to be made up by 2050.[13] This 2050 "emissions gap" increases despite reductions due to the Paris Agreement, because GHGs are stock pollutants that can remain in the atmosphere for hundreds or thousands of years. For this reason, UNEP, NOAA, the IPCC and many other highly credible organizations are urging significant action **now** to avoid the worst effects of climate change.

2.6 Chapter Conclusions

The evidence on global climate change that has been built up since the 1992 Earth Summit is unequivocal. Humans have been changing the climate of the Earth since the late 1700s and this process has particularly speeded up since the 1970s. A critical metric of climate change is the global average surface temperature, which is currently approximately 15°C/59°F and about 1.18°C above the 20th-century average.

The main reason for climate change is emissions of GHGs by people and especially CO_2, which mainly comes from burning fossil fuels. Paleoclimatic data suggest that the world has been in this temperature situation before and has previously seen such high GHG concentrations, but not for several hundred thousand or even millions of years.

Because GHGs stay in the atmosphere so long and therefore build up over time, a lot of future climate change is already baked in. The concern expressed by key scientific organizations, such as NOAA, the IPCC and many others, is that the world could be headed toward warming of 3°C+ compared with pre-industrial times by the end of the century. This situation would move the climate well outside of what humans – and many, many other species – have ever encountered, leading to numerous outcomes we do not want to think about, much less experience. Urgent and immediate action, which takes into account human aspirations, particularly in the Global South, is therefore very important.

The international community, under the auspices of the United Nations, has made some critical agreements to try to curb and eventually reverse climate change. The Paris Agreement, which operationalizes the UN FCCC through 2030, is in the process of reducing GHG emissions compared with what otherwise would have been the case, but current commitments are not enough to keep warming to even 2°C by 2100, which is itself not considered a safe level. Since 2021, the United Nations has therefore been calling on countries to increase their ambition and stay within a safe carbon budget. How to divide up that budget remains a critical and highly divisive issue.

Issues for Discussion

1. What do you see as the #1 takeaway from comparing information on temperatures and CO_2 emissions from the present and the distant past? What can we learn from the paleoclimate record?

2. How evocative do you find the data on climate change? Write down one paragraph that you think would help your friends understand the urgency of stopping global warming.

3. Some potential effects of additional climate change were discussed in this chapter. Do some quick additional research using Google or similar. What are some other climate impacts that can be expected if the world warms 2°C or more above pre-industrial times? What are two climate change impacts that most worry you?

4. In your view, which countries need to reduce their GHG emissions to slow and reverse climate change? What is your rationale? If you live in a country that you suggest should move toward net zero emissions, are you and others willing to make the necessary sacrifices?

5. The UN FCCC made the higher-income countries listed in Annex 1 responsible for addressing global climate change. Do you think that made sense back in 1992? Does it make sense now and do you think it was and is a good idea?

6. What do you think of the Paris Agreement? What do you see as the positive and negative features?

7. The carbon budget to keep the climate within safe limits is very tight. Considering everything you know about the situation, how would you suggest this carbon be allocated?

Notes

1 NOAA and the IPCC may have other methodological differences as well.
2 Just a quick reminder that percentage change, which is sometimes abbreviated as %←, is calculated as (new value - starting value)/starting value. In this case %← = (48 - 33)/33 = 0.46 = 46%.
3 The near term is especially important, because then we have fewer GHG molecules staying in the atmosphere for tens, hundreds and even thousands of years.
4 Names depend on the ocean where these extreme weather events originate.
5 Take a look at the text of the Framework Convention on Climate Change. It is reasonably readable and at least skimming this and other international climate change agreements helps us understand what the Parties were trying to accomplish. The texts of the UN FCCC, Kyoto Protocol and the Paris Agreement can be found at the following URLs: UN FCCC - https://UN FCCC.int/sites/default/files/convention_text_with_annexes_english_for_posting.pdf Kyoto Protocol - https://UN FCCC.int/resource/docs/convkp/kpeng.pdf Paris Agreement - https://UN FCCC.int/sites/default/files/english_paris_agreement.pdf
6 The OECD is a non-United Nations international organization that was launched in 1961 and is headquartered in Paris, France. The OECD seeks to "… shape policies that foster prosperity, equality, opportunity and well-being for all." There are currently 38 members, the newest of which is Costa Rica (2021). See https://www.oecd.org/about/ for more information.
7 This material on the Kyoto Protocol is from UN FCCC (undated a), which is perhaps the key authoritative information source on the Protocol. There are a variety of links focusing on all aspects of the Kyoto Protocol that are available from its web page.
8 For everything you ever wanted to know about NDCs, please see United Nations (undated). All NDCs submitted to the UN FCCC are available at https://UN FCCC.int/NDCREG
9 The principles underpinning the Paris Agreement were significantly informed by findings from The Harvard Project on International Climate Agreements summarized in Aldy and Stavins (2008).
10 Very few non-Annex 1 countries submit national communications so frequently. Please see https://UN FCCC.int/non-annex-I-NCs#:~:text=UN FCCC%20Nav&text=Non%2DAnnex%20I%20Parties%20are,and%20every%20four%20years%20thereafter
11 The UN FCCC Annex 1 countries have been required to submit annual GHG emissions inventories using UN FCCC methodologies for all main GHGs since the 1990s. See https://UN FCCC.int/ghg-inventories-annex-i-parties/2023
12 To see all country-level GHG data in table or map formats, in detail or in aggregate as countries have submitted them, please see https://di.UN FCCC.int/ghg_profile_annex1
13 This gap would be about 20% lower if countries requiring financial assistance received and fully utilized that support to reduce GHG emissions.

Further Reading and References

Aldy, J. and R. Stavins. 2008. *Designing the Post-Kyoto Climate Regime: Lessons from the Harvard Project on International Climate Agreements: An Interim Progress Report for the 14th Conference of the Parties, Framework Convention on Climate Change*. The Harvard Project on International Climate Agreements. Retrieved from https://www.belfercenter.org/publication/designing-post-kyoto-climate-regime-lessons-harvard-project-international-climate September 2024.
Alley, R. 2000. "The Younger Dryas Cold Interval as Viewed from Central Greenland." *Quaternary Science Reviews* 19 (1-5): 213-226.
Alley, R. 2014. *The Two-Mile Time Machine: Ice Cores, Abrupt Climate Change, and Our Future*. Princeton University Press: Princeton, NJ.
Böhringer, C. 2003. "The Kyoto Protocol: A Review and Perspectives." *Oxford Review of Economic Policy* 19 (3): 451-466.
ClimateWatch and World Resources Institute. 2021. *Historical GHG Emissions*. Retrieved from https://www.climatewatchdata.org/ghg-emissions?breakBy=countries&end_year=2020&regions=WORLD&source=Climate%20Watch&start_year=1990 February 2024.

Dow, G. and C. Reed. 2015. "The Origins of Sedentism: Climate, Population, and Technology." *Journal of Economic Behavior and Organization* 119: 56–71.

Finchon, M. 2013. "The Day Before Yesterday: When Abrupt Climate Change Came to the Chesapeake Bay." *Chesapeake Bay Quarterly: A Magazine from Maryland Sea Grant* 12 (4). Retrieved from https://www.chesapeakequarterly.net/V12N4/main1/ July 2024.

Gelles, D. 2024. "Scientists are Freaking Out about Ocean Temperatures: 'It's Like an Omen of the Future.'" *The New York Times*, Climate Forward, February 27, 2024.

Goodwin, J. and K. Kizzier. 2018. *Elaborating the Paris Agreement: National Greenhouse Gas Inventories*. Center for Climate and Energy Solutions, August. Downloaded from https://www.c2es.org/wp-content/uploads/2018/08/national-greenhouse-gas-inventories.pdf February 2024.

Harari, Y.N. 2015. *Sapiens: A Brief History of Humankind*. Harper Perennial Publishers: New York.

Helm, D. 2012. "The Kyoto Protocol Has Failed." *Nature* 491 (7426): 663–665.

IPCC. 2023. Summary for Policymakers. In: *Climate Change 2023: Synthesis Report. Contribution of Working Groups I, II and III to the Sixth Assessment Report of the Intergovernmental Panel on Climate Change* (Core Writing Team, H. Lee and J. Romero, eds.). IPCC: Geneva, Switzerland.

Lindsey, R. 2023. "Climate Change: Atmospheric Carbon Dioxide." *Climate.gov: Science & Information for a Climate-Smart Nation*. Retrieved from https://www.climate.gov/news-features/understanding-climate/climate-change-atmospheric-carbon-dioxide#:~:text=Before%20the%20Industrial%20Revolution%20started,was%20280%20ppm%20or%20less January 2024.

Lüthi, D., M. Le Floch, B. Bereiter, T. Blunier, J.-M. Barnola, U. Siegenthaler, D. Raynaud, J. Jouzel, H. Fischer, K. Kawamura and T.F. Stocker. 2008. "High-Resolution Carbon Dioxide Concentration Record 650,000–800,000 Years Before Present." *Nature* 453 (7193): 379–382.

Maamoun, N. 2019. "The Kyoto Protocol: Empirical Evidence of a Hidden Success." *Journal of Environmental Economics and Management* 95: 227–256.

Michon, S. 2023. "Does it Matter How Much the United States Reduces its Carbon Dioxide Emissions if China Doesn't Do the Same?" August 30. *Climate.gov: Science & Information for a Climate-Smart Nation*. NOAA. Retrieved from https://www.climate.gov/news-features/climate-qa/does-it-matter-how-much-united-states-reduces-its-carbon-dioxide-emissions February 2024.

Michon, S. and R. Lindsey. 2023. "What's the Hottest Earth's Ever Been?" *Climate.gov: Science & Information for a Climate-Smart Nation*. Retrieved from https://www.climate.gov/news-features/climate-qa/whats-hottest-earths-ever-been January 2024.

National Oceanic and Atmospheric Administration (NOAA). 2023. *Global Climate Report*. Retrieved from https://www.ncei.noaa.gov/access/monitoring/monthly-report/global/202313 July 2024.

National Oceanic and Atmospheric Administration (NOAA). Undated. "NOAA Index Tracks How Greenhouse Gas Pollution Amplified Global Warming in 2022." Retrieved from https://research.noaa.gov/2023/05/23/noaa-index-tracks-how-greenhouse-gas-pollution-amplified-global-warming-in-2022/ January 2024.

NOAA Global Monitoring Laboratory. 2023. "Trends in CO_2, CH_4, N_2O, SF_6." Retrieved from https://gml.noaa.gov/ccgg/trends/ December 2023.

Popovich, N. and B. Plumer. 2021. "Who Has the Most Historical Responsibility for Climate Change?" *The New York Times* November 12, 2021.

Rosen, A. 2015. "The Wrong Solution at the Right Time: The Failure of the Kyoto Protocol on Climate Change." *Politics and Policy* 43 (1): 30–58.

Stern, N. 2007. *The Economics of Climate Change: The Stern Review*. Chapter 3: "How Climate Change Will Affect People around the World," pp. 65–103. Cambridge University Press: Cambridge, UK. Based on presentation at Addis Ababa University, Addis Ababa, Ethiopia on January 29, 2007.

Stips, A., D. Macias, C. Coughlan, E. Garcia Gorriz and X.S. Liang. 2016. "On the Causal Structure between CO_2 and Global Temperature." *Scientific Reports* 6: 21691.

United Nations. Undated. *All about the NDCs*. Retrieved from https://www.un.org/en/climatechange/all-about-ndcs#:~:text=Simply%20put%2C%20an%20NDC%2C%20or,update%20it%20every%20five%20years February 2024.

United Nations Environment Programme (UNEP). 2023. *Emissions Gap Report 2023: Broken Record – Temperatures Hit New Highs, Yet World Fails to Cut Emissions (Again)*. Nairobi. Retrieved from https://doi.org/10.59117/20.500.11822/43922 February 2024.

United Nations Framework Convention on Climate Change (UN FCCC). Undated a. *What is the Kyoto Protocol?* Retrieved from https://UN FCCC.int/kyoto_protocol#:~:text=In%20short%2C%20the%20Kyoto%20Protocol,accordance%20with%20agreed%20individual%20targets February 2024.

United Nations Framework Convention on Climate Change (UN FCCC). Undated b. *The Paris Agreement: What is the Paris Agreement?* Retrieved from https://UN FCCC.int/process-and-meetings/the-paris-agreement February 2024.

US Environmental Protection Agency (USEPA). 2024. *Climate Change Indicators: Greenhouse Gases*. Retrieved from https://www.epa.gov/climate-indicators/greenhouse-gases January 2024.

US Environmental Protection Agency (USEPA). Undated. *Global Greenhouse Gas Emissions Data*. Retrieved from https://www.epa.gov/ghgemissions/global-greenhouse-gas-emissions-data January 2024.

Wing, S. Undated. *World with and without Polar Ice Caps over 500 Million Years*. Received directly from the Smithsonian Institution July 2024.

World Bank. Undated. "GDP (Current US$)" in *World Development Indicators*. Retrieved from https://data.worldbank.org/indicator/NY.GDP.MKTP.CD February 2024.

3 Economic Fundamentals, Ecosystem Services and Value

What You Will Learn in this Chapter

- The nature and causes of economic scarcity
- About the opportunity cost of decisions
- The many ecosystem services of environmental assets
- The value of ecosystem services
- How to work with linear functions
- The supply-demand model and the predicted equilibrium of the model
- The importance of market prices as indicators of value and scarcity

---

3.1 Recap and Introduction to the Chapter

In Chapter 1, we distinguished between environmental *assets*, which are stocks like forests, clean air, the Columbia River and the Appalachian Mountains, and the ecosystem services those assets produce, such as clean water, carbon sequestration and recreational opportunities. The Millennium Ecosystem Assessment (MEA), which was published in 2005 and involved more than 1300 experts, sought to categorize the ecosystem services humans receive from the natural world, and identified provisioning, regulating, cultural and supporting classes of ecosystem services. The study also evaluated the status of those services and found that many if not most ecosystem services were declining.

A critical environmental asset discussed in the MEA and elsewhere is a stable climate. As we learned in Chapter 2, human emissions of greenhouse gases into the atmosphere since the industrial revolution have steadily depleted this asset. CO_2 is the most important greenhouse gas and is emitted whenever carbon-based fuels are burned. It is not the most potent greenhouse gas, but it is very long-lived and the most pervasive.

As of 2023, the CO_2 concentration was over 420 parts per million and we are starting to see critical ecosystem services, which are derived from our stable climate, erode. Examples of climate-based ecosystem services that are at risk include moderate sea levels (limited flooding along coasts); sufficient rainfall, especially in tropical regions (e.g., limited droughts

DOI: 10.4324/9781003308225-4

in agricultural areas); and forests that are resilient to fires (e.g., limited super-fires in the western US). All these ecosystem services and many more are in the process of being eroded by our warming and increasingly unstable climate.

To use economics to analyze and solve environmental problems, such as climate change, we need to know some key ideas that have stood the test of time. This chapter discusses critical economic concepts that are accepted by virtually all economists, environmental, ecological, natural resource or otherwise. Scarcity, value and choices in economics are specific in meaning and central to any meaningful analysis. We therefore start there. To understand behavior, we also need to think about decision-making, which moves us into the realm of incentives and the roles of self-interest, gains from trade, prices and markets, which are widely accepted, but about which reasonable people can differ in their opinions.

It is too strong to call most of these notions facts, but they are more basic than, for example, "models" or "theories" that require testing. They are certainly much beyond mere "beliefs." That said, these ideas, which go back to Adam Smith, David Ricardo, Alfredo Pigou, Alfred Marshall and even Karl Marx, do represent a particular perspective and focus. It is therefore possible, certainly in analyzing environmental problems, to be on a useful, but completely different, totally non-economic wavelength. Such, perhaps, is the nature of interdisciplinarity and most environmental problems.

3.2 Scarcity and the Need to Make Choices

In normal conversation something is scarce if it is hard to get or there is not enough to go around. People then have to search for the item, line up or use other means to try to get it. In economics, though, it is much easier for something to be scarce than the normal way the word is used. Something is *scarce* in economics if **everyone** cannot have all they want of something for **free**. In other words, something is scarce if we try to give it away and we run out before everyone gets as much as they want.

The idea is important, because if everyone cannot have all they want for free, a bunch of choices need to be made. For example, who gets the scarce item and how much each person gets must be determined in ways that seem reasonable. That is, the choices that are made have to be set within the norms of the society we are talking about, and it is totally possible that different societies could have different norms regarding how scarce things are allocated.

Challenge Yourself

Can you think of something that is not scarce?

If you think about it, most things fulfill this fairly liberal criterion. The computer I am typing on now as I write this chapter and computers in general are certainly scarce, as is the Costa Rican coffee I am drinking. Everyone cannot have all they want of them for free, because we would quickly run out. The gas you may have burned or the bike you rode to work are also scarce. If we gave them away, we would run out. If you are now at work, not considering the issue of why you are reading this book at work, your time is scarce; I am confident that everyone cannot have all of your time – and the skills that go with it – for free. Indeed, if you go through a mental list, it is hard to identify something that is not scarce and therefore need not be rationed.

What causes scarcity? Clearly, bicycles, gasoline, computers and your time are things that people want. Scarce things are therefore *desired* by people. Stuff nobody wants cannot be scarce. Second, to be scarce, goods, services and the resources that make them up cannot be produced by magic technologies that require no work. One way to circumvent scarcity is therefore to invent technologies that require no scarce inputs. For example, if we were to be able to magically grow grapes and then use an ultra-advanced, work-free technology to ferment those grapes, we could produce non-scarce wine.

Equation 3.1

Desirable + Costly = Scarce

Of course, the problem is that wine is made using a bunch of inputs that are themselves scarce and it will probably always be that way. Choices must therefore be made about inputs like fertile land, clean water, energy, labor, winemaking knowledge etc. Science fiction stories often portray the future as being a time when humans get things they want without using anything up, but at the foundation of such a reality, scarcity would have to be overcome.

We notice that in the wine example a number of the resources needed to make wine are actually ecosystem services from the environment that carry the baggage of choices. Land, water and fossil fuels, like most environmental resources, have many, many uses. Those services are needed to grow grapes and produce wine. Ecosystem services are therefore given up when environmental assets are used to produce wine. Economists typically call the package of foregone options *opportunity cost*.

Challenge Yourself

What spillover effects does suburban sprawl have on ecosystem services of air and water resources?

Land growing pinot noir grapes in the Willamette Valley in Oregon may not be very far from the major urban area of Portland and therefore could easily be used for housing, shopping and other built-environment uses. Those areas are also famous for growing trees, such as Douglas fir, hemlock and other tall coniferous varieties, and probably that land was originally forested. These alternate uses are given up when vineyards are developed. The same is true for the clean water from underground aquifers or the Tualatin River that is used to produce grapes, and the energy, likely derived from fossil fuels, that drives the tractors and other machinery.

Box 3.1 The Opportunity Cost of Using Rivers for Irrigation

John Loomis of Colorado State University and several colleagues examined the ecosystem services provided by the South Platte River watershed in Colorado. They note that the watershed area simultaneously supplies several services, including recreation, irrigation water supply, habitat for fish and wildlife, natural water purification, wastewater dilution and erosion control. Their analysis indicates that irrigation water supply is dramatically interfering with the other ecosystem services and today "the river is operated as a plumbing system with about 500 irrigation ditches."

They estimate that partially restoring the neglected services by purchasing conservation easements along 45 miles of the South Platte River would be valued at $19–$70 million. They find that even the lowest value would be sufficient to purchase the necessary water rights and conservation easements. Reducing the use of water from the South Platte River for irrigation therefore makes economic sense.

Source: Loomis et al. (2000).

That environmental assets have multiple uses is not unusual. Non-environmental resources have similar features and indeed it is rare that resources are so specialized that they can only do one thing. For example, your car can be used to make money driving Über, for your own personal transport or to deliver packages. A bedroom can be used for sleeping, an office, a TV room or to rent to a boarder. Human skills are highly fungible across uses and it is not at all unusual for people to switch careers. Labor can be used for virtually anything.

But environmental assets are different from other types of resources in at least three respects, which can magnify their scarcities and increase the opportunity costs we incur when choices are made. First, environmental assets are often complex and, depending on human management, can simultaneously produce many ecosystem services.[1] Forests, for example, not only produce timber, but they also filter water, capturing particles that would otherwise remain suspended in water; forested areas therefore have cleaner water than places that are deforested.

Furthermore, lands downstream from forests have less erosion and are not as prone to flooding during extreme rainfall events, because land with lots of vegetative cover slows the movement of water and absorbs it. Protecting downstream lands from erosion and floods is therefore an important ecosystem service provided by forests.

Forests are ecosystems that provide habitat for animals and plants, which may or may not be of direct benefit to humans, but people often like healthy ecosystems where animals and plants have a place to live. This feature therefore can and should be viewed as an ecosystem

Photo 3.1 Eroded Slope in Nepal
Source: USDA Agricultural Research Service.

service forests provide to humans. Of course, enthusiasts like bird watchers and other eco-tourists get direct enjoyment from healthy ecosystems.

As discussed in the previous chapter, climate change is one of the most important challenges humanity faces in the 21st century. It turns out that forests capture carbon from the atmosphere in their wood as they grow, helping offset carbon emissions from fuel burning and other sources. Forests therefore help stabilize the climate and indeed can grow faster when there is more CO_2 in the atmosphere.

The second key difference between human-made and environmental assets is that many ecosystem services can simultaneously be enjoyed by a number of people. For example, a large forest can reduce erosion on many farmers' downstream plots and can protect several hundred houses from flooding. It can purify the water for perhaps thousands of people and sequester carbon, which benefits the whole planet. We will return to these issues later in this chapter and get into the details in Chapter 8.

Finally, scarcity of environmental assets and the choices we make about ecosystem services can have spillover effects on other environmental assets. An obvious and particularly simple example is our use of fossil fuel environmental assets. Processing and burning these resources yield important energy-related ecosystem services like the ability to drive tractors, trucks and have light at night. Of course, utilizing this ecosystem service by its nature degrades environmental assets and reduces their capacity to provide other ecosystem services. For example, burning fossil fuels degrades air quality, reducing important services like supporting human respiratory health. It can also degrade land and water through acid rain, if fuels are acidic. Areas with high rainfall and fossil fuel emissions therefore generally have overly acidic waterways, soils and even swimming pools.

Box 3.2 Ecosystem Services from Sustainable Wine Production

Using land as a vineyard can generate multiple ecosystem services, but few appear to conflict. While the main ecosystem service is soil fertility to grow grapes, depending on the management system chosen, other services can be produced. For example, sustainable viticulture and enology can help maintain bee populations, reduce erosion and increase water quality compared with other land uses.

   Winkler et al. (2017) and Winkler and Nicholas (2016) analyzed ecosystem services provided by vineyards. They especially identify disease control, such as control of fungal disease and insect pests, as critical ecosystem services produced when viticulturists maintain biodiversity in vineyards. Using key informant interview data, they find that in California's Napa Valley, both vineyard owners and residents see important cultural heritage values in vineyards. They especially note regional traditions associated with grape cultivation and pride in the *terroir* of the region, which is the combination of natural features that gives wine grapes their character. They also discuss carbon sequestration and scientific discovery as important ecosystem services provided by sustainable vineyards.

Sources: Winkler et al. (2017); Winkler and Nicholas (2016).

Challenge Yourself

What ecosystem services are produced by the natural resources a) clean air, b) coastal areas of the Indian Ocean?

As we know from the previous chapter, burning fossil fuels is the #1 way we degrade our stable climate, which provides an amazing number of important ecosystem services. Here are a few: A stable climate implies lower sea levels, which means less coastal flooding, which in Chapter 2 we learned is increasingly occurring. It also means there are fewer extreme weather events, such as heat waves, meaning we need to buy fewer costly cooling technologies, such as air conditioners, and use them less often.

   Climate stability also offers humans more predictable and probably better weather, producing a variety of benefits. For example, a variety of sources, including IPCC (2023), suggest that Sub-Saharan Africa, which in many areas is subject to drought, will experience more frequent and deeper droughts as the atmosphere warms. This will make agriculture, which is the main source of income for over 65% of people on the continent, less reliable as an income source.

In sum, environmental assets – and the ecosystem services they produce – are scarce and therefore subject to the same opportunity costs as decisions about other resources. Unlike non-environmental resources, though, environmental assets can often simultaneously generate a variety of different benefits, which may accrue to hundreds, thousands or even billions of people at the same time. These features imply that opportunity costs can be very sensitive to our choices around environmental assets. Bad decisions can have big implications, because ecosystem services benefitting many, many people at the same time can be wiped out.

3.3 Ecosystem Service Values

Ecosystem services are clearly valuable, but what makes ecosystem services valuable and how do we know their values? Perhaps more importantly, how can values be compared when we need to make decisions?

Value is a measure of how scarce is a good or service, so desirability and cost also drive value. First and foremost, to be valuable an ecosystem service must be desirable. An ecosystem service nobody wants is not scarce or valuable. Of course, ecosystem service desirability can and does vary across individuals, societies and contexts. People living in places with a lot of different birds, such as Ethiopia, which has one of the highest concentrations of endemic species in the world, for example, may not view showy birds as any big deal, whereas in the US spotting a bald eagle or California condor can make someone's day. In fact, if exotic birds are very numerous, people may view them as nuisances!

Air quality is typically measured using the concentration of pollutants per cubic meter of air volume. Particulate matter like dust that is less than 2.5 microns in diameter ($PM_{2.5}$) can be especially harmful to people, causing respiratory irritation and transporting compounds that cause illness, including cancer, deep into lungs. According to the World Health Organization, average annual $PM_{2.5}$ exposure in Ethiopian cities is two to three times that of urban areas in Japan, the US and Germany.[2] Much of this may be outside residents' control and they therefore have no choice, but one also does not hear much complaining about air pollution in Addis Ababa. Most people seem to have other things on their minds.

Ecosystem service values can change over time. Beijing has experienced over 20 years of legendary smog that a 2013 paper reported reduced average life expectancy by 5.5 years. Starting in 2011, though, many residents had had enough. Protests erupted and people voted with their feet by enjoying so-called "Clean Air Tourism" in coastal provinces like Fujian.[3] By June 2013, the government issued new rules aimed specifically at reducing urban air pollution. Part of this change was probably due to increased desirability of ecosystem services like visibility, respiratory health and comfort provided by clean air.

And what about individual differences in ecosystem service values? Of course, just as societies differ, some individuals are very attuned to ecosystem services and worry a lot about air pollution, climate change, polluted water and species extinction. Some people, for example, always drink out of glass containers, because they worry that other cups will leach toxic materials into their drinking water. Other people could not care less and others are annoyed even to discuss or do anything positive for the environment. For example, as reported by several news agencies, including Forbes, and posted to YouTube, the conservative commentator Glenn Beck in August 2013 declared that he would fire any employee caught using efficient light bulbs or recycled products. He further added that "...global warming is a pile of crap... a load of socialist, communist crap" (Kanellos, 2013).

Incomes also cause ecosystem service values to differ. People and societies with higher incomes have the capacity to have more of most things, including ecosystem services. We will come back to this idea below in the context of non-ecosystem goods and services.

It is very important to note that levels of ecosystem services that people already have affect individuals' values. When air is very polluted, the elderly and children are likely to particularly suffer from respiratory illnesses. Starting from such a low baseline, all else equal (especially income), any initial improvements in air quality are likely to be particularly welcomed and highly valued. In contrast, if the air is already very clean, any further improvement may not even be noticed.

Polluted water is closely linked with diarrheal diseases that can particularly affect children under five years old. Such diseases may even be fatal due to extreme dehydration, but as surface water quality improves – and availability of water supply infrastructure increases – the incidence of such diseases declines and child mortality falls. Initial steps to improve water quality and eliminate potentially life-threatening diseases are therefore viewed by all as critical. Additional improvements probably will be less valued, however, and there may be disagreements about whether such investments are worth the cost.

Suppose we focus on this last potential determinant of value and order a particular set of ecosystem service improvements from highest to lowest value and plot the information on a graph. On the horizontal axis is the level of the ecosystem service. This measure could be acres preserved for habitat, reduction in diarrheal disease cases due to less water pollution, days of good visibility due to reduced air pollution etc. On the vertical axis is the benefit derived from the improvement by people in the society. The origin of the graph is where none of the ecosystem service is provided and no additional benefits are generated (0, 0).

Our construction would likely give us something like Fig. 3.1. The graph may not be continuous as I've drawn it, in which case it would have "jumps." It may be more or less steep and closer or farther away from the origin, but it seems reasonable that it will slope downward, because near the origin there is very little of the ecosystem service and as we move to the

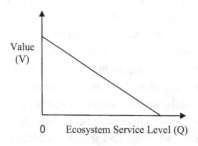

Figure 3.1 Marginal Benefit from Ecosystem Services

right there are lots of those services. Because this graph tells us the value of **increased** services, it is typically called a *marginal benefit* (MB) function.

Underpinning the placement of the graph are the preferences of people in the society, potential substitutes for ecosystem services and incomes. For example, how passionate is the average person about the ecosystem service? If people are very concerned about maintaining the service, all else equal, the graph will be farther to the right. The marginal benefit of a service that is very unique, with few or no substitutes, would also all else equal be placed farther away from the origin. Finally, people with higher incomes, all else equal, might be expected to desire more of any ecosystem service than those with lower incomes, pushing the graph away from the origin.

To write something like Fig. 3.1 as a linear equation (i.e., a line), it would be of the form Value = a_1 - b_1*(Ecosystem Service Level), where both a and b are positive numbers. As it tells us how the X (horizontal) axis variable affects the Y (vertical) axis variable, this equation would be called a *function*. We will be using such equations throughout this book, with value (V) sometimes being monetary and sometimes not, and Q could be a non-marketed ecosystem service or market good quantity. The generic form of this function is therefore V = a_1 - b_1*Q, where the subscript "1" indicates in this case that it is related to users. The position and slope of the function are determined by the *parameters* "a_1" and "b_1." The part of the equation called "a_1" is the intercept and tells us where the function touches the Y axis. The "a_1" term therefore indicates where on the graph our linear function is placed. Increasing "a_1" shifts the marginal benefit function up or to the right. Reducing it shifts the curve down or to the left. Changing the parameter "b_1" changes the slope of the line, with a larger value meaning a steeper slope.[4]

Box 3.3 How to Graph a Function using an Equation

Suppose we have a *linear* function of the form V = c + dQ, where V stands for the *total* value (for simplicity, in dollars) of an ecosystem service and Q is the level or "quality" of that ecosystem service. This equation says that if Q is increased by one unit, V increases by d dollars. Suppose that Q is a reduction in the number of summer days *E. coli* bacteria concentrations are over health limits in Lake Erie, which is the Great Lake bordering the US states of Ohio, Michigan, Pennsylvania and the Canadian province of Ontario. *E. coli* bacteria can cause diarrhea, abdominal pain, fever and vomiting, and are associated with sewage pollution in lakes and rivers. Regulators may sometimes close beaches when *E. coli* exceeds safe levels.

Suppose that V is the total value per person from being able to safely swim for an extra day during the summer. The *parameter* c is $10, meaning that without controlling pollution, people get $10 in value, which could be from being able to walk along the lake without swimming. The parameter d is $40, meaning each additional person-day at a Lake Erie beach is worth $40.* Our function is therefore V = $10 + $40Q. Let's plot the function for a total of five extra swimmable days during the summer, pairing each value of V with the number of additional days people can swim. We therefore simply.

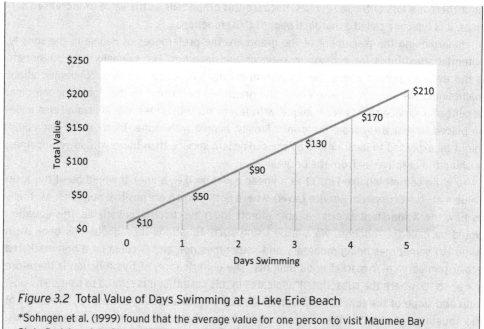

Figure 3.2 Total Value of Days Swimming at a Lake Erie Beach

*Sohngen et al. (1999) found that the average value for one person to visit Maumee Bay State Park beach in Ohio in 1997 was about $25. Accounting for inflation, but assuming users' valuation of trips to the park stays constant over time, this estimate is roughly equivalent to $47 in 2023 dollars.

The relationship between value and the second aspect of scarcity, that something must be given up to provide it, also deserves some thought. Within the world of services that people enjoy, ecosystem services may be unique in that they can often be provided with limited explicit human effort or diversion of human-made resources. One can only get a massage, repair your car or motorcycle or have a haircut, if resources like labor, skills, energy and buildings are diverted for that purpose, but ecosystem services derived from clean air, wildlife, a walk in the woods and stable climates are largely provided directly to humans by nature. That said, we definitely incur opportunity costs when choices are made about the use of natural resources. Ecosystem services are therefore costly.

But let us think about the value of goods and services other than ecosystem services. For example, which is more valuable, a Mercedes-Benz car or a Snickers bar? A Mercedes or a bag of basmati rice?

Before reading on, please think about which goods are more valuable, under what circumstances and why.

Of course, the Mercedes-Benz is on average more valuable. Why? Because it provides more and higher-quality services that, at least under present circumstances (especially how much we have available now), people find more desirable. On average, who wouldn't prefer a Mercedes to a bag of rice? It also, though, takes more resources to make. Among other things, Mercedes-Benz cars require a lot of sophisticated engineering, metal, plastic, rubber, computer equipment and glass. Candy bars and rice have much simpler production processes and require less scarce inputs. The same comparison can be made for services. Which is more valuable, a haircut or a consultation with a cardiologist?

Notice that the values of goods and services such as cars, grains and haircuts, in contrast to many ecosystem services, are the result of collisions between input resource scarcity and desirability. Typically, particularly in higher-income countries, these values are determined by *markets*, which are institutions that facilitate trade. Scarcities of resources to produce private goods and services are reflected, perhaps imperfectly, in the costs of production, which are in monetary units. Desirability is not fundamentally in monetary units, but if we are talking only about private goods and people have a lot of choice about what they can buy, desirability is determined by what else people can do with their money, along with their incomes. Desirability of private goods and services is therefore expressed by people's *willingness to pay*, which is in monetary units.

These two sides of the equation, sometimes called *demand* and *supply*, interact to determine private good and service prices, which offer expressions of value. These values are in money terms and perhaps as a result, "value" is often erroneously presumed to be monetary. For example, the first two definitions of value in the 2024 Merriam-Webster dictionary are the following:

1: the monetary worth of something: MARKET PRICE

2: a fair return or equivalent in goods, services, or money for something exchanged

3.4 Prices, Markets and Ecosystem Services

Why do prices say something about value? What exactly do they tell us and how do prices relate to ecosystem services? To answer these questions, we need to know something about market behavior and price formation. The desirability side of the market starts from the recognition that, just like for valuation of ecosystem services, people are different. For example, some people love to have new technologies and therefore they buy any new program, app or high-tech device as soon as it comes out. Other people could not care less, and some are annoyed by the range of options.

For relatively homogeneous products and services like massages and pizza, the history of enjoying these goods and services matters. If you have only one massage per year, it means you will look forward to and highly value your upcoming massage. Someone who gets a massage each week probably after a while will view the experience as pretty routine. Some people never get massages, because they can't afford them or don't like massages. Because of variation in preferences, incomes and experience, people are different.

Desirability values of marketed goods are quite directly represented by willingness to pay, because it seems reasonable to suppose that people want to use their money wisely and are not willing to pay more for a good or service than its intrinsic value. Preferences and alternative uses for money are important determinants of willingness to pay. Incomes and wealth are also important, because people who have money to burn, all else equal, should have higher willingness to pay. Dependence of willingness to pay on incomes and wealth is therefore fundamental.

Suppose, as we did for ecosystem services, we order people from highest to lowest value – now represented by willingness to pay – and plot this representation of value on a graph. We would get something like Fig. 3.1, except it would be for the marginal benefit of a private good or service. From the perspective of desirability, ecosystem services are therefore very similar to private goods. In fact, there is no reason we cannot analyze them in the same way.

Challenge Yourself

What private goods and services are so unique they have almost no substitutes?

On the supply side, suppose we are analyzing relatively homogeneous private goods and services, so we can meaningfully talk about "the" good made by different producers, which are typically called *firms*. Examples of such goods and services need not be limited to commodities like specific types of coal, wheat and gold, but can also include the many, many goods and services that are quite similar. Pizza, soft drinks, tax advising, yoga classes and beer would be good examples of close, but imperfect, substitutes. Highly unique products, like very specialized software or unique drugs, are not what we are talking about, but for many products and services, offerings of different providers have many more similarities than differences.

In our quest to analyze price formation in such markets, costs to produce market goods and services are important. In general, producers have different *marginal costs (MC)*, with some having low costs and others high costs. In fact, as producers try to ramp up production, costs may rise as they, for example, hire more people and need to train them. Ordering these additional costs from lowest to highest, we can plot this information on a graph. The result is something akin to Fig. 3.3. We can once again represent the figure as a linear equation of the form $v = a_2 + b_2 Q$, where a_2 and b_2 are positive numbers that are different than for buyers, which is why we give them the subscript "2," in this case for the producer side of the market. Note that the sign before b_2 is positive rather than negative, so the function is upward-sloping.

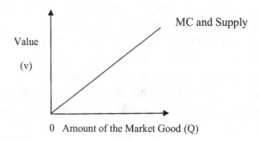

Figure 3.3 Marginal Cost to Supply Market Goods

To analyze behavior and price formation, we now must make some minimal assumptions about what people are trying to accomplish by participating in such markets. Let's keep it simple. On the desirability/benefit side, suppose that people are trying to get a good deal. They will therefore buy a product or service if they value it more than the price they have to pay for it, given what else they can do with their money. If they value it a lot and it costs them little, so much the better. On the other hand, those who produce private goods and services are trying to make money. They will therefore sell goods and services as long as their products cost less to produce than the prices they receive in the market.[5] Suppose they try to earn as much money as they can.

In our behavioral model, the motivations of producers and users appear to be exact opposites! Producers want users to overspend, even as buyers try to get great deals. How are these divergent interests reconciled? To answer this question, let's plot our private good version of Fig. 3.1 and Fig. 3.3 on the same graph, add a hypothetical price – any price – and ask ourselves how producers and users might respond. Let us focus in on price P_1 in Fig. 3.4.

Implementing our simple behavioral assumption that users try to get bargains, we compare users' marginal benefit to price P_1.[6] We see that all users up to quantity A on the horizontal axis have values (plotted on the vertical axis) greater than P_1 and presumably would buy the product or service. On only the "last" unit of the good (the Ath one) do users "break even." On all units up to point A, users get good deals. This is especially true for the first users who for whatever reason really covet the good or service.

At price P_1, reading off the marginal cost function, which in market situations is often called *supply*, because it predicts how much will be produced at various prices, on all units up to the Bth one producers earn profits. Producers are therefore predicted to be willing to provide B units of the good or service. Especially the low-cost, highly efficient producers do very well. Only on the last unit is profit zero, where profit equals price minus the MC.

Of course, quantity B is greater than quantity A, which means that at price P_1 those who might buy the product want fewer than are produced by firms. This situation is clearly not stable, because at price P_1 producers are producing products nobody wants. Market pressure is therefore put on suppliers to reduce prices to avoid over-supply. As price falls, less efficient, high-cost producers are priced out of the market and the quantity produced declines. Users with lower marginal values (i.e., those who all else equal desire the product less) are tempted to buy the product or service and the number of users increases.

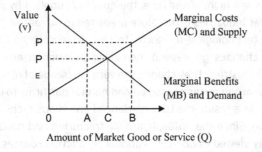

Figure 3.4 Price of Market Goods and Services

This dynamic stops and there will be no more pressure for price change when the price declines to price P_E and the quantity is equal to C. The price P_E is called the *equilibrium* price. At this price, the quantity users want to buy is the same as the quantity produced.

Notice that this *supply-demand* model predicts that goods and services will have only one price. This price may change due to changes in the *marginal benefit* of users and/or changes in producer costs, but at any point in time there will be one price. What are the characteristics of this price and what does it tell us about value? First, we see that this stable price is predicted to be a lower-bound measure of user value, because users only buy if they get a good deal. Indeed, the level of the price is expected to be exactly equal to the marginal benefit from the **last** unit of the good or service bought in the market. The equilibrium price is therefore predicted to give critical insight into user benefits, which is the desirability side of scarcity and value. Because the equilibrium price P_E is simultaneously the *marginal cost* of production, the price also sheds important light on the scarcity of resources needed to make available the good or service. In fact, the equilibrium price is just the highest production cost of units exchanged in the market!

In sum, market prices can provide very important information about value, which is a metric for scarcity. High prices for Mercedes-Benz cars yell out to us loud and clear that a lot of people like that kind of car and furthermore some pretty valuable resources have gone into making them. The positive, but much lower, prices of 10-kg bags of rice and Snickers bars tell us we value those products and they are scarce, but they are much less scarce than Mercedes-Benz cars.

How do we react to the information prices give us about scarcity? High prices, of course, all else equal cause people to buy less. We therefore tend not to use so many resources in that way and instead conserve resources for other purposes. For example, the high cost of a cardiologist visit tells us we should not go to such specialists for each and every ailment. Instead, we see less specialized and perhaps less scarce medical practitioners for non-life-threatening problems and go to cardiologists only when we really need to. High prices therefore steer our behavior in resource-conserving directions. On the other hand, low prices tell us that resources are not very scarce, and we should use more.

As we've seen, high prices make production profitable for producers with high costs who at lower prices would not find it profitable to sell. Over time, though, high prices also attract investment to the industry and spur technical change. High prices mean high profits, which pulls investable resources from other parts of the economy. In essence, high prices scream out to producers, "Hey, the economy wants more of this and less of that. Come and invest here where the profits are high!" Over time, the quantity supplied by producers who previously produced in other industries is therefore predicted to increase.

As sectors boom, technological advances to take advantage of high profitability may be stimulated. Such changes go beyond increasing production and sucking investment resources from other industries. They change the very landscape of production. For example, decades of high hydrocarbon prices caused oil and natural gas firms to investigate new ways to extract oil and gas. As a result, the US experienced enormous technical change in oil and natural gas production. Since the 2010s, hydraulic fracturing and horizontal drilling, which have been traditionally viewed as non-conventional production sources, have revolutionized these industries and made the US a net exporter of both natural gas and oil, a trend that is projected to continue through 2050 (US EIA, 2023).

If price tells us about market value, we can substitute price (P) in for value in our marginal benefit and marginal cost equations. Though buyers *want* low prices and sellers *want* high prices, the supply-demand model predicts that, ultimately, they will have to agree on one market price, which in equilibrium simultaneously equals the marginal benefit to buyers and the marginal cost to sellers.

In terms of our equations, finding the equilibrium price (P_E) and quantity (Q) is therefore just a matter of recognizing that in equilibrium there will only be one price, and quantity supplied will have to equal quantity demanded. Operationally, we set the marginal benefit and marginal cost functions equal to each other and solve for Q and then P. We can get solutions, because we have two unique equations (marginal benefit and marginal cost) and two unknown variables (price and quantity). The basic setup and method using linear equations are shown below.

Equation 3.2 Equilibrium Price and Quantity Equation

P = marginal benefit = $a_1 - b_1 Q$ = marginal cost = $a_2 + b_2 Q$

Let us recall that we can add, subtract, multiply and divide both sides of any equality and the equation will still be equal.

Step 1: subtract a_2 from both sides.

$a_1 - a_2 - b_1 Q = b_2 Q$

Step 2: add $b_1 Q$ to both sides and combine terms.

$a_1 - a_2 = (b_1 + b_2) Q$

Step 3: divide both sides by $(b_1 + b_2)$ and voila!

$Q = (a_1 - a_2)/(b_1 + b_2)$

As the four parameters (a_1, a_2, b_1, b_2) are just positive numbers, we can find equilibrium Q. Once we know the equilibrium Q, we just plug that number back into either the demand or supply equation[7] to find the equilibrium price.

Box 3.4 Finding Equilibrium Price and Quantity with Linear Supply and Demand

Suppose the MB (i.e., demand) for phone charging cords is given by MB = $12 - 2Q_D$ and the supply can be represented as MC = $3 + Q_S$, where the subscript D stands for the consumer side (demand) and the subscript S means the producers (supply). How can we use these two equations to solve for the equilibrium Price (P) and Quantity (Q)? The first step is to recognize that in equilibrium the quantity demanded (Q_D) has to equal the quantity supplied (Q_S), otherwise there would be a shortage or surplus. Second, ultimately the producers and consumers must agree on an equilibrium price. Third, in equilibrium P = MC = MB. These critical steps are logic rather than math.

Recognizing that P = MC = MB, we set the two equations equal to each other and solve for Q. That means 12 − 2Q = 3 + Q. We first subtract 3 and add 2Q to both sides of the equation, which we can do, because adding the same thing to each side of an equality still implies equality (i.e., if we add 1 to each side of the equality 4 + 3 = 6 + 1, the result would still be equal). This gives us 9 = 3Q. Dividing both sides of the equation by 3 gives us Q = 3, which is our equilibrium quantity. To get the equilibrium price, we just plug Q = 3 into either price equation (which we can do, because in equilibrium they are equal) and find that P = 6. The equilibrium price is therefore 6 and the quantity is 3.

3.5 Complications in Using Prices as Indicators of Value

A key complication with using price to infer value is the dependence of willingness to pay (i.e., the desirability side of value) on incomes, which makes drawing strong conclusions difficult. This is particularly true when income differences are very large. Haircuts are a classic example of a service that is not tradable across international boundaries. Often such goods and services, where prices are locally generated, are called *non-tradables*. As it is impossible to import a haircut, local prices are unaffected by those in other countries. In US cities a haircut costs $20 and up, which suggests that the marginal benefit of a haircut of average quality is at least $20.

Over the years, I got many haircuts in Addis Ababa, the capital of Ethiopia. The last time was a 45-minute virtuoso performance by Elias at Ras New Fashion Barbers, which cost $1.80 plus tip. Whenever I go to cities in the Global South, I try to get a haircut, because the quality is high (much better than a $20 haircut in the US) and the cost is low. In other words, I get a great deal.

The $1.80 price tells us that haircuts are less scarce in Ethiopia than in the US. But what else can we infer? Does the low price mean people value grooming less in Ethiopia? No. More likely is that key resources used to produce haircuts, mainly labor, are cheaper than in the US. Average wages in Addis Ababa for barbers are about $8.00 per day and many do not make even that much. Does that mean Ethiopian barbers are less valuable than US barbers? Probably not. It only means that labor is relatively abundant in Addis Ababa.

Challenge Yourself

Can you think of policies to send people the message that non-priced ecosystem services are indeed valuable?

It is also the case that many economic and social features are so different from the US that comparisons become difficult, and it is hard to draw strong conclusions about value. For example, perhaps half of all adults are not working, because they cannot find jobs (Gebeyaw, 2011). Saying that haircuts are valued at 1/15th of their value in the US therefore probably misses the main point that it is very difficult to get work in Addis Ababa, which drives down wages. Through little or no fault of Addis Ababa residents, incomes are low and that feeds into the prices barbers can charge.

The dependence of willingness to pay on incomes and wealth is an analytical nuisance, but the biggest complication with depending on price to know value is that valuable and potentially crucial ecosystem services have absolutely no prices. What information does no price give us? No price is the same as a price that equals zero, which tells users that ecosystem services must not be scarce. This means that unpriced services are not desirable and/or any resources that go into producing such services are not valuable.

This observation is especially important for analyzing ecosystem services and improving their use, because so many ecosystem services are not priced. For example, what is the price to us as users to get fresh, breathable air? Stable climate? Fewer extinctions? Clean beaches? The answer is generally zero. We do not pay directly for those services and no prices tell us - erroneously, of course - that such ecosystem services are not scarce.

Many ecosystem services do not have prices simply because they are not traded in markets where prices are formed. The messages we can get from no prices for ecosystem services are therefore that those services are not desirable, we should use the environmental assets in some other ways and/or getting these services is costless. Of course, these conclusions are ridiculous. Clean air, for example, is very desirable because lives may be several years shorter without it. Of course clean air is valuable!

It is also not true that getting these ecosystem services is costless, because we have to give up stuff to get them. In fact, people spend billions every year to stop using the environment to absorb our pollution, which is one of the competing uses of breathable air. These costs may be indirectly included in prices if they are passed on to users, for example in electricity bills, but such costs may be small and not very transparent. Other ecosystem services like enjoyable beaches, big salmon runs due to good habitat and toxic waste-free areas require direct cleanup, as well as prevention. Often these costs are completely unpriced.

Some ecosystem services are "privatized" and have prices. For example, timber and fish have prices and we therefore know something about their values and how scarce they are. Campgrounds and private lakes allow us to experience nature for a fee, as do national and state/provincial parks in some countries. These services are priced. In general, wildlife areas are publicly provided, but many countries also have private parks that operate like businesses. The 160,000-acre Sabi Sand Game Reserve in eastern South Africa, for example, provides habitat for thousands of species of plants and animals. It allows people to view those animals and experience life in such an environment. It is also the oldest private game reserve in South Africa. The cost to stay at the Sabi Sand Game Reserve is 6000-9000 South African Rand or $600-$900 per night with full luxury board and game tours.

Costa Rica also has a number of private wildlife areas, including the 1000-acre Lapa Rios Eco-Lodge. Visitors can enjoy the environment with full board for about $400 per night.

Photo 3.2 Yosemite National Park Tioga Pass Entrance
Source: Author.

Perhaps the most famous private reserve in Costa Rica is the Montverde Cloud Forest Biological Reserve. It and four other reserves are administered by the non-profit Tropical Science Center. Foreigners can enter the reserve for $17 or can stay in the La Casona Lodge for $73 per night. Prices are cheaper for locals.

Why are ecosystem services like timber and fish always traded, while others like nature viewing and beaches only sometimes have prices? Why are others, such as climate stability, ocean health and air quality, never marketed or priced? All ecosystem services are valued (to varying degrees) by people, but only services that businesses are interested in providing are traded and therefore priced. These ecosystem services, combined with other inputs like labor, machines and human skills, can be sold at a profit just like all other business services.

As we will discuss in detail in Chapter 8, the key difference here is that providers of marketed ecosystem services can make people pay to enjoy their services. Conversely, people who do not pay for the services cannot have them and they can be *excluded*. It is critical to note that the ability to exclude those who don't pay is a technical property of some ecosystem services. It is also a bit difficult to manufacture. For example, it is pretty much inherent in many parks that they can have gates and therefore can charge entrance fees. They therefore fulfill an *excludability* requirement. Fish and timber are also fully excludable just like other services, such as accountancy, restaurant and auto mechanic services. If someone refuses to pay, they cannot expect to get their fish or have their car fixed. If someone were to take fish without paying, it would be regarded as theft.

Such is not the case for most of our critical ecosystem services, which by their very natures are shared by perhaps millions of people. Suppose, for example, a power producer decided to invest $50 million to reduce its carbon dioxide emissions into the atmosphere and therefore help contribute to climate stability. This step would be welcomed by all and probably would indirectly benefit the company via an improved reputation etc. It is very difficult to envision how the company could **directly** financially benefit from such a step, however. Companies certainly could not charge for such services, as they do for electricity, because it seems impossible to exclude nonpayers.

Or suppose you invented a machine to stop climate change. I don't mean producing greenhouse gas-free energy like wind power, which is excludable, but a machine to directly reduce the average surface temperature of the planet. How could such services – independent of governments – be marketed and assure that most of the people who benefit pay? Free riding would likely be enormous! You could expect a Nobel Prize, foundation awards and lucrative government contracts, but you probably would do no purely private sector business like a private timber producer or a commercial fisher. The benefits are too widespread and once they are created (i.e., the climate is stabilized) there is no way to keep freeloaders out.

Many, many other ecosystem services have this same non-excludability property. A city that depends on good land cover to have good water quality enjoys a largely non-excludable ecosystem service. Whale watching eco-tourism is also non-excludable (though being able to get on a whale-watching boat is not), because whales have a wide range and migrate long distances. When every year the people and Government of Tanzania preserve habitat roughly the size of Denmark for animals in Serengeti National Park, they generate a variety of ecosystem services. Most of those accrue to those living outside the country who value animals like the lion, African elephant and ostrich. Those people get non-excludable ecosystem service benefits via media like YouTube but may never visit Tanzania.

No prices for non-excludable ecosystem services is the worst possible price problem, but there are other cases where prices give wrong signals. Subsidies, which are government policies that deliberately distort prices, will be discussed in Chapters 11 and 12. Monopolies and other forms of market power can jack up prices, causing them to exceed scarcity values. Bad information to producers and consumers can also cause prices to poorly reflect values. For example, if a product is hazardous, but the effects are kept secret from buyers, this will inflate the price compared to what it would have been if the hazards were known. For example, the highly toxic pesticide diazinon used to be widely sold in garden centers for residential use. Over time, though, the toxic effects on people became clear and eventually the chemical was banned for home use. Probably if those effects had been known all along, the bottom would have fallen out of the market much earlier, providing a better indicator of the value of using diazinon at home.

3.6 Chapter Conclusions

Environmental assets like clean air, coastal zones and forests represent the closest thing to true magicians that any of us will ever meet. They simultaneously produce numerous ecosystem services that may be shared by thousands or even millions of people. These services are scarce and valuable, because they are desirable to most people, and something needs to be given up to get them. Though everyone does not equally value ecosystem services, people in general have a willingness to pay for even non-marketed ecosystem services, and society will indeed need to pay (i.e., give up something) if they want them.

Many of the ecosystem services we most care about have no private sector supply. We know from our supply-demand model that market prices come from the collision of producers and consumers in markets, and without private sector producers we simply cannot have markets. Without markets, we still have value – not all value is generated in markets – but we do not have market prices.

Lack of prices creates a variety of problems, including not knowing values, which complicates decision making. More fundamentally, though, unpriced ecosystem services may be treated as if they are free, which tells markets that they are not valuable. Though unpriced ecosystem services are hardly the only price distortion, zero prices cause under-investment (or no investment) and overuse, and are perhaps the ultimate price distortion.

As will be discussed in detail in Section III, **Solving Environmental Problems around the World**, fortunately a variety of institutions, approaches and instruments are available to help us understand and take into account the value of unpriced ecosystem services. In the remainder of this section, though, we begin to investigate ecosystem service values and why valuable ecosystem services might be lost. The next chapter discusses how we might estimate the monetary value of unpriced ecosystem services and the natural resources that produce them.

Issues for Discussion

1. Are ecosystem services scarce? How do you know?
2. Can you think of any ecosystem service that is not scarce?
3. What is the relationship between scarcity and opportunity cost? What are some opportunity costs people incur when they make decisions about the use of environmental assets?
4. Mr. Market says the private sector can provide everything society needs. The government should therefore get out of the way of business. What is your response?
5. Are there non-ecosystem services that are non-excludable? What are they and who typically provides them?

Practice Problems

1. Researchers at the University of California, Davis estimate that taking walks in forests improves human health by reducing blood pressure and anxiety. They conduct research on those who take walks in forests and those that do not and find that marginal benefit (v) = 20 – 4*Q, where v is the value of an additional walk and Q is the number of walks. For simplicity, suppose we can convert v to monetary units like dollars.

   a. For each additional walk, by how much does the marginal benefit decline?
   b. What is the maximum v possible to walkers?
   c. After how many walks are people worn out and v is driven to 0?
   d. At what level of v is Q at a maximum, assuming v cannot be less than 0?
   e. At what level of v is Q at a minimum?
   f. Suppose Q = 2, 3 and 4, what are the predicted values of v?
   g. Suppose v = 4, 8 and 12, what are the predicted values of Q?

2. Suppose that MC/supply and MB/demand are as in Fig. 3.4, but the government dictates a price that is less than P_E. Why is this price not an equilibrium? If not, which quantity is greater, quantity supplied or quantity demanded? If that price is not an equilibrium, can you discuss the expected mechanism for price to move toward equilibrium and quantity supplied to equal quantity demanded?

3. The market for computer hard drives is competitive. The marginal cost of producing hard drives (supply) is given by marginal cost = 50 + 3*Q and marginal benefit = 300 - 2*Q.

   a. What is the predicted equilibrium P and Q in the market for hard drives?
      Hint: Please use the method presented in Box 3.4.

   b. Suppose the price somehow gets stuck at $100. Is there a shortage or surplus? If yes, which is greater, quantity supplied or demanded, and by how much (i.e. how much is the shortage or surplus)? **Definition:** a shortage is the amount by which quantity demanded exceeds quantity supplied and a surplus is the amount quantity supplied exceeds quantity demanded.

   c. Answer the same questions as in 3b for a price of $300.

4. Suppose China is a key maker of hard drives for the US market. There is a trade war between the US and China, and the US imposes a tax called a *tariff* of $25 on each hard drive imported from China into the US. This policy drives up the cost of selling hard drives by $25 **at all levels of production** (i.e. the marginal cost function shifts up by $25). Incorporate this development into your calculation of equilibrium price and quantity in question 3a.

5. **Challenge Problem:** The popularity of cloud computing has increased dramatically since 2015. This trend continues, reducing the quantity demanded for hard drives by 50 hard drives at all price levels (i.e., the demand for hard drives shifts down/left by 50 hard drives). Incorporate this development into your calculation of equilibrium price and quantity in question 3a.

6. **Challenge Problem:** Suppose the effects in Practice Problems 4 and 5 happen at the same time. Please calculate the equilibrium price and quantity.

Notes

1 This topic is examined in more detail in Chapter 8.
2 https://data.who.int/indicators/i/87345F3/F810947
3 http://www.marketplace.org/topics/world/chinese-province-offers-clean-air-tourism-suffering-urbanites
4 A variety of functional forms could be used, especially if we want the graph to be curved rather than a straight line. One example is a quadratic function, which would take the form $V = a_1 - b_1*Q - b_2*Q^2$, where if b_2 is positive, the curve will bend downward (convex), and if b_2 is negative, it will bend upward (concave).
5 There are a couple of other details on the producer side that are not critical for our purposes that can be found in any microeconomics textbook.
6 When analyzing a market situation, the marginal benefit function is often called *demand*, because the function predicts how much users will buy at various prices. It slopes downward, because as they get more, people get tired of market goods and services. This feature of demand is universal enough that it is known as the *law of demand*.
7 We can do this because the equations are equal.

Further Reading and References

Acharya, R.P., T. Maraseni and G. Cockfield. 2019. "Global Trend of Forest Ecosystem Services Valuation – An Analysis of Publications." *Ecosystem Services* 39: 100979.

Gebeyaw, T. 2011. "Socio-Demographic Determinants of Urban Unemployment: The Case of Addis Ababa." *Ethiopian Journal of Development Research* 33 (2): 79-124.

IPCC. 2023. Summary for Policymakers. In: *Climate Change 2023: Synthesis Report. Contribution of Working Groups I, II and III to the Sixth Assessment Report of the Intergovernmental Panel on Climate Change* (Core Writing Team, H. Lee and J. Romero, eds.). IPCC: Geneva, Switzerland.

Kanellos, M. 2013. "Glenn Beck Vows to Fire Employees For Buying Fluorescent Bulbs, Recyclables." *Forbes*, August 27, 2013. Downloaded from http://www.forbes.com/sites/michaelkanellos/2013/08/27/glenn-beck-vows-to-fire-employees-for-buying-fluorescent-bulbs-recyclables/ July 15, 2014.

Loomis, J., P. Kent, L. Strange, K. Fausch and A. Covich. 2000. "Measuring the Total Economic Value of Restoring Ecosystem Services in an Impaired River Basin: Results from a Contingent Valuation Survey." *Ecological Economics* 33 (1): 103-117.

Naime, J., F. Mora, M. Sánchez-Martínez, F. Arreola and P. Balvanera. 2020. "Economic Valuation of Ecosystem Services from Secondary Tropical Forests: Trade-Offs and Implications for Policy Making." *Forest Ecology and Management* 473: 118294.

Sohngen, B., F. Lichtkoppler and M. Bielen. 1999. "The Value of Day Trips to Lake Erie Beaches." Ohio Sea Grant College Program Technical Bulletin OSHU-TB-039.

US Energy Information Administration (US EIA). 2023. *Annual Energy Outlook 2023*. Retrieved from https://www.eia.gov/outlooks/aeo/ March 2024.

Winkler, K. and K. Nicholas. 2016. "More than Wine: Cultural Ecosystem Services in Vineyard Landscapes in England and California." *Ecological Economics* 124: 86-98.

Winkler, K., J. Viers and K. Nicholas. 2017. "Assessing Ecosystem Services and Multi-Functionality for Vineyard Systems." *Frontiers in Environmental Science* 5: 15.

4 Estimating the Benefits of Environmental Assets and Ecosystem Services

---

What You Will Learn in this Chapter

- About the economic benefits of ecosystem services
- How to estimate the economic benefits of natural resources and ecosystem services using revealed preference methods, such as avoided cost, hedonic pricing and travel cost approaches
- Using experimental methods, such as contingent valuation and choice experiments, to estimate the economic benefits of natural resources and ecosystem services

---

4.1 Recap and Introduction to the Chapter

In Chapter 3, we emphasized that marketed and non-marketed ecosystem services are valuable in much the same way that other services are valuable. After all, what is the big difference between relaxing in a meadow under a clear sky on a beautiful summer day and going to a day spa?

But knowing that value *exists* and suspecting that it may be large is a far cry from figuring out an average person's actual value! Yet another step is to compare non-marketed ecosystem service benefits – in monetary terms – against the costs of getting those benefits. Most ecosystem services are not traded in markets and are fundamentally non-monetary. While the monetary costs associated with environmental improvements may be reasonably easy to estimate, benefits are therefore less obvious.

In Chapter 3, we considered a framework for analyzing environmental problems using benefits and costs. We left aside the issue of estimating the benefits of ecosystem services that are not transacted in markets. In this chapter, we focus on why we might want to estimate ecosystem service benefits, how to conceptualize those benefits and how we might estimate them in monetary terms.

DOI: 10.4324/9781003308225-5

4.2 Why Estimate Environmental Benefits?

Around the world, there are important human-wildlife conflicts. In sub-Saharan Africa, large mammals, such as the elephant, rhino and hyena, can make life difficult in rural areas by eating crops or livestock and damaging structures. The same is also true in South Asia, where elephants in Sri Lanka, wild pigs in Nepal and monkeys basically throughout the sub-continent can be serious nuisances for humans (e.g., see Bandara and Tisdell, 2003).

In the US, there have for a long time been major controversies related to the Gray Wolf. Hunted under an organized program of species eradication, by the 1940s it was hard to find gray wolves in the lower 48 US states. In 1974, they were listed as "endangered" under the US federal Endangered Species Act of 1973 and subsequently reintroduced. Endangered means a species is under significant risk of extinction throughout all or most of its range and therefore must be protected.

Federal wolf policy has vacillated significantly since the beginning of the millennium. After 35 years of protection, in 2009 the species was removed from the endangered category in Montana and Idaho and was actively hunted; the Gray Wolf was not removed from the list in Wyoming, where state law considers the species "vermin" (Morell, 2010).[1] The administration later proposed delisting the Gray Wolf throughout the US, with the exception of the Mexican Gray Wolf in the Southwestern US, but this was put on hold by courts until 2017. In April 2017, the administration of President Donald J. Trump delisted the Gray Wolf throughout the Northern Rockies, including Wyoming.

In March 2019, the Trump Administration proposed removing all federal protections on the Gray Wolf throughout the US, with the exception of the Mexican Gray Wolf, and on September 29, 2020 the US Fish and Wildlife Service finalized the change in management rules under the Endangered Species Act, removing all federal protections. This change ceded management of the species to the US states and indigenous peoples (Brown et al., 2020).

A chorus of conservation groups immediately condemned the rule change and vowed to continue the fight in court, saying that the Gray Wolf offers important ecosystem services. Many still see the Gray Wolf as "vermin" and some may even ask if Gray Wolves have *any* value, because they prey on farm animals. Sheep produce meat and wool, but what does the Gray Wolf give people except headaches? Does anybody really care about the Gray Wolf?

John Krutilla argued in his seminal 1967 article, *Conservation Reconsidered*, that just because people do not actively "use" nature does not mean they do not value it. Perhaps people sufficiently value the *existence* of the Gray Wolf that based on a strict cost-benefit analysis, their willingness to pay would be greater than the losses to ranchers and others? Though listings under the Endangered Species Act are not based on cost-benefit analysis, high existence value can create strong constituencies for protection. Estimating the existence value of the Gray Wolf - if analysts could figure out how to do it - might help convince policymakers that despite limited or no market value, people highly value protecting this species.

Regulations may also require estimating environmental benefits. Under a presidential executive order dating to the 1990s, any regulation that is expected to have economic effects of at least $250 million (a small amount in the approximately $27 trillion US economy) must be subjected to some degree of cost-benefit analysis. This order also applies to environmental regulations, which means that environmental benefits must somehow be estimated and weighed against estimated costs.

Governments have a variety of competing investment priorities. Around the world, governments are in charge of critical social services like public health, support for low-income people and education. They are also typically responsible for investing in and maintaining important infrastructure like roads, water, bridges and electrical supply. Local governments especially take leading roles in public safety, such as police and fire protection. And then there is the environment.

Governments use public resources to protect air and water quality, maintain open space, protect habitat for other species and provide natural recreation areas like parks. The costs of government environmental investments are part of public budgets and are easy to see, but how much do constituents get from those budget outlays? Is it worth it to hire 20 people to work all summer to remove invasive species from a major city park? Would those resources be better spent elsewhere? To answer such a question, it is necessary to estimate the environmental benefits constituents get from such a policy.

A final potential reason we may want to estimate the monetary value either of ecosystem services or of environmental assets is legal liability. When environmental assets are damaged, flows of ecosystem services that are shared by local communities or even whole countries are reduced. What is the monetary value of the ecosystem services *lost* as a result of damaging assets?

Box 4.1 How Do You Put a Price on Marine Oil Pollution Damages?

Ships, marine terminals, offshore oil rigs and harbors can spill oil into bays, gulfs, oceans and seas, damaging marine ecosystems and affecting people's livelihoods and enjoyment of those systems. Under a variety of US federal legislation, owners and operators of marine vessels and facilities can be held liable for damages and must "make the environment and public whole" following a spill.

But how much do those responsible for oil spills owe? For example, how much did BP owe when on April 20, 2010 the Deepwater Horizon oil rig exploded and sank off the Gulf of Mexico, spilling 4 million barrels of oil into the Gulf?

Federal government agencies who are the trustees of marine resources have understandably struggled with answering this question. As a first cut, agencies have valued damages at so-called "replacement cost," which is the amount required to restore ecosystems to their previous states. Assessing costs in complex marine ecosystems can be difficult and the value of *replacement* is likely to be far less than the *damage*, particularly when ecosystem effects are uncertain and far-reaching.

Source: Boyd (2010).

Answering such a question could have very practical implications. Under civil legal codes in many countries, as well as US states, in addition to potential criminal liability, those who damage common assets are responsible for civil penalties. How much should companies pay, for example, for an accidental release of a heavy metal like arsenic into the atmosphere? Such a release could be toxic to humans and other species, but what is the value of those damages? What should they be required to pay those who were damaged?

In sum, estimating the monetary value of ecosystem services, either for the purpose of understanding what we gain from improvements or for measuring damages, can be very useful for advocacy, regulatory analysis, investment policy and determining civil liability, among other possible reasons. Of course, for some purposes, such as evaluating safe min-imum standards, valuation does not *need* to be monetary, and sometimes physical units (e.g., species conserved or lives saved) or even indices (e.g., percentage risk reduction) are more than sufficient.

4.3 Types of Environmental Benefits

In Chapter 1, we discussed the Millennium Ecosystem Assessment typology of ecosystem ser-vices as 1) provisioning; 2) regulating; 3) cultural; and 4) supporting. We now discuss a more "economicsy" categorization, which focuses on the degree to which ecosystem services are marketed. This approach preceded the Millennium Ecosystem Assessment by perhaps a few decades, continues to be widely used and, as discussed in the next section, is often the organizing framework for actual valuation studies.

Figure 4.1 presents a diagram of environmental benefit categories. It can be used to con-sider both environmental assets (e.g., lakes, forests and rivers) and individual ecosystem services that flow from assets. At the far left of the diagram are *direct use values*, which are provisioning services actually extracted from the environment. Examples include timber cutting, oil drilling, shellfish harvests and hunting. There are two distinguishing features of direct use values. First, taking them actually changes and typically degrades environmental assets. For example, when oil is pumped or gold is mined from underground deposits, those assets have less materials remaining in them, reducing the value of those deposits.

Second, direct use values are often marketed. As we will see in the next subsection, having market prices to directly value ecosystem services can be quite handy. If we want, for example, to value the potential timber in a forest, we can simply use the average market price for the type of wood that is in the forest multiplied by the estimated board-feet or meters. In general, with direct use values we don't need to use the complicated valuation methods that have been developed by economists and others, because we have market prices to work with.

Indirect use values come from services that humans use, but it is not necessary to actu-ally extract anything in order to enjoy those services. An example is nature photography or hiking. We *use* the mountain, forest or river when we do these types of recreation activities, but we don't meaningfully change those natural resources by using them. When we enjoy a view of a beautiful sunset from one of the beaches on the west coast of Florida or on the island of Zanzibar, off the coast of Tanzania, we are using the environment to get valuable benefits, but we are not changing the beach or the sky by looking at the sunset and beach.

When we attempt to estimate indirect use values, we need to take a significant step away from directly using market prices. Though there may be some market information that can be tapped to infer ecosystem service values, there is nothing directly applicable. For example,

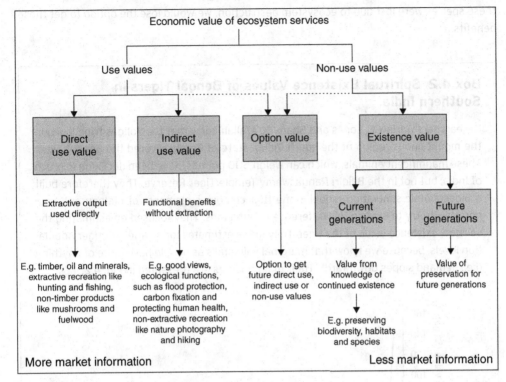

Figure 4.1 Economic Benefits of Ecosystem Services

*There are a variety of such taxonomies and so presentations by other authors may vary somewhat from this one.

unlike timber and mushrooms, which have market prices that make valuation easy, there is no price for sunset views. We know, though, that real estate in places that are on beaches and have water views is typically more expensive than properties without such amenities. We therefore may be able to utilize data from real estate markets to infer ecosystem service values.

Option value recognizes that a lot is not known about what the natural environment offers humans and indeed values may change over time. Preserving nature therefore maintains options that we might want to utilize in the future. For example, a variety of pharmaceutical drugs and naturopathic medicines have been taken from natural ecosystems. Examples include atropine, which is a drug to control spasms and seizures; ephedrine, which some of us might take to control allergy problems; and the former antimalarial drug, quinine. Many more pharmaceuticals have been derived from or inspired by natural molecular structures. Examples include chloroquine, which is a modern antimalarial drug; and aspirin, which is made completely synthetically, but was originally derived from the willow tree.

Numerous naturopathic medicines are taken directly from plants, including gingko, Saint John's wort and valerian (Houghton, 2001). These are direct use values from plants that keep us healthy. Many personal care products are also directly used or derived from plants, but if these species were lost due to ecosystem destruction, we would lose the *option to* get those benefits.

Box 4.2 Spiritual Existence Values of Bengal Tigers in Southern India

Researchers Adrian A. Lopes and Shady S. Atallah knew that the Soligas tribe living in the mountainous region of the south Indian state of Kerala revered the Bengal Tiger. These magnificent animals, which can weigh 500 pounds, have been declining in much of India, but not in the Biligiri Rangaswamy Temple Tiger Reserve. They therefore built a bioeconomic simulation model – one that combines elements of biology and economics – to try to explain this difference in outcomes. They focused on estimating the spiritual existence value of the tiger. They made estimates for a variety of tiger population levels, because we know that marginal value falls as people have more of anything (i.e., demand slopes downward). Their main findings are presented in Fig. 4.2.

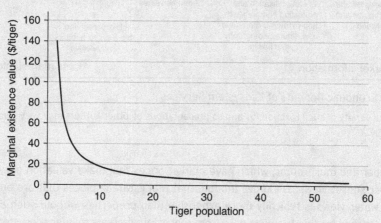

Figure 4.2 Marginal Value of Tigers as a Function of the Tiger Population
Source: Lopes and Atallah (2020).

Furthermore, there is increasing evidence that ecosystem preservation and especially biodiversity offer important indirect use values. Smith et al. (2014) found that between 1980 and 2013 there were over 12,000 infectious disease outbreaks involving over 200 diseases, with the frequency of such outbreaks significantly increasing over time. The World Economic Forum, which is a private economic organization based in Switzerland, notes that this increase in epidemics and perhaps even the COVID-19 pandemic are linked to loss of biodiversity.[2]

Jabr (2020) summarizes a variety of evidence that human disruption of ecosystems may be contributing to zoonoses – the spreading of diseases from other animals to humans. He notes that as much as 75% of infectious diseases come from other animals, including SARS, Ebola, West Nile and Zika viruses. COVID-19 is probably also a zoonotic disease and the pandemic, which started in 2020, may have been prevented by avoiding disruption of animals (likely bats) and their habitats.

The Inter-American Development Bank, which is one of four regional development banks associated with the United Nations, also raises the possibility that ecosystem disruption may be linked to the COVID-19 pandemic.[3] Preserving biodiversity and habitat therefore offers the possibility – the *option value* – to avoid zoonotic disease transmission.

Existence value acknowledges that people may not need to actually *use* environmental assets at all to get ecosystem service benefits from them. Most people get benefits from good-quality environmental assets, such as wildlife in terrestrial ecosystems, marine species and gigantic trees in Sri Lanka, the Pacific Northwest and elsewhere. These values may be in terms of benefits to current generations or future generations, in which case they might be called *bequest values*. If there is uncertainty regarding future existence values, they may be thought of as option (existence) values.

As existence value is a non-use ecosystem service value that merely requires caring about assets and the services they provide, the concept has been broadly applied to a variety of services and assets. For example, as discussed in Box 4.2, Lopes and Atallah (2020) developed a model to estimate the existence value of the Bengal Tiger to indigenous groups living in a tiger reserve in Southern India. The Soligas tribe worships the tiger, so the existence value estimated is spiritual value.

Species preservation and recovery offer particularly important types of existence values. Richardson and Loomis (2009) conducted a *meta-analysis* – a study of over 30 studies – of how people value threatened, endangered and rare species. They discussed research that analyzed human valuation of species we may have heard about, like sea otters, salmon, the Gray Wolf, Bald Eagle, Humpback Whale, Gray Whale and the Whooping Crane. Also covered were species with which we may not be familiar, such as the Monk Seal, Riverside Fairy Shrimp, Silvery Minnow, Striped Shiner and the Steller Sea Lion.

How big are the values? They found that a 50% increase in sea otter populations was on average worth about $88 per household per year in 2007, which when updated for inflation is approximately $114 (2023 US dollars). This means that Richardson and Loomis (2009) estimated that an average US household would be willing to pay about $114 (in 2023 prices) to improve the welfare of sea otters by taking steps to increase otter populations by 50%. As there are about 130 million households in the US (US Census Bureau, 2021), that means a 50% increase in sea otter populations could have a total US value of about $15 billion!

But how were the valuation studies analyzed by Richardson and Loomis (2009) actually done? What were the specific methods used and what is the menu of options available to value environmental assets and ecosystem services? Answering these questions is the subject of the remainder of this chapter.

4.4 Estimating Ecosystem Service Benefits

As we discuss the *how* of estimating ecosystem service values, we must also remind ourselves *why* we need to go to the often extraordinary lengths required to estimate values. No ecosystem services except provisioning/direct use ecosystem services have actual market prices attached to them, so we are essentially forced to figure out and implement clever, assumption-filled, often imperfect and generally resource-intensive and time-consuming methods, to estimate ecosystem service values. As we evaluate these methods, let us remember that if there were more straightforward methods, we would certainly use them!

There are two approaches. The first uses observed market data to estimate ecosystem service values. As people actually spent real money in market transactions, the markets "reveal" something about people's valuations and preferences (assuming people only pay up to their valuations). These methods are therefore called *revealed preference* or market-based valuation methods.

For many ecosystem services, though, there are simply no markets to latch onto. For example, what market could we possibly use to estimate the spiritual value of Bengal Tigers? In such situations, the only possibilities are to construct a model of human valuation that has nothing to do with markets. Such a model could be a bioeconomic model, such as the one used by Lopes and Atallah (2020), but another method would be to use behavioral experiments specifically crafted to figure out people's values. Because preferences and values are not linked to market behaviors, but instead are "stated," these methods are often called *stated preference* methods.

4.4.1 Revealed Preference: Estimating Benefits Using Market Information

Replacement Cost Benefit Valuation

The replacement cost method of ecosystem service valuation is perhaps the most straightforward, because it simply undertakes a thought experiment of the following form: If particular ecosystem services were to be reduced or disappear, what would it cost to replace them?

Replacement cost relies on estimates of the cost of replacing ecosystem services to estimate their values. We can think of this process as a group of decisionmakers, informed by relevant experts, such as ecologists, economists and engineers, sitting down and budgeting out a public project to replace lost ecosystem services. For example, one way to value clean spring water in a community would be to estimate how much water is used and the cost of getting that water from another source.

A lot of attention has been paid recently to the loss of insect pollination services, due to declines in populations of species, such as wild bees. VanBergen et al. (2013) estimate that 75% of crops and between 78% and 94% of flowering plants depend on animal pollinators to produce seeds. Species used for food by humans, such as tree fruits and grains, are part of this dependence.

Allsopp et al. (2008) estimated the cost of replacing the pollination services currently provided for free to the fruit-production sector in the Western Cape Province of South Africa. Apples, pears, peaches and plums are examples of fruits grown in Western Cape Province, which are pollinated by animals, and especially insects. If bees and other species were not able to pollinate fruit trees, they assumed people would have to either dust trees with pollen or pollinate them by hand. They used the estimated cost of replacing these services as the value of insect pollination, and found that values were between $68 million and $394 million per year in $2020.[4]

The replacement cost valuation method has the clear virtue that it is understandable and, with the right data, anyone can make these calculations. The problem with the method is that we need to presume - albeit based on expert knowledge - what would otherwise have happened if some or all of an ecosystem service were lost.

There is indeed a serious question as to whether the replacement cost method even estimates benefits in the normal sense in which we have been using that word. As we discussed in Chapter 3, benefits - marginal, total or otherwise - are on the user or demand side of the equation. Using information from the supply side - what it might cost to *replace* services - to estimate demand-side ecosystem service values, is a bit ... confusing.

Productivity and Avoided Cost Benefit Valuation

Ecosystem services often help us avoid costs or they may improve human productivity in important ways. Clean air is a critical natural resource supporting human respiratory health. When the concentrations of air pollutants like $PM_{2.5}$ are reduced, it reduces the incidence of respiratory health problems and even human mortality. This important value is not directly priced in any market, but markets that are *related* to respiratory health exist and can perhaps be used to help estimate the value of such services.

For example, it costs money to treat asthma and other respiratory conditions that can be aggravated by air pollution. These costs are *avoided* if the incidence of respiratory health problems is reduced. Health care markets may therefore help us get at least partial estimates of the respiratory health benefits from reducing air pollution.

Voorhees et al. (2014) used the productivity/avoided cost method to estimate the public health benefits in Greater Shanghai, China from achieving China's $PM_{2.5}$ air quality standards. They started with information from epidemiological studies, which connect changes in air pollution concentrations with health outcomes, such as chronic acute bronchitis, asthma attacks, lower respiratory infections and cardiovascular problems. They determined that for both adults and children these health effects can require outpatient doctor visits, hospitalizations and even admission to emergency departments.

Such interactions with the health care system are costly. The authors cleverly isolated the effects of $PM_{2.5}$ emissions from other drivers of health problems to estimate the *reduction* in health system costs that can be expected if Shanghai were to achieve China's so-called "Class II" air quality standards. They estimated that achieving those standards would reduce hospital admissions by 5400-7900 per year. As of 2014, the estimated reduced Shanghai health care system costs ranged from $11.3 million to $21 million per year, with just under half of the benefits made up by avoided hospital admissions.

Econometrics and Valuing Ecosystem Service Benefits

Like the replacement cost method, using avoided costs and improvements in productivity to value inherently non-marketed ecosystem services has the clear benefit that they are understandable to most people. The method only involves estimating how many incidences of human benefits a policy would generate and multiplying by the avoided monetary cost or productivity benefit that each case generates. It is all addition and multiplication, which is good work if you can get it!

The problem with this approach is that these methods *presume* relationships between ecosystem service improvements and human benefits. There is almost an engineering aspect to the avoided cost and productivity methods in which we define – rather than estimate – what is gained from environmental improvements.

Unfortunately, if we want to truly estimate the monetary value of environmental benefits, more complicated methods are needed. Indeed, all the rest of the environmental valuation methods – hedonic pricing, travel cost, contingent valuation and choice experiments – require a branch of statistics called *regression* analysis or what in economics is typically referred to as *econometrics*. Econometrics looks at relationships between variables. Typically, there is an outcome variable of interest, which is called the *dependent variable* or left-side variable. The variation in this dependent variable is then explained by many *independent variables*, which are hypothesized to affect the outcome variable.

Econometrics involves statistically testing these hypotheses to see if indeed there are relationships – either correlation or perhaps causality – between the independent and dependent variables. Often, indeed, we are most interested in *causality*, meaning that changes in the independent variables actually *cause* changes in the dependent variable.

Challenge Yourself

Can you name three variables other than work experience that may help explain variation in income levels?

Figure 4.3 gives an econometric model using years of work experience and annual income ('000s of $US) as an example. The graph plots combinations of years of experience and annual income as dots. For example, one person has no experience and earns $20,000 per year. Another person has 15 years of work experience and earns $48,000. Although two people with ten years of experience earn different salaries ($29,000 and $49,000), in general there is an increasing relationship between years of experience and income.

The line on the graph that runs through the data is a "best-fit" line, which broadly speaking minimizes the sum of the distances between the line and the data points. Such best-fit functions are called *regression lines*. Though the fit is far from perfect, the regression line is strongly upward-sloping, indicating that as years of work experience increase, on average, so does annual income.[5]

Like replacement cost and productivity/avoided cost methods, the two econometrics-dependent revealed preference valuation methods – hedonic pricing and travel cost – rely on closely linked markets to value non-marketed ecosystem services. The difference, though, is that the relationships between the linked markets and ecosystem services are no longer *assumed* to be correct. Instead, they are statistically tested.

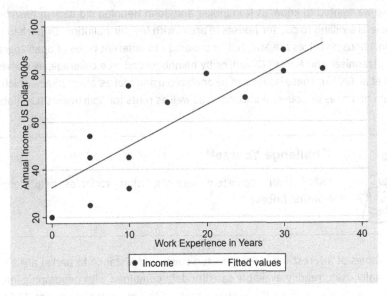

Figure 4.3 Regression Analysis Example

Hedonic Pricing

One interesting application of hedonic pricing taps into the wealth of data from real estate market transactions to value environmental amenities like parks and clean air. It can also be used to value non-environmental amenities such as good schools, proximity to coffee shops, availability of public transport and public safety. As long as public amenities can be measured and it is reasonable to suppose that property values are related to them, hedonic pricing can be used.

Box 4.3 Seminal Research on Air Pollution and Residential Property Values

In their seminal (i.e., first-ever) work, Ron Ridker and John Henning estimated how sulfate (SO_2, SO_3, H_2S and H_2SO_4) and particulate air pollution affected the median (i.e., 50th percentile) house price in St. Louis, Missouri. In the 1960s, sulfate air pollution and the acid rain it caused were a serious problem in the US, as it is currently in many countries.

   The authors found that sulfate air pollution explained a bit over 1% of the variance in median house prices. They noted that if low-sulfur fuels had been used back then (common in the US today), St. Louis property values would have been about $83 million higher – roughly $800 million in 2022 dollars. They were the first authors to establish that air pollution drives down property values.

Source: Ridker and Henning (1967).

Suppose we wanted to know, as Ron Ridker and John Henning did back in 1967, how much more people are willing to pay for houses in areas with less air pollution. Or perhaps we want to know how housing prices are affected by proximity to different types of open space, as was done by Lutzenhiser and Netusil (2001), or by neighborhood tree coverage, as was estimated by Netusil et al. (2010). These issues can be analyzed using real estate markets, which produce a lot of data on prices of houses and condos, as well as rents for apartments (i.e., flats).

Challenge Yourself

Other than ecosystem services, what variables help determine housing prices?

Our *variables of interest* (air pollution, tree cover and distance to parks) are all measurable, especially using readily available satellite data combined with geographic information system (GIS) software, such as the systems sold by the prominent company ESRI and the non-profit, open-source, QGIS.[6] These tools can identify greenery, such as tree coverage, and calculate and map distance to neighborhood features like parks and schools.

Data on air pollution across a landscape can be *overlaid* on maps of housing prices to look at relationships. Of course, air pollution data are not measured everywhere in cities, even in higher-income countries, and actual monitoring of air pollution may be quite spotty in low-income settings. Fortunately, some pollutants can be estimated using satellite imagery, which avoids the need for earthbound air pollution monitoring stations.

While our interest as analysts might be in open space, tree cover or air pollution, a lot of other variables affect housing prices. **Before reading on, please think for a minute about some of the factors that affect house prices. These factors should come to you quickly.**

Perhaps house size, lot size, number of bathrooms, neighborhood, school quality and heating system type came to mind. All these factors must be included in a model of housing prices. If key features that happen to be correlated with our environmental variables, are omitted from hedonic models, we could wrongly attribute effects (+ or -) to environmental variables that are really due to something else. For example, we could confuse the effect on house prices of more bathrooms with being close to a park, simply because big houses tend to be near parks!

Equation 4.1 Sample Hedonic Price Equation Specification

House price = a_0 + a_1%tree cover + a_2*size of house + a_3*size of land + a_4*number of bathrooms + a_5*fireplace + etc. + error

---

Suppose, like Netusil et al. (2010), we are interested in better understanding the value of tree canopy. A linear version of our hedonic price model might look something like Eq. 4.1, where all the "a"s are parameters to be estimated and the parameter a_1 tells us how much the data suggest the average house price increases when the percentage of tree cover in a neighborhood increases by 1%. For example, if a_1 = 2000, it means that a 1% increase in tree cover increases house prices on average by $2000. Tapping into a housing market, we therefore now know that investing in tree cover is valued at $2000 per 1% of tree cover, which is pretty valuable information for a city trying to figure out how much to invest in street trees!

Hedonic studies have also been used to value amenities and even *disamenities* in lower-income countries. Cities in low- and middle-income countries can be pretty messy. Air quality is often hazardous, garbage may be openly dumped in streets and sewage can make its way to rivers and streams via open canals that run through neighborhoods. Such public aspects of cities represent costs to those who live in or visit those places, but how much does an open sewer reduce human welfare on a day-to-day basis? What about garbage in the streets? Hedonic pricing to the rescue!

Box 4.4 Indicative Data Used in Hedonic Price Analysis by Nepal et al. (2020)

	HH1	HH2	HH3	HH4	HH5
Dependent Variable					
Sale Price (Million NPR)	8.42	5.52	5.53	7.51	9.22
Select Independent Variables					
House is in Hills (1/0)	1	1	0	1	0
Garbage Collection? (1/0)	0	0	1	1	0
No. Rooms	7	5	9	6	4
Land Size (Sq. ft.)	4442	2102	1467	7239	6351
Toilet in House? (1/0)	1	1	1	0	0
Piped Water in House (1/0)	0	1	1	1	1
Distance to Bus Stop (km)	0.52	1.86	0.78	2.20	0.33
Distance to Paved Road (km)	3.32	2.11	1.04	1.53	4.72

Yes coded as 1, no as 0. Data are indicative only. HH = household.

Nepal et al. (2020) analyzed just these questions in urban areas of Nepal. Nepal is tucked between India and the Himalayan mountains, and some of its cities have serious problems with uncollected and uncontrolled garbage. Nepal has purchasing power parity (PPP)-adjusted gross national income (GNI) per person of about $4726 in 2023 dollars (World Bank, undated), which is just high enough for the World Bank to classify it as a lower middle-income country.[7]

PPP stands for *purchasing power parity*, which is a price adjustment that accounts for differences in overall price levels across countries. As we discussed in Chapter 3, prices of goods and services tend to be similar across countries if they are traded, but some services like haircuts cannot be traded and therefore prices may vary dramatically across countries. The PPP price adjustment, which was developed by the World Bank,[8] adjusts for the reality that people can buy *non-tradables* more cheaply in some countries (usually low-wage) than others.

The authors found that in 2014/2015 the average house in urban areas in the hills of Nepal had an owner-assessed value of 7,640,000 Nepali Rupees (NPR). The average house price in the lowland region of the country, which is called the *Terai*, was about half the hill price. The exchange rate from NPR to $US in December 2014 was NPR 102/$US, meaning that the average house sold for about $75,000.

They transformed the reported house prices into natural logs.[9] Mathematically, when running regressions with a dependent variable as a natural log, the estimated coefficients (i.e., the "a"s in Eq. 4.1) tell us the *percentage* change due to a one-unit change in the independent variables.

Using data such as the example in Box 4.4, the authors found that the existence of garbage collection services increased house prices a lot! Having those services increased prices by 30%-45% compared with what they otherwise would have been without trash pickup. They also found that an additional square foot of land area increased prices by 30%-40% and an indoor toilet had roughly the same effect. They concluded that expanding municipal waste collection is easily worth the cost, because residents seem to value garbage pickup so highly. They suggested that their results armed municipal officials with arguments for introducing fees to clean up Nepal's cities.

Hedonic pricing is also used to value avoided mortality risk (i.e., risk of death), such as from automobile accidents or pollution exposure. In this case, the closely linked market that researchers tap into – this time to value safety – is the labor market within which workers earn wages and salaries. Of course, different jobs have different mortality risks, and we can use econometrics to try to estimate the increase in wages and salaries workers get when they accept a greater risk of death. Using this information, it is possible to generate an estimate of the value of reducing mortality risk to the point that one person's life would be saved. This value, perhaps unfortunately, has come to be known as the *value of a statistical life*, or VSL (Viscusi and Aldy, 2003; Cameron, 2010).

Suppose researchers find using regression analysis that on average workers require an additional $1000 in current-year lifetime wages to accept a 1 in 10,000 risk of death. This result means that for every 10,000 workers who accept the additional $1000 in wages, there is a *certainty equivalent* of one death. What is the value of this certainty equivalent or VSL?

It is $\frac{\$1000}{(1/10,000)}$ = $1000 * 10,000 workers = $10 million.

The VSL is very important for analyzing public investments that reduce human mortality. The US Department of Transportation (US DOT) posts full guidelines for these cost-benefit analyses on its website (US DOT, 2022a), including guidance on the VSL analysts should use. As of 2022, US DOT was using a VSL of $11.8 million to value mortality risk reductions (increases) (US DOT, 2022b). Such a large VSL means that when a project reduces human mortality, estimated benefits can be very high.

Travel Cost

The travel cost method is used to value visits to recreation sites. It taps into the opportunity costs visitors typically incur when they visit recreation sites and uses that information to infer the value of recreation experiences. These opportunity costs include time – by its nature, recreation of any type takes time, implying opportunity costs – and money, which means people have given up other goods and services that could have been purchased.

The money side of travel costs usually includes transportation (e.g., car or airplane), perhaps entry fees, maybe hotels and possibly restaurant meals. The opportunity cost of time is what visitors would have otherwise done with that time. The monetary value of time (e.g., a day) is generally estimated using a relatively small portion (usually 30%–50%) of people's wage rates (Chavas et al., 1989; Yuan and Wang, 2019), because people tend to vacation when it is convenient; working and recreation therefore only partially conflict.

Box 4.5 Willingness to Pay for a Glimpse of the Tiger

Indrila Guha and Santadas Ghosh (2009) used a travel cost method to estimate the average value of a trip to the Sundarban Tiger Reserve, which is a mangrove forest and UNESCO World Heritage Site in India near the border with Bangladesh. Tourists mainly from the Indian state of West Bengal, but also other areas of India and internationally, book different tour packages to see the Bengal Tiger, which lives in the mangrove forest. The authors noted that Sundarban is the only place in the world where Bengal Tigers live in mangrove forests.

The tour package chosen and the distance from the Reserve, which the authors classified as "zones," were used to estimate visitors' willingness to pay for a trip. About 65,000 visitors came to Sundarban Tiger Reserve each year, generating about $377,000 in benefits. The authors estimated that the value generated was high enough that fees to enter the Reserve could easily be increased from the current very low levels to help fund the park.

Source: Guha and Ghosh (2009).

Do you have a park or other outdoor recreation site near your house that you and others like to visit? Perhaps it has no access fee and you can even get there by walking. No accommodations or airplane tickets or even gasoline are needed to get there and travel does not take a lot of time. Think for a minute about how many times you went there during the last few years or maybe even over your lifetime. Probably a lot, because you enjoy the place (it is desirable), but it is not that costly. Bottom line, using our methods from Chapter 3, we might conclude it is a good deal. If you go there a lot, we might also say that you're way down your marginal benefit function (i.e., MB is low), but you still go a lot, because the cost is also low.

Contrast the number of times you've visited your local recreation area with a higher-profile, world-class natural recreation site. Depending on where you live, maybe we are talking about the Western Ghats in India, Yosemite National Park in the US, Mount Kilimanjaro in Kenya, Kruger National Park in South Africa, Mount Fuji in Japan, the Galapagos Islands in Ecuador, the Swiss Alps or the Black Forest in Germany. These are places that attract visitors from all over the world and people lay down serious money to experience them. All these natural areas require special arrangements and people may not go many times or even at all during their lifetimes. Why do most people visit infrequently or not at all when those places are so amazing? Of course, the answer is that visiting them is expensive. Most people would need to buy airplane tickets, pay big entrance fees, stay in hotels, take a lot of time off work etc. to visit.

But what about people who live in the Indian state of Kerala and can go out to the Western Ghats for a weekend? What about South African visitors who could do the same in Kruger National Park or Swiss people who actually live in the Alps? Of course, these folks are likely to visit such places very often. Why? Because the costs are low and the sites are amazing.

I'm reminded of when we lived in Southern California when our kids were young. One year we bought annual passes to Disneyland in Pasadena, California, which were only available to people who lived in the region. It was about an hour's drive from Redlands, California, where we lived, so we would go whenever we felt like it, usually when we thought the place would not be swarming with thousands of people coming from all over the world to visit the famous, original Disneyland!

The basic idea of travel cost is to use differences in visitors' total costs of visiting a recreation site to estimate the demand, with the expectation that on average how much people visit depends on how much it costs them. If you would like to be an environmental valuation expert, consider focusing on travel cost, because data are collected on-site by actually surveying visitors using specially developed questionnaires that allow collection of travel costs and other needed data – and nobody recreates in unpleasant places. Being a travel cost expert means you are always visiting great recreation areas!

Equation 4.2 Sample Travel Cost Equation Specification

Number of times visited the specific recreation site = a_0 + $a_1$total travel cost + a_2*age + a_3*gender + a_4*income + a_5*experience with such recreation sites + etc. + error

---

The trick is to use the survey questionnaire to get all travel cost information and then add up those details into an aggregated "travel cost" for each visitor. As for most environmental valuation, the travel cost model is estimated using econometrics. Equation 4.2 provides an example of a potential linear *specification* (i.e., a particular form) of the model. Think of visits as the quantity and travel cost as the price. The estimated parameter a_1 therefore tells us how much quantity demanded falls when total travel cost increases by one monetary unit (e.g., €1.00). As always, in such regression models, we need to include variables that affect the quantity that might be correlated with the cost. Examples of such variables in travel cost models include income, age, gender and experience with the type of recreation site analyzed. Including all the relevant variables can help us get *unbiased* estimates of a_1 and demand.

What we want to end up with is a demand function, so after we estimate the "a"s, for all but a_1 we substitute the average values from the data into the travel cost equation to get the average effect of all those variables on visitation. For example, suppose we are analyzing a recreation site in France. Average annual income in our sample is €40,000 and using econometrics we estimate that a_4 = 0.0000003. This means that a €1.00 increase in income is estimated to increase visitation by 0.0000003*€40,000 or 0.012 times. We can do this for all variables other than travel cost and then just lump those numbers into a_0, the vertical intercept (sometimes called the *constant term*), because we are not interested in their independent effects.

Equation 4.3 Example Estimated Travel Cost Equation

Number of times visited the specific recreation site = 3.22 - 0.00044*Total travel cost

---

Suppose we followed this procedure for all the "a"s other than a_1, yielding a function such as Eq. 4.3. This equation says that the estimated maximum number of visits (i.e., when travel cost approaches 0) of those in the sample was 3.22. The equation also says that an extra €100 in travel cost reduces visitation by 0.044 times. Of course, to estimate the total willingness to pay for a visit or an average number of visits, we want the travel cost as the dependent variable. Subtracting 0.00044*Total travel cost and number of times visited from both sides and dividing through by 0.00044 gives Travel Cost = 7318 - 2273 * Number of Times Visited, which is the demand equation. It is plotted as Fig. 4.4. The predicted travel cost for one visit is therefore (€7318 - €2273) * 1 Visit = €5040.

As shown in Fig. 4.4, the total value of one trip has two parts represented by areas A and B. Area A is total travel cost of one trip (€5040*1) and Area B is the "extra value" from taking one trip to the recreation site. This area, which is typically called *consumer surplus*, is the value over-and-above what is actually paid in travel costs to visit the site. With our linear specification, we can calculate Area B using the triangle formula $\left(\frac{1}{2} \text{Base*Height}\right)$. The total value, estimated using willingness to pay, is therefore €5040*1 + (€7318 - €5040)/2 = €6179. A visit to the site therefore turns out to be valued at €6179, which seems quite valuable!

Travel cost, hedonic pricing, productivity/avoided cost and even the relatively simple replacement cost method use people's actual market decisions to estimate the value of environmental assets and ecosystem services. These methods use market data - health care,

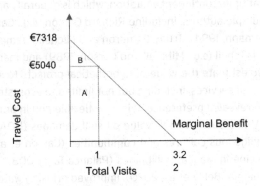

Figure 4.4 Example Marginal Benefit for a European Recreation Site for One Person

equipment, wages, agricultural prices, travel costs, real estate – to "reveal" humans' under-lying preferences for natural resources and ecosystem services. As decisions have actually been made, as long as the econometrics (or arithmetic in the case of replacement cost and productivity/avoided cost methods) are good, the analysis provides at least a partial esti-mate of value.

But what if the ecosystem service or environmental asset to be valued has absolutely no links to markets? For example, if you would like to value the survival of rhinos in southern Nepal or the set of ecosystem services provided by wetlands in Quebec, there may be no markets. Under such circumstances, the only choice is to use survey-based methods in which environmental valuations are derived using experimental methods.

4.4.2 *Stated Preference: Using Experiments to Estimate Benefits*

When there are no linked markets or analysts want to try to be more comprehensive in their valuation estimates, one must use one of two experimental methods. *Contingent valuation* and *choice experiments* are also sometimes known as *stated preference* methods, because we infer valuations based on answers to survey questions. In contingent valuation and choice experiment studies, analysts craft surveys that bring participants, which are typically called *respondents*, into their hypothetical, experimental worlds, with the goal of estimating their values.

Contingent valuation is used when estimating the value of a whole project, such as an improvement in a forest, perhaps with many components. Such projects improve environ-mental assets and thereby increase key ecosystem services. Choice experiments are very helpful when valuing the individual components (typically called *attributes*) of a project vis-à-vis each other and can also be used to estimate total project value. The challenge with choice experiments is that the value of perhaps a maximum of six project attributes can be estimated.

Contingent Valuation

The contingent valuation method was first proposed as a possibility by S.V. Ciriacy-Wantrup in 1947 for valuing what he called "extra-market goods" generated by soil conservation investments. Since that time, contingent valuation, which is generally abbreviated as CV, has been developed by multiple authors, including Richard Carson (e.g., Carson, 2000), Michael Hanemann (e.g., Hanemann, 1994), Trudy Cameron and Michelle James (e.g., Cameron and James, 1987), Robert Mitchell (e.g., Mitchell and Carson, 1989) and many others.

CV uses a survey to estimate the value of hypothetical projects to improve one or more services or value losses of services, including but not limited to ecosystem services. Because it is *stated* rather than *revealed* preference, it is very flexible and in principle can value any-thing. For example, CV has been used to value oil spill damages (NOAA, undated), preser-vation of American indigenous cultures and communities (Carson et al., 2020), availability of a dengue fever vaccine in Manila, Philippines (Palanca-Tan, 2008), soil conservation on communal lands in Ethiopia (Belay et al., 2020), improved drinking water in Kenya (Brouwer et al., 2015), green energy in China (Xie and Zhao, 2018), urban cultural sites in Iceland (Cook et al., 2018) and PM$_{2.5}$ air pollution reductions in Nanjing, China (He and Zhang, 2021), among many other applications.

Box 4.6 Valuing Dengue Fever Vaccines in Manila, Philippines

Dengue is an illness spread by day-biting mosquitos, which causes high fevers, and especially affects children. In the Philippines, about 1 in 100,000 children die of dengue fever each year. In 2006, Rosalina Palanca-Tan conducted a CV study to estimate the demand for what was at that time a hypothetical possibility – a dengue vaccine. She found that on average, households in Manila were willing to buy the necessary two doses of the vaccine at a price of $38–$45 per dose in 2022 dollars, even if the vaccine had only a one-year efficacy.

As average household income in Manila is only about $8000/year (PSA, 2019), the study showed that people highly valued the development of a dengue vaccine. Fortunately, in 2017 a vaccine for dengue fever was developed and it is available under the trade name Dengvaxia.

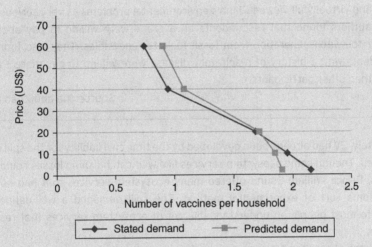

Figure 4.5 Marginal Value of a Dengue Vaccine as a Function of Number of Vaccines
Source: Palanca-Tan (2008).

CV researchers implement an experiment in which respondents respond to questions that are rolled out using a predetermined procedure. When well done, CV studies value environmental or other changes that are well-defined, specific and understandable to respondents. Carefully defining what is valued helps avoid a key potential problem with CV, which is that respondents cannot relate to or understand what is being valued, and the whole valuation enterprise becomes purely hypothetical.

On March 24, 1989, the oil tanker Exxon Valdez, which was owned by the Exxon Oil Company (now ExxonMobil), ran aground in Prince William Sound off the coast of Alaska, spilling 11 million gallons of crude oil, affecting 1300 miles of coastline and killing an estimated 250,000 seabirds and many other animals (NOAA, undated). The whole US mourned the loss of fish productivity, eco-tourism and existence value ecosystem services. After the mourning was over, the country turned to the question of civil liability. What exactly did Exxon owe the country because of this environmental damage?

Box 4.7 Willingness to Pay for Green Electricity in Tianjin, China

Electricity generation is a major source of air pollution, including greenhouse gases, in most countries. China is an especially important user of electricity, absorbing over one-quarter of all electricity produced. A critical way to reduce the environmental footprint of electricity generation is to switch from fossil fuels to renewable, low-carbon energy, such as hydropower, wind and solar.

Bai-Chen Xie and Wei Zhao examined the willingness to pay for renewable, green energy in Tianjin, China, which is located in northern China not far from Beijing. They sampled 468 households and found that most respondents were aware of green energy and virtually all viewed China's environmental problems as very serious.

The authors found that respondents on average were willing to pay about $56 per year to switch generation from fossil fuels to renewables. They also found that respondents with a history of respiratory disease were willing to pay more – all else equal – than other participants.

Source: Xie and Zhao (2018).

Fortunately, CV had already been developed by the time civil liability for the spill needed to be estimated. Though some ecosystem services like wild-catch fishing losses could be linked to markets, Prince William Sound offered many ecosystem services that had no markets, requiring some sort of experimental method. The spill damaged a well-defined, identifiable and to most people an understandable set of ecosystem services that respondents could value.

CV was part of a suite of economic valuation tools that helped courts set the damages Exxon needed to pay. In 2022 dollars, in addition to criminal penalties, Exxon paid the US government and others about $1.9 billion to fund remediation and restoration efforts. The case and the horror of that large oil spill in an environmentally fragile marine ecosystem also ultimately led to the Oil Pollution Act of 1990, which established the NOAA Damage Assessment, Remediation and Restoration Program (NOAA, undated).

The heart of a CV experiment is a valuation question where respondents say under what circumstances they would be willing to contribute to improving a good or service.

There are three forms of the CV valuation question. First, though not considered best practice, analysts may use so-called "open ended" questions of the form, "How much would you pay for X?" Even if respondents take the exercise seriously, this approach is cognitively demanding and therefore may yield a wide variety of estimates. Second, researchers can present respondents with sequentially higher and higher prices (so-called *bid values*), asking them if they would be willing to pay that value to fund the project.

The most common and well-respected method, though, is for researchers to simply ask a valuation question, which is of the form, "Would you pay $X per period (e.g., week, month, year or in total) for the project?" This bid value is one of several (e.g., 6-12) that attempt to represent the range of what people would be willing to pay.

Of special importance is that the single bid value presented to each respondent must be randomly assigned. Random bid assignment, combined with random sampling of respondents and a sufficiently large sample size, are what help ensure that the sample represents the overall population, allowing us to draw inferences from the sample.[10]

In 2010, CV was again called upon to provide estimates of oil spill damage, this time when the BP Deepwater Horizon oil rig failed in the Gulf of Mexico off the coast of the US, releasing 134 million gallons of crude oil into the Gulf (more than ten times that from the Exxon Valdez) and causing widespread damage (Bishop et al., 2017). Researchers framed the bids as payments for insurance, which would ensure that if such spills occurred, they would not cause damage. Respondents said yes or no to randomly assigned one-time bids of $15, $65, $135, $265 and $435, which were derived from formal group discussions called *focus groups*.[11] The researchers found that US willingness to invest to avoid such damages in the future was at least $17.2 billion, which was just a bit below the $20.8 billion in ecological and economic damages estimated by the court (Bishop et al., 2017).

Box 4.8 Implementation of CV Studies

1. Carefully define the natural resource and ecosystem service or set of ecosystem services to be valued.
2. Define a "project" to be valued and the change in ecosystem services associated with that project.
3. Define the population of people that would be affected by the project.
4. Conduct focus groups to help map out the questionnaire and identify the range (i.e., highest and lowest reasonable values) of willingness-to-pay values.
5. Develop a *scenario*, which describes the project to respondents, including the *status quo* if the project were not implemented.
6. Carefully define all aspects of how the project will be funded. This part of the study is called the *payment vehicle*.
7. Pretest the questionnaire to be sure it is understandable and works for people.
8. Conduct a pilot study on a small sample to better understand implementation issues.
9. Administer the survey either in person or online.
10. Gather the data and test for biases, such as effects of particular interviewers on estimated values.

Inspired by Johnston et al. (2017).

The demand curve, like the one shown in Fig. 4.5, is then estimated using economet-rics (Cameron and James, 1987).[12] Once the valuation question is answered, the experi-mental portion of the study is concluded and the rest of the survey includes questions about respondent characteristics, demographics, income and anything else the analysts would like to ask.

The steps in developing a CV study, which are presented in Box 4.8, are well-known and highly scripted, particularly until the respondent answers the valuation question. Best practices have been developed and refined over almost 40 years (Johnston et al., 2017). The structure of most CV questionnaires is also quite standard and generally includes the following elements:

1. Greeting, explanation of purpose and any disclosures. Request for participation.
2. Respondent's experience with the natural resource and/or ecosystem services being valued.
3. Questions about preferences (often based on 1-5 scales) for features of the natural resource and/or ecosystem services valued.
4. Introduce the project, provide a detailed scenario, including the payment vehicle.
5. Ask the valuation question!
6. Ask open-ended valuation and other valuation-related follow-up questions (e.g., what is the maximum you would be willing to pay? Why were you unwilling to pay $X?).
7. Questions regarding respondent and household characteristics and socioeconomics.
8. Any other questions, including questions related to potential biases and whether respondents were truly paying attention when answering the survey questions.
9. Conclusion, thanks and discussion of any follow-up (e.g., promise to send a summary of the results).

Given the long history of CV research, the technique is considered reputable. As noted by Johnston et al. (2017), though, CV results are only as good as the experiments crafted by researchers. Following best practices for effectively bringing respondents into researchers' experimental universes and taking steps to avoid biases can ensure that CV studies truly estimate willingness to pay and value.

Choice Experiments

Choice experiments (CE) are a lot like CV studies, but they allow researchers to compare individual components of projects. As CV studies estimate overall project values and CE estimates the relative values of project components, CV and CE studies can be conducted at the same time using many of the same survey questions.

CE is sometimes (perhaps wrongly, e.g., see Louviere et al., 2010) also called conjoint analysis, and originally comes from the marketing literature. My colleague Sahan T.M. Dissanayake likes to use the example of valuing a pizza. Suppose you had no idea of the scar-city value of pizza add-ons like mushrooms, extra cheese, onions and peppers and wanted to find the relative willingness to pay for those aspects of a large pizza.

As is the case with CV studies as well, respondents are presented with a scenario; in this case perhaps they are told they can have either pizza. The toppings are then presented

Photo 4.1 Which Pizza Do You Want?
Source: Ivan Torres – https://unsplash.com/photos/pizza-with-berries-MQUqbmszGGM.

Photo 4.2 Which Pizza Do You Want?
Source: Fernando Andrade – https://unsplash.com/photos/pizza-with-cheese-and-tomato-_P76
trHTWDE.

Option	Forest restoration	Educational opportunities	Trails improvements for recreation	Signs at entry points and more parking	Interpretive and navigation signs within park	Cost per adult per year	I would choose ⬇
Option A	50 additional acres	25 more discovery hikes and start school programs for 200 children	16 miles	Better signage at entry points to the park	Improved signage	$50	❏ A
Option B	150 additional acres	Only start educational programs for school children	8 miles	20 more parking spaces	Improved signage	$30	❏ B
Option C	No change					No cost	❏ C

Figure 4.6 Sample Choice Set from Study of Forest Park Improvements

and compared with pizzas that have other toppings, and just cheese. Respondents are then presented with different prices and asked to choose which pizza they want.

Using two econometric methods – conditional logit or multinomial logit – believe it or not, it is possible to estimate the value of mushrooms versus green pepper versus extra cheese, etc. Sahan, I and our graduate student Jake Kennedy[13] used this method to estimate the value of upgrades to a large city park in Portland, Oregon called Forest Park. We examined different upgrades to the park the city was considering, in the hope that we could provide guidance on where to best allocate resources. As shown in Fig. 4.6, those upgrade *attributes* included forest restoration (e.g., invasive species elimination), educational opportunities for children and adults, trail improvements, parking and signage and interpretive/navigational signage. Each package of attributes had a price attached to it and there was always a no-upgrade option available, so respondents were not forced to choose among upgrade options they did not want.

The *attributes* (e.g., more navigational signs or not) and *levels* of attributes (e.g., 50 acres restored versus 150 acres restored), along with the cost per adult per year, were randomly varied using statistical rules. Each respondent then made approximately six of these *choice set* choices. Using these data, we will be able to discern the marginal benefit (in $US) associated with an additional acre of restoration, an additional mile of trail upgrade, increased signage etc.

4.5 Chapter Conclusions

Ecosystem services are valuable even if they are not traded in markets, because they are desirable, and it takes work to create or preserve them. But how valuable? In money terms, I mean. Knowing that ecosystem services are valuable is quite different from estimating values in monetary terms. Economists and others have developed methods for estimating the value of ecosystem services. The easiest approaches rely directly on market values and either estimate direct use (i.e., provisioning) values or rely on assumptions about replacement costs. More sophisticated methods, which focus on productivity improvements or costs that are avoided by environmental assets and the ecosystem services they produce, rely on closely linked markets. For example, with the help of natural science and health professionals, combined with a few assumptions and computational techniques, it is fairly straightforward to estimate the health care cost improvements or additional work/school days that can be expected from air or water quality improvements. Such methods are nice, because in terms of mathematics they don't require much more than arithmetic and are understandable to most people.

More sophisticated and powerful revealed preference methods require statistics, and especially regression analysis. Hedonic pricing and travel cost have the advantage that analysts don't need to assume people's decisions as ecosystem services change, but instead can estimate them. The downside is that results become more difficult to interpret, putting a potentially significant communication burden on analysts, if they want their results to be policy relevant.

Many ecosystem services have absolutely no links to markets. For example, the existence and option values highlighted by Krutilla (1967) have no revealed behaviors, particularly if the project analyzed is prospective (i.e., looking forward). Under such circumstances, experimental methods must be used to estimate environmental values. These methods have the advantages that they can value a wide variety of ecosystem services and use sophisticated and well-respected empirical methods.

As Johnston et al. (2017) note, however, experimental methods are only as good as the research designs that underpin them. The same is also true for revealed preference methods. While CV and choice experiments have very well-developed and respected methodologies, if poorly constructed they can become hypothetical exercises that inadequately estimate monetary valuations. Paying close attention to the procedures developed during the last 40-plus years can help ensure that CV and choice experiments actually estimate the values analysts hope to measure.

Issues for Discussion

1. Can you name three environmental problems you would really like to solve and know the value of those solutions in monetary terms? Please suggest which valuation techniques might be most appropriate and defend your answers.
2. Discuss two important environmental problems that you would like to solve, but for which you think it would be pointless to estimate monetary values. Why do you think monetary values are beside the point for these problems?
3. The chapter is hardly glowing in its endorsement of the replacement cost method as a way to value ecosystem services. Do you think this pessimism is justified and are there positive features that could outweigh any disadvantages?
4. How serious do you think hypothetical bias is when using experimental valuation methods? Do you think it is possible to make respondents take the exercises seriously

enough to actually estimate their inherent values? Please give examples from other realms in which hypothetical exercises were or were not taken seriously.

Practice Problems

1. The inspectorate of the Ministry of Environment in Slovakia discovers that a large underground oil tank has been slowly leaking oil into an underground aquifer for the past 20 years. This behavior violates Slovak civil law.

    The aquifer is used for drinking water by rural residents and health authorities estimate that concentrations of oil in the aquifer have reached the point where the water is undrinkable. Local officials estimate that residents in the area use 50,000 liters per day at a cost of €1000. They have developed the potential approaches to replacing the water lost due to the contamination in the table below and are open to combining replacement approaches. The inspectorate would like to replace the lost water as *cost-effectively* as possible (i.e., at the lowest possible cost).

Approach	Maximum Water Production (Liters/Day)	Total Cost per Day (Including All Costs) at Maximum Production
Build a new pipeline to bring water from the Tatra Mountain foothills	14,000	€420
Deliver water by tanker truck	22,000	€880
Tap an aquifer that is farther away	88,000	€7040
Buy bottled water on the open market	1,000,000	€60,000

    The environmental authority would like to levy a fine on the owners of the underground storage tank which is in line with the damages they caused, plus a penalty of 50% to deter such behaviors in the future. Using the replacement cost method, what is your best estimate of the penalty the authority should require? Having limited data is standard in real-world analysis and so perhaps you wish you had a bit more information. What specific assumptions about replacement costs did you need to make to reach your estimate?

2. You are using the avoided cost approach to analyze the costs of $PM_{2.5}$ emissions from large lorries (i.e., trucks) in a neighborhood in Delhi. You comb the literature and find that a 3 microgram per cubic meter ($\mu g/m^3$) increase in particulate emissions on average increases asthma rates by 44 cases. Treating an asthma case costs about 9,000 Indian Rupees ($\$1.00 = $ INR 78). Average $PM_{2.5}$ concentrations in the neighborhood are 220 $\mu g/m^3$ during winter and the literature suggests that 1/3 of this concentration is due to lorries. You conduct a vehicle count in your target neighborhood and find that 6000 trucks on average pass through the neighborhood each day. The city government proposes a project to reroute trucks through less populated neighborhoods, reducing truck traffic by 50%. What is your estimate of the benefit of the policy in INR and US dollars?

3. Suppose that hedonic analysis suggests that the willingness to pay to reduce the risk of premature death from an environmental hazard by 1/100,000 is $70. A policy is expected to reduce the average risk of death from 6/100,000 to 2/100,000 in a population of 10 million. What is the implied value of a statistical life? What is the estimated total value of the policy?

4. Box 4.8 and the list in Section 4.4.2 of this chapter provide the outlines of a guide for doing a contingent valuation experiment. The Appendix to this chapter also includes a sample questionnaire used to value a park in Southern California. Now it is time for you to try. Identify a well-defined environmental problem you would like to solve. This issue can be local, regional, national or global, but you must be able to describe it clearly and it should be potentially relevant for others.

    a. Go through as many steps as you can from Box 4.8, including defining the problem you would like to solve as a "project," developing a project scenario, a payment vehicle to pay for that project, an open-ended payment question and other questions typical on contingent valuation questionnaires.

    b. Develop a range of five payment values that you think represent the highest and lowest values your friends might pay.

    c. Survey at least ten of your friends and note their responses either on paper questionnaires or in electronic form.

    d. Enter the data into Excel or another spreadsheet program. Calculate average values and, if possible, estimate correlations between your variables using more sophisticated statistical methods.

    e. Write up at least two pages that describe your "project," method and results.

Notes

1 Between 1999 and 2010, wolves in Wyoming killed 418 sheep (out of a 2010 estimated total of 375,000) and 474 cattle out of 1.3 million (Anselmi, 2017).

2 https://www.weforum.org/agenda/2020/03/biodiversity-loss-is-hurting-our-ability-to-prepare-for-pandemics/

3 https://blogs.iadb.org/sostenibilidad/en/what-is-the-link-between-covid-19-and-the-ecological-and-climate-emergencies/

4 The authors report results in 2005 US dollars, but producer prices in South Africa fell approximately 11% between 2005 and 2020. Interestingly, in 2012 producer prices fell by roughly 50% (https://tradingeconomics.com/south-africa/producer-prices).

5 This example is a simplification of the relationship, because the model has only one independent variable. Several variables affect incomes and if some omitted variables were correlated with work experience, we might attribute effects to work experience that really were due to those other variables. We therefore need to include all independent variables we think could reasonably affect income.

6 https://www.esri.com/en-us/home and https://qgis.org/en/site/

7 GNI is the total income earned by a country and includes both domestic production and incomes earned abroad.

8 Of course, some services like call center technical assistance or even tax preparation can be internationally traded. See http://pubdocs.worldbank.org/en/332341517441011666/PPP-brochure-2017-webformat-rev.pdf for a how-to guide on the PPP adjustment.

9 Any data can be converted to natural logs. For example, the natural log of 2 = 0.693 and the natural log of 117 = 4.7.

10 Often, however, bids are randomly assigned, but the sample is a non-random "convenience" sample. Without a random sample, drawing population inferences is not possible.
11 Focus groups are critical to the two experiment-based valuation methods. Only through focus groups can researchers know the relevant issues to include, define the range of bid values and even settle on a way to frame how the payment would hypothetically occur.
12 Logit and probit econometric models are the most common.
13 Who also provided valuable assistance with this chapter!

Further Reading and References

Allsopp, M.H., W.J. de Lange and R. Veldtman. 2008. "Valuing Insect Pollination Services with Cost of Replacement." *PLoS ONE* 3 (9): e3128.
Anselmi, J.J. 2017. "Wyoming's War on Wolves." *JSTOR Daily*, May 10, 2017. Downloaded from https://daily.jstor.org/wyomings-war-on-wolves/ April 10, 2020.
Bandara, R. and C. Tisdell. 2003. "Comparison of Rural and Urban Attitudes to the Conservation of Asian Elephants in Sri Lanka: Empirical Evidence." *Biological Conservation* 110 (3): 327-342.
Belay, G., M. Ketema and M. Hasen. 2020. "Households' Willingness to Pay for Soil Conservation on Communal Lands: Application of the Contingent Valuation Method in Northeastern Ethiopia." *Journal of Environmental Planning and Management* 63 (12): 2227-2245.
Bishop, R., K.J. Boyle, R.T. Carson, D. Chapman, W.M. Hanemann, B. Kanninen, R.J. Kopp, J. Krosnik, J. List, N. Meade et al. 2017. "Putting a Value on Injuries to Natural Assets: The BP Oil Spill." *Science* 365 (6335): 253-254.
Boyd, J. 2010. "How Do You Put a Price on Marine Oil Pollution Damages?" *Resources*, Summer 2010.
Brouwer, R., F.C. Job, B. van der Kroon and R. Johnston. 2015. "Comparing Willingness to Pay for Improved Drinking-Water Quality Using Stated Preference Methods in Rural and Urban Kenya." *Applied Health Economics and Health Policy* 13 (1): 81-94.
Brown, M., J. Flesher and J. Mone. 2020. "Trump Officials End Gray Wolf Protections across most of US." *Associated Press*, October 30. Downloaded from https://apnews.com/article/election-2020-joe-biden-donald-trump-michigan-elections-30b039ee99901e14a1b0e3d3f63f5b6a October 30, 2020.
Cameron, T.A. 2010. "Euthanizing the Value of a Statistical Life." *Review of Environmental Economics and Policy* 4 (2): 161-178.
Cameron, T.A. and M. James. 1987. "Efficient Estimation Methods for 'Closed-Ended' Contingent Valuation Surveys." *The Review of Economics and Statistics* 69 (2): 269-276.
Carson, R.T. 2000. "Contingent Valuation: A User's Guide." *Environmental Science and Technology* 34 (8): 1413-1418.
Carson, R.T., W.M. Hanemann and D. Whittington. 2020. "The Existence Value of a Distinctive Native American Culture: Survival of the Hopi Reservation." *Environmental and Resource Economics* 75 (4): 931-951.
Centers for Disease Control and Prevention (CDC). Undated. *About a Dengue Vaccine*. Retrieved from https://www.cdc.gov/dengue/vaccine/index.html 16 April 2022.
Chavas, J.-P., J. Stoll and C. Sellar. 1989. "On the Commodity Value of Travel Time in Recreational Activities." *Applied Economics* 21 (6): 711-722.
Ciriacy-Wantrup, S.V. 1947. "Capital Returns from Soil-Conservation Practices." *Journal of Farm Economics* 29 (4 Part II): 1181-1196.
Cook, D., K. Eiríksdottir, B. Davídsdottir and D.M. Kristofersson. 2018. "The Contingent Valuation Study of Heiðmörk, Iceland – Willingness to Pay for its Preservation." *Journal of Environmental Management* 209: 126-138.
Guha, I. and S. Ghosh. 2009. "A Glimpse of the Tiger: How Much are Indians Willing to Pay for it?" SANDEE Working Paper No. 39-09. South Asian Network for Development and Environmental Economics (SANDEE), Kathmandu.
Hanemann, W.M. 1994. "Valuing the Environment through Contingent Valuation." *Journal of Economic Perspectives* 8 (4): 19-43.

He, J. and B. Zhang. 2021. "Current Air Pollution and Willingness to Pay for Better Air Quality: Revisiting the Temporal Reliability of the Contingent Valuation Method." *Environmental and Resource Economics* 79 (1): 135-168.

Houghton, P. 2001. "Old Yet New—Pharmaceuticals from Plants." *Journal of Chemical Education* 78 (2): 175-184.

Jabr, F. 2020. "Out of the Wild." *New York Times Magazine*, June 21, 2020.

Johnston, R.J., K.J. Boyle, W. Adamowicz, J. Bennett, R. Brouwer, T.A. Cameron, W.M. Hanemann, N. Hanley, M. Ryan, R. Scarpa et al. 2017. "Contemporary Guidance for Stated Preference Studies." *Journal of the Association of Environmental and Resource Economics* 4 (2): 319-405.

Krutilla, J. 1967. "Conservation Reconsidered." *American Economic Review* 57 (4): 777-786.

Lopes, A.A. and S.S. Atallah. 2020. "Worshipping the Tiger: Modeling Non-Use Existence Values of Wildlife Spiritual Services." *Environmental and Resource Economics* 76 (1): 69-90.

Louviere, J., T.N. Flynn and R.T. Carson. 2010. "Discrete Choice Experiments are Not Conjoint Analysis." *Journal of Choice Modelling* 3 (3): 57-72.

Lutzenhiser, M. and N. Netusil. 2001. "The Effect of Open Space on a Home's Sale Price." *Contemporary Economic Policy* 19 (3): 291-298.

Mitchell, R.C. and R.T. Carson. 1989. *Using Surveys to Value Public Goods: The Contingent Valuation Method*. Resources for the Future: Washington, DC.

Morell, V. 2010. "Gray Wolves Back on the Endangered List." *Science*, August 6, 2010. Downloaded from https://www.sciencemag.org/news/2010/08/gray-wolves-back-endangered-list April 10, 2020.

Nepal, M., R. Rai, M. Khadayat and E. Somanathan. 2020. "Value of Cleaner Neighborhoods: Application of Hedonic Price Model in Low Income Context." *World Development* 131: 104965.

Netusil, N., S. Chattopadhyay and K.F. Kovacs. 2010. "Estimating the Demand for Tree Canopy: A Second-Stage Hedonic Price Analysis in Portland, Oregon." *Land Economics* 86 (2): 281-293.

NOAA. Undated. *Exxon Valdez*. Damage Assessment, Remediation and Restoration Program. Retrieved from https://darrp.noaa.gov/oil-spills/exxon-valdez 16 April 2022.

Palanca-Tan, R. 2008. "The Demand for a Dengue Vaccine: A Contingent Valuation Survey." *Vaccine* 26 (7): 914-923.

Philippines Statistics Authority (PSA). 2019. "Annual Family Income is Estimated at PhP 313 Thousand, on Average, in 2018." Retrieved from https://psa.gov.ph/statistics/income-expenditure/fies/node/144731 April 16, 2022.

Richardson, L. and J. Loomis. 2009. "The Total Economic Value of Threatened, Endangered and Rare Species: An Updated Meta-Analysis." *Ecological Economics* 68 (5): 1535-1548.

Ridker, R.G. and J.A. Henning. 1967. "The Determinants of Residential Property Values with Special Reference to Air Pollution." *The Review of Economics and Statistics* 49 (2): 246-257.

Smith, K.F., M. Goldberg, S. Rosenthal, L. Carlson, J. Chen, C. Chen and S. Ramachandran. 2014. "Global Rise in Human Infectious Disease Outbreaks." *Journal of the Royal Society Interface* 11 (101): 20140950.

US Census Bureau. 2021. "Historical Households Tables." Downloaded from https://www.census.gov/data/tables/time-series/demo/families/households.html 30 January 2022.

US Department of Transportation (US DOT). 2022a. *Benefit-Cost Analysis Guidance for Discretionary Grant Programs*. Office of the Secretary, US DOT, March 2022 (Revised). Downloaded from https://www.transportation.gov/sites/dot.gov/files/2022-03/Benefit%20Cost%20Analysis%20Guidance%202022%20%28Revised%29.pdf 26 March 2022.

US Department of Transportation (US DOT). 2022b. *Departmental Guidance on Valuation of a Statistical Life in Economic Analysis: Effective Date March 4, 2022*. Office of the Chief Economist. Assistant Secretary for Transportation Policy, US DOT. Retrieved from https://www.transportation.gov/office-policy/transportation-policy/revised-departmental-guidance-on-valuation-of-a-statistical-life-in-economic-analysis 27 March 2022.

Vanbergen, A. and The Insect Pollinators Initiative. 2013. "Threats to an Ecosystem Service: Pressures on Pollinators." *Frontiers in Ecology and the Environment* 11 (5): 251-259.

Viscusi, W.K. and J.E. Aldy. 2003. "The Value of a Statistical Life: A Critical Review of Market Estimates Throughout the World." *The Journal of Risk and Uncertainty* 27 (1): 5-76.

Voorhees, A.S., J. Wang, C. Wang, B. Zhao, S. Wang and H. Kan. 2014. "Public Health Benefits of Reducing Air Pollution in Shanghai: A Proof-of-Concept Methodology with Application to BenMAP." *Science of the Total Environment* 485-486: 396-405.

World Bank. Undated. *GNI per Capita, PPP (Constant 2021 International $) - Nepal*. Retrieved from https://data.worldbank.org/indicator/NY.GNP.PCAP.PP.KD?locations=NP 26 September 2024.

Xie, B.-C. and W. Zhao. 2018. "Willingness to Pay for Green Electricity in Tianjin, China: Based on the Contingent Valuation Method." *Energy Policy* 114: 98-107.

Yuan, L. and S. Wang. 2019. "Recreational Value of Glacier Tourism Resources: A Travel Cost Analysis for Yulong Snow Mountain." *Journal of Mountain Science* 15 (7): 1446-1459.

APPENDIX
Example of a Contingent Valuation Questionnaire
Proposed State Park in the San Timoteo Canyon near Redlands, California

(Prepared with Rafat Fazeli)

Section 1: Perceptions and Uses of the San Timoteo Canyon

1. Do you know where San Timoteo Canyon is located?

    ___ Yes
    ___ No (**go to #7**)
    ___ Prefers not to answer

2. Would you please describe where San Timoteo Canyon is located?

    ___ Right
    ___ Wrong
    ___ Prefers not to answer

3. Have you ever visited San Timoteo Canyon?

    ___ Yes
    ___ No (**go to #7**)
    ___ Prefers not to answer

4. When was your last visit?

    ___ Within Last Week
    ___ Within Last Month
    ___ Within Last 6 Months
    ___ Within Last Year
    ___ Within Last 2 Years
    ___ Within Last 5 Years
    ___ Within Last 10 Years
    ___ Prefers not to answer

5. How regularly do you visit the canyon?

___ Daily
___ Twice a Week
___ Weekly
___ Bi-Weekly
___ Monthly
___ Twice a Year
___ Yearly
___ Other
___ Prefers not to answer

6. When you visit the canyon what activities do you participate in? Please list all that apply.

___ Hiking
___ Running
___ Fishing
___ Biking
___ Horse Back Riding
___ Bird Watching
___ Off Road Vehicle Use
___ Camping
___ Drive
___ Other
___ Prefers not to answer

7. In general, what outdoor activities do you participate in? Please list all that apply.

___ Hiking
___ Running
___ Fishing
___ Walking
___ Mountain Biking
___ Horse Back Riding
___ Bird Watching
___ Off Road Vehicle Use
___ Camping
___ None
___ Other
___ Prefers not to answer

8. In general, how often do you participate in outdoor activities?

___ Daily
___ Twice a Week
___ Weekly
___ Bi-Weekly

___ Monthly
___ Twice a Year
___ Yearly
___ Prefers not to answer

9. Do you feel your household would benefit from a state park near Redlands?

___ Yes
___ No
___ Prefers not to answer

10. If a state park were created near Redlands, would the amount of outdoor activities your family participates in increase?

___ Yes
___ No
___ Prefers not to answer

11. How many different state parks have you visited during the past year?

___ Zero
___ 1
___ 2
___ 3
___ 4-6
___ 7-9
___ 10+
___ Other
___ Prefers not to answer

12. How many visits to state parks have you made during the past year?

___ Zero
___ 1
___ 2
___ 3
___ 4-6
___ 7-10
___ 11-15
___ 16+
___ Other
___ Prefers not to answer

Section 2: Relative Importance of Values Generated by the San Timoteo Canyon

Description of San Timoteo Canyon and the Proposed State Park

Now I would like to introduce you to San Timoteo Canyon and the proposed state park. San Timoteo Canyon is located southeast of Redlands between Calimesa and Loma Linda. At the bottom of the canyon is San Timoteo Canyon Road. The canyon provides a natural barrier between Redlands and Riverside County. Currently, most of the land in the canyon is privately owned except a small portion that has been designated the Norton Younglove Preserve.

Historically, the Canyon has been an important area for Native American tribes and is the burial site for the Cahuilla Chief Juan Antonio, who died in 1862. The oldest schoolhouse in Riverside County is also located in the canyon **(show picture)**, and Chief Juan Antonio is buried behind this schoolhouse. The canyon has been a passageway for people throughout time, including Native Americans, the Spanish, missionaries, miners in search of gold, the Wells Fargo stages and the first continental railroad to Southern California. Also located in the canyon is an Old Spanish cemetery.

San Timoteo Canyon not only is of historical interest, but also is an important habitat area. A variety of habitats exist within the canyon, including the dense live oak forest, which is one of the last closed canopies in Southern California. Creekside habitat also flourishes on the canyon floor **(show picture)**.

San Timoteo Canyon is home to several important vegetation communities. Twenty species of plants in the canyon are designated as under threat by the state or federal governments. Many animal species inhabit the canyon, such as the Gray Fox **(show picture)**, Mule Deer, Orange California Tree Frogs **(show picture)**, and endangered species such as the Kangaroo Rat **(show picture)** and the California Gnatcatcher **(show picture)**. The federal government has designated the San Timoteo Canyon as critical habitat for the California Gnatcatcher. One important aspect of the canyon is that it provides the only wildlife corridor between the lowlands of the valley and the highlands of the San Bernardino Mountains.

A variety of recreational activities are available in the canyon, including road running, horseback riding and hiking through the creekside forest. Trail running and mountain biking through the hills, as well as fishing and bird watching, are also common.

The proposed San Timoteo Canyon State Park is a vision of several local governments, environmental groups and historical preservation societies. It would cover 18,200 acres in the base and hills of the eastern part of the canyon near the Oak Valley freeway exit. It would be directly next to and across San Timoteo Canyon Road from the Oak Valley Housing development which, when completed, will consist of over 4,000 houses. The park would straddle San Bernardino and Riverside Counties, which are two of the most rapidly growing counties in the country. The park would be available for passive recreational uses, such as biking, hiking, horseback riding, historical sightseeing and bird watching, but housing development would not be allowed. Some state funds will most likely be given to purchase about 2000 acres of land from interested sellers. Enough money is therefore probably available to purchase a core area of the proposed State Park, including the El Casco Lake, but more funds are needed to *create and maintain the full 18,200-acre park. These funds would have to be generated locally from surrounding communities.*

Would you like me to repeat any part of the description of the canyon and proposed park or would you like to see any of the pictures again?

From the previous description of the canyon and your own experience, please rank the following characteristics of the canyon on a scale from 1 to 5, 1 being of absolutely no importance and 5 being extremely important.

1.   Historical values are:

___

___ Prefers not to answer

2.   Wildlife and Habitat Conservation is:

___

___ Prefers not to answer

3.   Endangered Species protection is:

___

___ Prefers not to answer

4.   Providing a natural boundary for the city of Redlands is:

___

___ Prefers not to answer

5.   Recreational Opportunities are:

___

___ Prefers not to answer

Section 3: Willingness to Pay to Establish and Maintain the Proposed Park

Taking into consideration the characteristics of the canyon and your own experiences, I would like to ask you how much you as a representative of your household would be willing to pay as a yearly fee for the San Timoteo Canyon State Park to be established and maintained. This annual fee would come in the form of an increase in your utility bill that now includes water, waste collection and recycling. The park fee would simply be another item on your bill. All communities near the canyon will experience increased utility charges to fund the park. I would like to be clear that we are not asking for a donation and your utility bill will not actually be increased as a result of your answers. But please do not let the hypothetical nature of these questions influence your answers. An accurate reflection of your willingness to pay will help city and state officials understand your views on the creation of the park.

Any money from the park fees charged on utility bills, plus the state funds, would be used for the purchase of an 18,200-acre park land area and for maintenance of the park. Again, please remember that you are completely anonymous and your answers will be kept confidential. Every household has different needs and levels of income and wealth. You and your household also have a variety of goods and services - clothes, gifts, entertainment, transportation, food, drink etc. - available to you on which you can potentially spend your money. There is no right or wrong answer; we really want to know how you value San Timoteo Canyon. Please simply answer yes or no to the following two questions. **Would you like me to repeat any part of this description?**

Yearly and Bi-Monthly (Every-Other-Month) Fees for All Household Types

Household Type	0	1	2	3	4	5	6	7	8	9	10	11	12	13
Yearly Fee	$1	$20	$40	$60	$80	$100	$120	$140	$160	$180	$200	$220	$240	$260
Every Other Month Fee	$.20	$3	$7	$10	$13	$17	$20	$23	$27	$30	$33	$36	$40	$43

ENUMERATOR: **Please note the household type of the respondent and choose the appropriate yearly and bi-monthly fees before asking the following question.**

1. Would you be willing to have your utility bill increase by $___ a year, or $___ per bi-monthly payment in order to create the San Timoteo Canyon State Park?

    ___ Yes (**go to #2**)
    ___ No (**go to #3**)
    ___ Prefers not to answer

ENUMERATOR: *Please increase the yearly cost by one category ($20 per year and $3 every other month).*

2. Would you be willing to have your utility bill increase by $___ a year, or $___ per bi-monthly payment in order to create the San Timoteo Canyon State Park?

    ___ Yes (**go to #4**)
    ___ No (**go to #4**)
    ___ Prefers not to answer

ENUMERATOR: *Please decrease the yearly cost by one category ($20 per year and $3 every other month) compared with Question 1.*

3. Would you be willing to have your utility bill increase by $___ a year, or $___ per bi-monthly payment in order to create the San Timoteo Canyon State Park?
    ___ Yes
    ___ No
    ___ Prefers not to answer

4. What would be the maximum increase in your utility bill you would be willing to pay per year to create the San Timoteo Canyon State Park?
    $_____ per year
    _____ Prefers not to answer

5. If people with property in the canyon were willing to sell enough land to increase the size of the park to 27,000 acres, which is about 50% larger than previously described, what would be the *maximum* increase in your yearly utility bill you would be willing to pay to create this larger park?

    $_____

    _____ Prefers not to answer

ENUMERATOR: If the answer to Question #4 was more than zero, answer this question, otherwise go to Question #7.

6. What was the *main* reason you would be willing to pay the amount you decided on?
    ___ Conserving habitats.
    ___ Protecting animals.
    ___ Maintaining Redlands' borders.
    ___ Increased recreational opportunities.
    ___ Preserve historical sites.
    ___ Other
    ___ Prefers not to answer

7. Why would you be unwilling to increase your utility bill to create the San Timoteo State Park?

    _____

    _____

Section 4: Respondent and Household Information

Now I would like to ask about you and your household.

1. Which of the following age brackets do you fall within?
    ___ 20-29 years
    ___ 30-39 years
    ___ 40-49 years
    ___ 50-59 years
    ___ 60-69 years
    ___ 70+ years
    ___ Prefers not to answer

2. How many people are in your household living in this house?

    ____
    ___ Prefers not to answer

3. What is the highest educational degree you have achieved?
    ___ No degree
    ___ High School Diploma
    ___ Associate's Degree

___ Bachelor's Degree
___ Master's Degree
___ Ph.D.
___ Vocational Degree
___ Prefers not to answer

4. Which income range does your household fall within?
    ___ $15,000 or under
    ___ $15,000-$40,000
    ___ $40,000-$60,000
    ___ $60,000-$80,000
    ___ $80,000+
    ___ Prefers not to answer

5. Do you rent or own this home?
    ___ Rent
    ___ Own
    ___ Prefers not to answer

6. How would you best describe your ethnic background? Please choose only one category.
    ___ Native American
    ___ Hispanic
    ___ South Asian origin
    ___ African American
    ___ East Asian origin
    ___ European origin
    ___ Other
    ___ Prefers not to answer

7. What political party would you affiliate yourself with?
    ___ Democrat
    ___ Republican
    ___ Other political party
    ___ No political party
    ___ Prefers not to answer

8. Do you consider global warming a serious issue?
    ___ Yes
    ___ No
    ___ Prefers not to answer

9. Do you think pollution is a problem in Southern California?
    ___ Yes
    ___ No
    ___ Prefers not to answer

10. With what gender do you identify?

    ___ Male
    ___ Female
    ___ Non-binary or gender non-conforming
    ___ Other

Section 5: Concluding Questions

1. Could we please have your phone number for a single follow-up call to check on the quality of this interview?
Phone number _____ No _____

2. Thank you very much for your time and thoughtful answers. Would you like to receive a summary of the study results by mail?

    ____ Yes
    ____ No

5 Consumption

5.1 Recap and Introduction to the Chapter

In Chapter 3 we discussed why things are valuable from an economic standpoint and in Chapter 4 we learned about methods for estimating the economic benefits of ecosystem services in monetary terms. All ecosystem services are, of course, valuable, but we found the value particularly of non-excludable ecosystem services may not be immediately apparent to us, because they do not have prices. We will discuss this topic in more detail in Chapter 8, when we delve deeply into the issue of *Why are Environmental Assets So Difficult to Manage?*

With non-excludable ecosystem services, we might mistake – or act like we mis- take – extremely valuable ecosystem services like a stable climate or biological diversity for low-value or worthless services simply because they are not traded in markets. Most envir- onmental assets produce several ecosystem services, which often conflict, so there is the dis- tinct possibility that markets could conflict in basic ways with some ecosystem services. But what exactly is the culprit here? One possibility, which has been eloquently covered by a var- iety of authors, is overconsumption, especially by higher-income countries. The idea behind overconsumption is that high levels of consumption require provisioning ecosystem services, including waste disposal services, that degrade natural resources. Sustainable Development Goal #12 in fact focuses on responsible production and **consumption**.

DOI: 10.4324/9781003308225-6

What are authors exactly saying about overconsumption and what does it mean? How might we know if we are overconsuming and whether we might just be exporting our environmental problems to other countries? Moreover, if we or others are implicating human consumption in the environmental problems about which we are concerned, let us try to understand the connection between consumption and environmental assets. We therefore now turn our attention to this task.

5.2 Perspectives on Overconsumption

That "economies," "markets" and "globalization" destroy the environment is what may jump first to our minds when we wrestle with why we are damaging the planet. In fact, we may take it as obvious that the #1 way to save the world is to roll back markets and/or shrink our economies. This point that economies need to shrink to save the planet was made back in 1972 in the famous book titled *The Limits to Growth* (Meadows et al., 1972) and more recently in *Prosperity without Growth: Economics for a Finite Planet* (Jackson, 2009). If so-called "degrowth" is accepted, it is a very, very short step to conclude that to save the world we have to reduce our consumption.[1]

Box 5.1

"Globally, burgeoning consumption has diminished or cancelled out any gains brought about by technological change aimed at reducing environmental impact."

Source: Wiedmann et al. (2020).

There is no shortage of overconsumption critiques and Googling "Overconsumption and the Environment" will yield over 3 million hits. Back in 1991, writing in *The Futurist*, Alan Durning noted that "Skyrocketing consumption is the hallmark of our era" and in the same issue Andrew Bard Schmookler summarized a conventional wisdom that one hears today:

> The materialistic appetite of Western civilization serves as the engine of our environmental destructiveness. It is, therefore, important to understand why it is that, having so much, we are still fast devouring the earth in our hunger for more.
>
> (Schmookler, 1991)

A bit more recently, Erik Assadourian commented on the 2010 Worldwatch Institute *State of the World* report that he directed. He noted that "Until we recognize that our environmental problems, from climate change to deforestation to species loss, are driven by unsustainable habits, we will not be able to solve the ecological crises that threaten to wash over civilization." The article further notes that the Worldwatch Institute says "...the cult of consumption and greed could wipe out any gains from government action on climate change..." (Goldenberg, 2010). What is necessary to reduce human impacts on natural resources has often been referred to as an economic "paradigm shift," with reduced consumption at its core (Daly and Farley, 2004).

Challenge Yourself

Where in your life could you change your consumption habits in order to reduce your environmental impact?

The UCLA geography professor Jared Diamond (author of *Guns, Germs and Steel* and *Collapse*) in a 2008 New York Times editorial challenges us to evaluate our consumption, but also puts forth a critique especially of Americans' consumption. He alleges that reducing our consumption won't require real sacrifice, because "Much American consumption is wasteful and contributes little or nothing to quality of life." He notes that the average American consumes 32 times more than a Kenyan.

This assertion that the things we buy do not make us happy is a common one. We can find it in a variety of places, including in SERI (2009), which challenges us to develop new economic approaches "...that focus on well-being instead of increased production and consumption," and from philosophers, such as Mark Sagoff of George Mason University. Sagoff lays out this argument in detail in a 1997 article in *The Atlantic* that argues (among other things) that once basic needs are met money does not make people happier. Similarly, Jackson (2009, p. xiii) notes that "One study after another has shown in recent years that the tie between more stuff and more happiness has broken down."

Challenge Yourself

Under what circumstances do you think it is appropriate to judge the consumption decisions of others or for others to comment on your consumption patterns?

A particularly straightforward expression of the view that increasing consumption in higher-income countries often does not bring more happiness is in a 2007 YouTube video called *The Story of Stuff* (www.storyofstuff.org). In the video, Annie Leonard says that Americans' primary identity is as consumers and that consumption is the way we demonstrate our value. We are on a "work-consume treadmill" that is pushed by business and has led to average Americans consuming two times more than 50 years ago, says Leonard.

5.3 What is Consumption?

In economics, "consumption" refers to the goods and services that are used (i.e., consumed) by people. Think about what you spend your money on. What did you buy during the past week? What do you hope to buy this year? All this is consumption. For example, you may have had a pizza yesterday. This is part of your consumption. Perhaps you enjoyed a drink from a coffee shop or purchased a television set or a screwdriver so you can put up a picture in your home. That is consumption, but so is the movie you saw, the music you downloaded from iTunes, your subscription to a gaming website and working out in a gym. Anything that you voluntarily lay out perfectly good money for - physical or not - that you use is consumption.

Conversely and to avoid double-counting, all the *intermediate* goods and services that are produced in the economy that lead up to consumption are not consumption. For example, the web developer services to build the gaming web page are not consumption, but the massage you got last week is consumption. The milk in the coffee or tea you bought and the cheese in the pizza are intermediate inputs and are therefore not consumption.

Consumption is also a category in the so-called *national accounts* that keep track of national economies. Virtually all countries use the same system of national accounts. Looking at consumption and the possibility of overconsumption from the perspective of the national accounts is potentially helpful, because it can help us think about linkages with other *macroeconomic* measures like income and output. Let us take the US as an example using information from the Bureau of Economic Analysis in the Department of Commerce, which is responsible for reporting on the national accounts.[2]

Challenge Yourself

Can you name three things you consumed this week? In what respect were they "consumed?"

At the highest level of aggregation, we can look at economic activity in terms of either economic output or the incomes people get from that production. Of course, these two approaches have to be closely related, because the way most of us get our incomes is by providing goods and services to others. For example, we can evaluate the economic activity of a McDonalds restaurant in terms of the value of all the Big Macs, milkshakes etc. sold. We can also look at it based on the way the income earned by the McDonalds is divided up between the owner, the workers, building owner and others; incomes and production are not exactly equal, but they are closely related.

The total value of production in a country is called the *gross domestic product* (GDP) and in 2023 the US GDP totaled a whopping $27.1 trillion. GDP is the total value of *final* goods and services produced in an economy. National income was about $26.80 trillion, with the difference due to some details, including depreciation of the machines and buildings that make up *capital*. Capital consists of the "tools" used to produce final goods and services. With such a small discrepancy, GDP and national income are basically the same thing. This reality is noted in Eq. 5.1.

Equation 5.1

Total Output = GDP ≈ National Income

In the national accounts, GDP is broken down into several components based on how production is used. The first category is personal consumption, which is formally called *private consumption expenditure*. This use of GDP is the piece we are focusing on in this chapter. The second component of GDP is *private investment*. Investment is output that is used for capital (e.g., machines, vehicles, education and buildings) that in the future will produce economic output.

The third aspect of GDP is *government consumption expenditure and investment* at national, regional (e.g., state) and local levels. For example, expenditures on environmental protection, schools, roads, national defense and support for underprivileged groups are part of government expenditure. These are typically financed by taxes assessed on production, consumption and/or income. Government expenditure can be government consumption (e.g., housing for those without homes, drug treatment, food support) or investments that provide services over time (e.g., roads and buildings). We will discuss sources of government revenues in detail in Chapter 13. Finally, there are *net international exports*, which are the value of exports minus imports. Imports are netted out because the production that underpins imports occurs in other countries. As shown in Eq. 5.2, GDP is the sum of all these uses.

Equation 5.2

GDP = Consumption + Investment + Government Expenditure + Net Exports

How important is consumption in the overall economy? Table 5.1 gives the answer for the US for 2023. We see that consumption is very important indeed. A bit more than 2/3 of the $27.10 trillion in total US economic activity is classified as final goods and services that make up consumption. This is only slightly more than the world average of 62% and virtually identical to several other countries around the world. Most economic production is therefore used to produce final goods and services consumed by people around the world. To a first approximation, consumption basically **IS** production.

What then is the breakdown of goods and services consumption? For example, do Americans consume more IT services, education and car repairs or more lattes, shirts and automobiles?

Before reading on, please analyze the data in Table 5.1. What are three key takeaways from your analysis? Did any of these takeaways surprise you?

We see from Table 5.1 that in the US about 2/3 of consumption is services and 1/3 is goods. Therefore, about 44% (2/3*2/3) of our economic production is for services like music lessons, web design and haircuts. Another 23% comes from purchases of final goods like washing machines, hamburgers, vegetables, computers and tablets. Only about 1/3 of final goods production (8% of GDP) is for durable goods that last several years, such as automobiles, refrigerators, computers etc.

Private investment by businesses makes up another 18% of GDP, which is slightly more than government expenditure, the majority of which is at state and local levels. Production

Table 5.1 Makeup of 2023 Gross Domestic Product in the US

	$Billions	%GDP
Personal Consumption	$18,456	67.9%
Goods	$6170	22.7%
Services	$12,286	45.2%
Private Investment	$4807	17.7%
Net Exports of Goods and Services	-$804	-3.0%
Exports	$3019	11.1%
Imports	$3823	14.1%
Government Expenditures	$4703	17.3%
Federal	$1756	6.5%
National Defense	$986	3.6%
Non-Defense	$770	2.8%
State and Local	$2947	10.9%
Gross Domestic Product	$27,162	100.00%

Average annualized value of first three quarters.

Source: Bureau of Economic Analysis, available at http://www.bea.gov/iTable/index_nipa.cfm

for the federal government makes up less than 7% of GDP. Most private investment is made to produce goods and services for domestic consumption.

The trade sector is a very important part of GDP, with almost a 25% share if we add imports and exports (i.e., 11.1% + 14.1%); the US, despite its big population and large area, is an economy that is quite open to the rest of the world. The difference between exports and imports is negative, implying that in 2023 we imported about $800 billion more than we exported. This *trade deficit* is a relatively small portion of GDP (-3%); within the context of the overall US economy, exports from and imports to the US are approximately in balance.

Our journey through the 2023 national accounts of the US suggests a few points that should be considered when we think about overconsumption. The first, perhaps, is that in the US – and in virtually all countries in the world – private consumption expenditure makes up a very large percentage of the economy. Most private investment is also made to produce consumption goods and services. A goal to reduce consumption is therefore basically a call to shrink the economy. This point is a very important one.

Second, about two-thirds of consumption is services rather than goods, which likely have very different impacts on ecosystem services than, for example, manufactures. We will come back to this point in the next section of the chapter.

Finally, almost all economic output (most of which is used for consumption) comes back to us as income. Reducing consumption in the economy therefore means reducing incomes, because **personal consumption makes up almost 68% of national income** (18.5/27.2); decreasing consumption would have a very significant impact on incomes.

The reason for this tight linkage between consumption and income is the so-called *circular flow* of goods and services, factors of production ("land," labor and capital) and incomes between the three domestic economic actors: households, businesses and governments. Households buy goods and services from businesses, who make labor payments to households (providing incomes) and pay taxes to governments. Governments provide services and incomes to households and businesses and households pay taxes to governments. The economy is therefore an integrated system, which would have a lot of difficulty functioning if consumption, which is the end use for 2/3 of all economic activity, were to be substantially reduced. We very well may need a so-called "paradigm shift" to save the planet, but de-growth also means we will take big income hits in the process.

5.4 Natural Resource Impacts and Consumption

Consumption is made up of goods and services, which to varying degrees depend on the natural environment. The natural resource content of a car or a building is, of course, likely to be very high, because they are complicated and large physical objects. Cars require metal, rubber and plastics, all of which, to some degree, originate in the natural environment and buildings have a lot of concrete, metal and wood in them as well. Producing the intermediate inputs that go into cars and buildings and the final products themselves can also produce significant pollution. Lots of car and building production therefore can make heavy demands on natural resources.

Other components of consumption, such as services, could potentially have smaller natural resource footprints. For example, while movies, music and educational services rely on natural resources (e.g., because they require buildings), the provisioning service impacts could be less than in steel, cement, food, beer, leather or textile production, which directly use natural resources.

Challenge Yourself

Do you think it is possible for manufacturing businesses to have no pollution?

Any greenhouse gas (GHG) advantage of the service sector vis-à-vis goods appears to be significantly muted once the use of energy and manufactured products by services is included. Suh (2006) finds that services have much lower emissions per dollar of value created than goods, and direct GHG emissions from the service sector in the US are only 5% of total emissions. They also find, though, that when the emissions due to embedded goods and electricity are included, services make up about 38% of GHG emissions. These results are echoed by Roberts et al. (2021) for a sample of five countries (Australia, Germany, Italy, the UK and the US). They find that between 1990 and 2016, the service sector was responsible for between 17% and 24% of GHG emissions. These emissions are very significant, though services typically make up higher percentages of GDP.

Education is a key part of the service sector and as of 2015 over 640 universities serving hundreds of thousands of students had pledged to implement climate action plans leading to zero net carbon emissions (www.secondnature.org). Colby College in Maine and several others have indeed actually achieved this milestone. Portland State University has a climate action plan, which calls for carbon neutrality by 2040, as well as a number of 2030 goals related to campus buildings, waste and travel. Several government institutions in and outside the US – which largely provide public services – have also pledged to have no net carbon emissions. Can we imagine a zero net emissions automobile plant?

In the first part of this chapter, we discussed some perspectives on overconsumption, but what exactly is being overconsumed and why we should worry are sometimes unclear. For example, when Jared Diamond notes that an average American consumes 32 times as much as a Kenyan or when Mark Sagoff says that the wealthiest 20% of the world consumes 80% of the goods and services, what exactly do they mean? If this is just in terms of dollar value and we know that most of the monetary value of consumption is services that might have a relatively light environmental footprint, perhaps we shouldn't be too worried about consumption per se?

Measuring overconsumption in terms of money has another important complication, because there are two components of monetary value. First, there is the total quantity, which measures how much of something we use. There is also the price per unit of quantity. Services vary widely, so it is difficult to simply add up the "quantity "of services. Price is a different story, as we measure it in common monetary units, and services are relatively expensive in richer countries.

If you live in a high-income country, think about how much you might dread a major car repair that perhaps will cost you $80 per hour. How much of a special occasion would it be for you to get a massage at $90 per hour? If you own a house, do you worry about something going wrong that will require a lot of skilled labor and therefore cost a lot? Please consider these questions.

The cost of services varies mightily depending on countries' income levels. For example, as we discussed in Chapter 3, while a haircut costs at least $20 in the US and can easily cost $50, one can often get much higher-quality service for a fraction of those prices in the lower-income world. The same is true for most other services. While living in Ethiopia, for example, I had a manual transmission on my car completely rebuilt for $90. In the US, such a service would cost much more!

Personal consumption in the US is $18.5 trillion, which is about 18% of the estimated $101 trillion GDP of the whole world. About $12.3 trillion of that amount is services, which in some respects may have a light environmental touch. If "overconsuming" were to mean at least partially buying a lot of services and paying high prices for them, from a purely ecosystem services perspective perhaps we should not get too worked up about consumption?

5.5 Offshoring Environmental Problems

Most of what we consume that is produced domestically is services, but in the US, imports are huge at almost 14% of GDP. It is therefore possible that while Americans are taking classes, fixing each other's cars, taking hikes, going on bike trips and enjoying music lessons, we are importing our cars, refrigerators, air conditioners and other manufactures from other countries. Sometimes this type of pattern is referred to as *offshoring* environmental problems or exploiting *pollution havens*.

To some, this story may seem almost obviously true. Faced with tight environmental regulations and high costs at home, firms can simply migrate overseas to pollution havens and export the same goods – perhaps with worse environmental effects – back to the high-income world. This result is certainly possible, but a number of alternative tales could also be true. We know that many manufactures are durable and can last many years. Cars, fridges, air conditioners and other manufactures may easily last more than a decade and it may be the case that most people already have these durable goods.

There is also the issue of geographic perspective. So far in this chapter we have focused largely on higher-income settings and particularly the US. Perhaps, though, a global perspective that includes the lower-income host countries is most appropriate. In such a case, we need to think about how firms migrating from high- to lower-income countries affect average environmental outcomes. For example, maybe multinational firms bring with them environmental standards that upgrade environmental performance and, in the process, improve ecosystem services worldwide. This hypothesis is, of course, just that – a hypothesis – that can only be settled with real numbers.

The bottom line is that we need to determine whether countries offshore their pollution as their service sectors become more important. This is wholly a numbers question that must be *empirically* evaluated using actual data. Getting to the bottom of the issue turns out to be a bit difficult. Research is very much in progress and there is lots more to do. So far, though, in general the empirical evidence is at best mixed on whether the Global North mainly imports its pollution or if tough environmental laws drive out polluting firms. This conclusion broadly holds if we take the high-income industry-sending country or lower-income host country perspectives, and whether we look at the nature of high-income investments abroad or the composition of trade flows from lower-income countries to high-income nations.

As we discussed in Chapter 2, GHG emissions are generally attributed to countries when they *produce* emissions, but within the context of offshoring, when trade flows are significant, considering emissions based on consumption can also be important. Our World in Data is a UK non-profit linked to Oxford University. They used Global Carbon Project data, which are also used by NOAA (e.g., see Michon, 2023), to estimate consumption-based CO_2 emissions by country (Ritchie, 2019).

Of course, all countries both import and export goods and services. A country's *net* CO_2 emissions percentage can be calculated as [(CO_2 emissions from imports – CO_2 emissions from exports)/ (total CO_2 emissions)] *100. Countries can be net importers or net exporters of CO_2. You can think of net importing countries as "topping up" their domestic total CO_2 emissions inventory with imported emissions. Countries that are net exporters have a percentage of the total CO_2 emissions from their production on net end up in other countries to enable these countries' consumption. Therefore, a consumption-based view of climate change looks a bit different than one purely based on production.

Before reading on, what are some countries that you think might be net CO_2 importers? Or net CO_2 exporters?

The major petroleum-exporting countries in the Middle East, Venezuela, Iran, Russia and Kazakhstan, as well as coal exporters, Australia and South Africa, are net exporters of CO_2.

Kazakhstan, Venezuela and South Africa have net CO_2 exports over 30% of their total CO_2 emissions. Major goods exporters including China, India and Vietnam, as well as Canada, Argentina and Brazil, are also net exporters, but in the single digits as percentages of their total CO_2 emissions.

Net CO_2 importers include much of Europe, Eastern Africa and the US. The top importer is Switzerland at over 200%, but many countries have net imports over 100% of their total CO_2 emissions, especially in the lower-income world. Of note in the Global North are France (35%), the UK (47%), Sweden (77%), Germany (22%), Italy (28%) and Finland (36%).

Many countries in the Global South also top up their domestic emissions with imported CO_2 emissions, but total emissions in lower-income countries are typically quite low, implying such situations are due to low domestic production. The US adds to its emissions total of about 5300 million metric tons by about 11% because of net CO_2 imports (about 540 Mt), and the value for the UK is approximately 165 Mt, France is around 110 Mt and Peru is 8.4 Mt.

What about non-carbon pollution? Levinson (2010) used data to examine the composition of US imports over a 28-year period starting in 1972. He found that in 2000, even accounting for intermediate inputs, the US imported proportionally more goods that typically cause very little pollution (e.g., shoes, apparel and toys) and less dirty goods like paper and petroleum than it did in the 1970s. The composition of imports had therefore shifted over time in a greener direction. Furthermore, Levinson found that this shift was **bigger** than the green shift in domestic production, which suggests people in the US maybe just weren't using as much pollution-causing stuff as they used to. Of course, this analysis was in terms of proportions and over 28 years the **levels** of domestic production and imports had significantly increased.

In follow-up work, Levinson (2015) asked if the almost 70% reduction in air pollution emissions in 400 US manufacturing sectors between 1990 and 2008 was due to offshoring. Perhaps it will surprise some of us to learn that he found 90% of the reduction was due to reduced emissions intensity (i.e., higher output/pollution) rather than other causes. This result means that lower pollution was mainly due to higher production efficiencies.

Shapiro and Walker (2018) used the same data set as Levinson (2015) to further break down the cause of the enormous reduction in air pollution in US manufacturing. They noted that because consumption of manufactures in the US did not decline during the 18 years analyzed, the reduced emissions must be due to some combination of 1) changes in the composition of output; 2) environmental regulations; 3) improved productivity; or 4) offshoring. They found that most of the emissions reductions were driven by more stringent environmental regulations rather than any of the other explanations.

Eskeland and Harrison (2003) looked at the offshoring pollution hypothesis from the perspective of US foreign direct investments (FDI) in four countries in the Global South. They found no evidence that multinational companies were flocking to developing countries to take advantage of pollution havens. They concluded that FDI was driven by other factors, such as availability of appropriately trained workers, infrastructure, energy supplies etc. Cole and Elliott (2005) looked at Mexico and Brazil, which are middle-income countries with environmental regulations that may in some aspects be less stringent than the US, but where skilled workers, infrastructure and energy are probably not major constraints for foreign investors. They found that firms in industries with high pollution reduction costs were more likely to relocate, which may offer some evidence that US environmental regulations played a role in foreign investments.

In Japan, from 1998 to 2013, Cole et al. (2021) found some evidence of pollution offshoring. This was shown through a greater increase in the carbon content of imports over time than in the value of those imports. There was also a larger increase in pollution embodied in imports than in exports over the same time period.

Looking from the standpoint of the Global South, researchers have largely found that investments by firms originating in high-income countries raise the average standard of environmental protection. This was found, for example, in India by Dietzenbacher and Mukhopadhyay (2007) and for China by He (2006).

5.6 The IPAT Model

Identifying and evaluating the effect of overconsumption on the world turns out to be a bit complex. We therefore need some sort of analytical framework or model that helps us organize our thoughts so we can dial in on the key issues. Paul Ehrlich and John Holdren put forth such a model way back in 1971. They asserted that the impact (I) of humans on the environment can be expressed by a simple identity – a mathematical truth – that relates impact to population and impact per person. They suggested the relationship in Eq. 5.3.

Equation 5.3

Ecosystem Service Impact = $I = P * F$ where P is population and F = Impact per person or (I / population)

Please notice that $F = I/P$, so P is in both the numerator and the denominator of the right side of the equation. The Ps can therefore be cancelled, leaving $I = I$, which is what makes Eq. 5.3 an "identity" rather than a function. Its usefulness is merely that it helps us break the impact of humans on the environment into component parts so we can see what needs to be done.

This model has been extremely influential for analyzing how humans affect the environment and has spurred a variety of research and thinking. In 1972, Barry Commoner broke down the F term in Eq. 5.3 into production per person and ecosystem service impact per unit of production. The idea was that the impact per person really depended on how much is produced per person (i.e., A, the "affluence" of society) and the environmental impact per unit of production (i.e., "technology" T). This formulation came to be known as the IPAT model and is shown in Eq. 5.4.

Equation 5.4

$$\text{Population} * \frac{\text{Production}}{\text{Population}} * \frac{\text{Impact}}{\text{Production}} = I = \text{"Population"} * \text{"Affluence"} * \text{"Technology"}$$

Equation 5.4 is more sophisticated than Eq. 5.3, which focused our attention wholly on population growth. Equation 5.4 says that the degree to which our production damages the environment depends on at least two levers that can be pulled in addition to our mere existence. For example, in addition to depopulating, we can reduce our output per person (i.e., de-growth/reduce consumption) or we can reduce the impact of our production (i.e., we can improve our production technology to be less wasteful of natural resources and/or produce less pollution).

More recently, the I=PAT or IPAT model has been further expanded to add in materials consumption per capita. This was the model presented in the 1992 World Bank World Development Report,[3] which focused on development and the environment. It was also the basis for the empirical analysis of Waggoner and Ausubel (2002), who called their reformulation of the IPAT model ImPACT and expressed the identity as Eq. 5.5. The main difference with Eq. 5.4 is that they added in consumption (C) of natural resources per unit of production and redefined the technology (T) term as the ecosystem service impact (e.g., pollution) from using natural resources. It therefore adds another lever – natural resource conservation – that we can pull to reduce our impact on the environment

Equation 5.5

$$I = \text{Population} * \frac{\text{Production}}{\text{Population}} * \frac{\text{Natural Resources}}{\text{Production}} * \frac{\text{Impact}}{\text{Natural Resources}} = PACT$$

Let us push the IPAT model just a little further to highlight the insights already discussed that economic sectors have different impacts on the environment and also add that the location of economic activities matters for impact. These additions offer two more angles on our relationship with natural resources and are shown as Eq. 5.6.

Equation 5.6 "Economic Scale" * "Economic Structure" * "Technology" * "Siting"

$$P * \frac{Y}{P} * \frac{D}{Y} * \frac{I}{D} * \frac{d}{I} = d$$

Let us first combine Population*(Production/Population) in Eq. 5.5 into one term and call it "Economic Scale." Let us label population as P and production as Y. Economic scale is therefore (Y/P) * P. We can also perhaps define economic sectors, such as cement, petroleum refining, manufacturing, some power production and mining, that can have large impacts on ecosystem services as "dirty" and denote them by D. The "Economic Structure" term in the modified identity is the portion of the economy that is especially hard on the environment and is labeled D/Y. This reformulation is closely related to Natural Resources/Production in Eq. 5.5. Technology is the next term in Eq. 5.6. We just modify it to focus on the dirty economic sectors, which have greater impacts (I) on ecosystem services. "Technology" is therefore I/D.

Finally, in our expanded IPAT model we note that often damages can vary depending on siting decisions. For example, siting polluting facilities in sensitive environments, such as deserts, will have bigger impacts than the same plants in more adaptive environments like temperate regions. Similarly, placing facilities emitting lots of air pollution in valleys ringed by mountains likely will yield higher concentrations than in coastal areas and putting factories close to people increases health and aesthetic damages. Let us denote this component as "Siting," which will be the ecosystem service "damages" created because of impacts (I) or d/I.

You may already be thinking that several policies and feedbacks can influence each part of this identity. **Before reading on, using Eq. 5.6, please list three policies that could potentially reduce human impacts on the environment.**

As we will learn about in Chapter 12, energy subsidy policies that artificially reduce the price of fossil fuel energy are likely to have a variety of effects. Subsidies could expand the economy (economic scale), because they make energy, which is a key production input, cheaper. They could also shift the economic structure to promote "dirty," energy-intensive economic sectors and make renewable energy like solar and wind power less attractive. In the long term such subsidies may also twist research and development to favor (cheap) fossil fuel-intensive methods. Subsidies can therefore distort the technology of production to increase pollution because they make energy cheap.

Of course, standard environmental policies and public sector environmental investments could affect these same components of the identity. Holding businesses accountable for their pollution makes dirty sectors less profitable and causes them to grow more slowly or shrink. For example, partly due to US environmental regulations and policies, coal production and the coal-fired power production industry have shrunk dramatically during the last decade. Coal production in 2022 was basically half of the peak production, which occurred in 2008 (EIA, 2023a), and there is no sign of a revival.

Between 2010 and 2019, about 550 relatively intensively polluting coal-fired power plants were decommissioned in the US and power producers had plans to close another 17 by 2025. The average closed plant was 50 years old when it was decommissioned, meaning it was the old plants that were closed (EIA, 2019). Electricity production increased over time even as between 2010 and 2022 power production by coal plants fell about 60%. Renewable energy production increased threefold and outpaced coal as a source of energy (EIA, 2023b).

Public policies can also steer technological change. For example, important research and pilots are underway into so-called carbon capture and storage technologies that could allow fossil fuels to be burned, but then capture the carbon before it is emitted into the atmosphere. Public policies also routinely affect siting decisions. Zoning laws are indeed specifically designed to avoid excessive environmental damages and nuisances from private activities. That is why noisy, polluting factories are less often sited in residential neighborhoods. However, through an environmental justice lens, public policies and zoning laws can also have negative effects on lower-income and marginalized communities, where green space is less common and pollution from factories and vehicles may be greater.

Finally, of course as societies get richer (i.e., as Y/P increases) there are more resources available for everything, including environmental protection! Societies will have more capacity for public and private sector environmental investments and the only question is whether they will choose to use those resources to protect and enhance natural resources. It turns out that richer people, neighborhoods and societies often have less tolerance for environments without trees, open space, clean air and clean water. They may therefore insist that the government do a good job of ensuring that environmental assets are preserved. Such demands then affect public policies, the modified IPAT identity and damages. That higher incomes imply greater "demand" for goods and services is quite standard. Think about cars, movies, iPads and meat. Demand for all these goods and services increases as incomes rise. It turns out that ecosystem services are not that much different and as we showed in Chapter 4, it is even possible to estimate the demand for ecosystem services.

This process is sketched out in Figure 5.1, which attempts to illustrate the complex relationship between standard economic activity and ecosystem service damages, once we decompose the basic IPAT model to highlight key factors that make up "affluence" and "technology." One good thing about such a complicated system is there are a lot of policy levers to pull and public investments that could be made to perhaps avoid reducing personal consumption. For example, we can use policies, such as the air pollution regulations that Shapiro and Walker (2018) found to be so important for reducing air pollution in US manufacturing, to improve technology, optimize the economic structure and reduce damages.

5.7 The Environmental Kuznets Curve

The modified IPAT model developed in the previous section shows some ways ecosystem services and consumption (which is basically equivalent to income) are linked. A critical question, though, is how environmental assets and the ecosystem services they provide are affected by rising incomes. In other words, when all is said and done, how **do** rising levels of consumption actually affect ecosystem services? Like the pollution haven hypothesis, this is 100% a numbers question that needs to be empirically evaluated.

Starting with the seminal paper by Grossman and Krueger (1991) and followed quickly by Shafik and Bandyopaday (1992), there has been a virtual flood of research that has attempted to answer this question. Researchers have investigated the relationship between income and consumption per person and a variety of environmental metrics, including deforestation, sulfur dioxide, particulate and nitrogen oxide emissions and concentrations in the atmosphere,[4] CO_2 emissions, wastewater quality and solid waste per person.

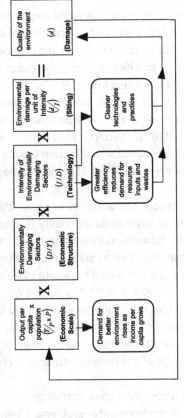

Environmental Damage, The Economy and Policy Feedbacks

Figure 5.1 The Expanded IPAT Model
Source: Adapted from World Bank (1992).

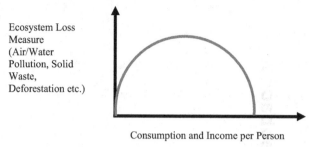

Consumption and Income per Person

Figure 5.2 Stylized Environmental Kuznets Curve

The two original papers, which looked across many countries at one point in time, found an inverted "U"-shape relationship, such as the one shown in Fig. 5.2. This is also the pattern postulated in the 1950s for income per capita and income inequality (a bad thing like pollution) by Nobel prizewinner Simon Kuznets. For this reason, the literature refers to this relationship as the "Environmental Kuznets Curve" (EKC).

The EKC pattern is, of course, not the only possible one. Indeed, for particular settings and ecosystem services the curve might be wholly downward-sloping or always have a posi- tive slope. Why then is the inverted "U" pattern of special interest? I would argue that it is of interest primarily because of a behavioral relationship lurking behind the bottom-left box in Fig. 5.1, which focuses on the potential effect of rising incomes on demand for ecosystem services.

The story goes that when people and societies are poor there is not much economic activity and little consumption, so environmental quality is excellent. As consumption and incomes rise, people are mainly focused on increasing their material well-being and don't care so much about the environment. At some stage (i.e., the "turning point") societies care more about ecosystem services and are willing to devote more resources toward their pro- tection. As summarized in Dinda (2004, p. 435), "As income grows, people achieve a higher standard of living and care more for the quality of environment they live in and demand for better environment induces structural changes in the economy that tends [sic[5]] to reduce environmental degradation." The EKC literature tests whether the relationship between income/consumption and measures of environmental quality is consistent with such a behav- ioral model.

The EKC has been studied across countries, across regions within countries and even across individuals. A variety of environmental problems have been examined, including aggregated measures of pressure on ecosystem services, such as ecological footprints (e.g., Caviglia-Harris et al., 2009). Given the complexity of the general relationship laid out in Fig. 5.1, it may not be particularly surprising that depending on the environmental issue examined, country analyzed and empirical methodology used, sometimes the EKC shows up and sometimes it does not.

Deacon and Norman (2006) looked at the relationship between GDP per capita and air pollution concentrations for 28 countries[6] over 20 years (1972–1992). Their findings are pretty similar to most other researchers' results. They find little if any evidence of the EKC and results are all over the map, including a number of cases where pollution concentrations

Table 5.2 Number of Country Observations Exhibiting Air Pollutant–GDP Relationships

Pattern as GDP/Person Increases	Air Pollutant		
	SO$_2$	Particulates	Smoke
Inverted "U"	4	2	3
Upright "U"	4	6	3
Always Increasing	2	1	2
Always Decreasing	15	5	5
Total Country Observations	25	14	13

Source: Deacon and Norman (2006).

are always falling as GDP/person increases and several where they are always increasing. Their results are summarized in Table 5.2.

Continuing the confusion, more recently Churchill et al. (2018) looked at data from 20 OECD countries over the period 1870–2014 and also found mixed results. Looking at the countries as a whole, they found supporting evidence for the EKC, using three of four different estimators. However, when looking at individual countries, there was evidence for the EKC in only 9 of the 20 countries. Özokcu and Özdemir (2017) investigated the validity of the EKC by looking at the relationship between income and CO$_2$ emissions in 26 high-income OECD countries and 52 emerging countries. They found no evidence supporting the EKC.

Other than this apparent jumble, what if anything can we take away from the over 25,000 academic articles that make up the EKC literature? First, some environmental problems in some places over certain periods of time exhibit EKC-type patterns. SO$_2$ concentrations, for example, have often been found to have such a relationship.

Second, water pollution that directly affects human health tends to decline monotonically[7] as incomes rise. As noted by Dinda (2004, p. 441), "…environmental problems having direct impact on human health (such as access to urban sanitation and clean water) tend to improve steadily with economic growth." Anyone who spends time in an urban or a densely populated rural area of a low-income country understands this point. Low-income countries – and especially urban areas – have some highly polluted rivers and streams. As incomes rise, clean water supply, sanitation and ultimately wastewater treatment are among the first environmental steps undertaken. This result was also highlighted back in 1992 by Shafik and Bandyopaday (1992).

Third, municipal solid waste exhibits always-increasing relationships with per capita income and consumption. The same is true for CO$_2$ emissions in low- to middle-income countries (Galeotti et al., 2006), but emissions have been falling for approximately the past decade in much of the Global North. Such findings suggest that the world will probably not produce or consume its way out of the climate crisis, and indeed the opposite may be true.

Finally, a lot of research articles have included policies, institutional quality and economic structure in EKC models. Most find that whatever the pattern (EKC, "U," downward-sloping or always increasing), good policies – particularly with regard to energy – push the curve downward toward the horizontal axis. For example, Caviglia-Harris et al. (2009) found that reducing energy consumption by 50% is required to bend the relationship between GDP per capita and the ecological footprint from a "U" to the more desirable inverted-"U" shape. In an influential early study, Panayotou (1997) found that SO_2 concentrations have a standard "U" relationship with GDP per capita, but with better public policies the "U" relationship becomes wholly downward-sloping. He concludes that high incomes and better air quality **can** coexist, but good public policies are what make it happen.

5.8 Chapter Conclusions

This chapter examined the relationship between consumption and ecosystem services. In economic terms, consumption is the value of final goods and services consumed by people and makes up most of economic output. Consumption is also roughly equal to total income, implying that reducing consumption means almost one-for-one reductions in incomes.

At least in higher-income countries like the US, most consumption is services, which probably on average have lighter environmental touches than manufactured goods. Evidence that the production of dirty manufactured goods gets offshored as services dominate local economies is weak, but by no means non-existent. No doubt, there is much more to come on that research front.

Successors to the IPAT model suggest there may be much more to loss of ecosystem services than overconsumption. Policies, technologies and the straightforward effect of increased incomes on demand for ecosystem services weigh in, complicating the picture, but also give us many ways to reduce pressure on the environment. We have said little about overpopulation, which is the subject of the next chapter, but perhaps population control policy could be part of the solution. This was indeed one of the key points made in the 2008 *New York Times* editorial by Jared Diamond (Diamond, 2008).

As a final note, Eduardo Porter in his column titled "Economic Scene" in the *New York Times* (Porter, 2015) cautions against latching on too hard to de-growth/reduced consumption strategies. He points out that Europe experienced no growth in income per person for the 500 years before the industrial revolution. For many regions, over long periods people's consumption levels actually shrank. Zero growth, he says, gave us Genghis Khan, conquest and subjugation and could bring "Mad Max" and "The Hunger Games" to real life. Solutions to important environmental problems that don't jeopardize human civilization, he asserts, need to be the focus of policies.

Issues for Discussion

1.   Watch *The Story of Stuff* at www.storyofstuff.org. Do you think that a work-consumption treadmill exists? If so, are you or your family members on it?
2.   How does consumption as a percentage of GDP vary around the world? What is the range? To answer these questions, use data from World Bank (2023).
3.   What is missing from the successors to the IPAT model? Is there anything you would like to add to Eq. 5.6?
4.   The quantity of services is difficult to measure. Can you suggest possible methods?
5.   Section 5.5 gives a brief introduction to the important issue of offshoring of environmental problems by higher-income countries. What, if anything, seems to be missing from the discussion? How would you set up a study to test for offshoring?

Notes

1 The concept is actually very old in economics and dates at least to 1920 when the Cambridge University economist Arthur Cecil Pigou published his classic book *The Economics of Welfare*.
2 All data and methodologies are online at http://www.bea.gov/iTable/index_nipa.cfm
3 Available back to 1978 at http://www.worldbank.org/en/publication/wdr/wdr-archive
4 These air pollutants are known to be harmful to human health, as well as the environment.
5 *Sic* denotes that a grammatical error was made.
6 Australia, Belgium, Brazil, Canada, Chile, China, Denmark, Egypt, Finland, Hong Kong, India, Indonesia, Iran, Ireland, Israel, Italy, Japan, Malaysia, Netherlands, New Zealand, Poland, Portugal, Spain, Thailand, UK, US, Venezuela, West Germany.
7 i.e., without ever reversing.

Further Reading and References

Brown, P.M. and L.D. Cameron. 2000. "What Can be Done to Reduce Overconsumption?" *Ecological Economics* 32 (1): 27–41.

Caviglia-Harris, J., D. Chambers and J. Kahn. 2009. "Taking the 'U' out of Kuznets: A Comprehensive Analysis of the EKC and Environmental Degradation." *Ecological Economics* 68 (4): 1149–1159.

Chernow, M. 2001. "The IPAT Equation and Its Variants: Changing Views of Technology and Environmental Impact." *Journal of Industrial Ecology* 4 (4): 13–29.

Churchill, S.A., J. Inekwe, K. Ivanovski and R. Smyth. 2018. "The Environmental Kuznets Curve in the OECD: 1870–2014." *Energy Economics* 75: 389–399.

Cole, M.A. and R.J.R. Elliott. "FDI and the Capital Intensity of 'Dirty' Sectors: A Missing Piece of the Pollution Haven Puzzle." *Review of Development Economics* 9 (4): 530–548.

Cole, M.A., R.J.R. Elliott, T. Okubo and L. Zhang. 2021. "Importing, Outsourcing, and Pollution Offshoring." *Energy Economics* 103: 105562.

Daly, H. and J. Farley. 2004. *Ecological Economics: Principles and Applications.* Island Press: Washington, DC.

Deacon, R.T. and C. Norman. 2006. "Does the Environmental Kuznets Curve Describe How Individual Countries Behave?" *Land Economics* 82 (2): 291–315.

Diamond, J. 2008. "What's Your Consumption Factor?" *The New York Times*, January 2, 2008.

Dietzenbacher, E. and K. Mukhopadhyay. 2007. "An Empirical Examination of the Pollution Haven Hypothesis for India: Towards a Green Leontief Paradox?" *Environmental and Resource Economics* 36 (4): 427–449.

Dinda, S. 2004. "Environmental Kuznets Curve Hypothesis: A Survey." *Ecological Economics* 49 (4): 431–455.

Durning, A. 1991. "Limiting Consumption: Toward a Sustainable Culture." *The Futurist* (July–August).

Ehrlich, P. and J. Holdren. 1971. "Impact of Population Growth." *Science* 171 (3977): 1212–1217.

Energy Information Agency (EIA). 2019. "More U.S. Coal-Fired Power Plants are Decommissioning as Retirements Continue." July 26. Retrieved from https://www.eia.gov/todayinenergy/detail.php?id= 40212 16 September 2024.

Energy Information Agency (EIA). 2023a. "Annual Coal Report." October 3. Downloaded from https://www.eia.gov/coal/annual/ 16 September 2024.

Energy Information Agency (EIA). 2023b. "Renewable Generation Surpassed Coal and Nuclear in the U.S. Electric Power Sector in 2022." December 27. Retrieved from https://www.eia.gov/todayinenergy/detail.php?id=61107 16 September 2024.

Eskeland, G.S. and A.E. Harrison. 2003. "Moving to Greener Pastures? Multinationals and the Pollution Haven Hypothesis." *Journal of Development Economics* 70 (1): 1–23.

Galeotti, M., A. Lanza and F. Pauli. 2006. "Reassessing the Environmental Kuznets Curve for CO_2 Emissions: A Robustness Exercise." *Ecological Economics* 57 (1): 152–163.

Goldenberg, S. 2010. "US Cult of Greed is Now a Global Environmental Threat, Report Warns," *The Guardian*, June 12, 2010. Retrieved from http://www.theguardian.com/environment/2010/jan/12/climate-change-greed-environment-threat 16 September 2024.

Grossman, G.M. and A.B. Krueger. 1991. *Environmental Impacts of a North American Free Trade Agreement.* NBER Working Paper 3914. National Bureau of Economic Research: Cambridge, MA.

He, J. 2006. "Pollution Haven Hypothesis and Environmental Impacts of Foreign Direct Investment: The Case of Industrial Emission of Sulfur Dioxide (SO_2) in Chinese Provinces." *Ecological Economics* 60 (1): 228–245.

Jackson, T. 2009. *Prosperity without Growth: Economics for a Finite Planet.* Earthscan/Routledge: London.

Levinson, A. 2010. "Offshoring Pollution: Is the United States Increasingly Importing Polluting Goods?" *Review of Environmental Economics and Policy* 4 (1): 63–83.

Levinson, A. 2015. "A Direct Estimate of the Technique Effect: Changes in the Pollution Intensity of US Manufacturing, 1990–2008." *Journal of the Association of Environmental and Resource Economists* 2 (1): 43–56.

Meadows, D.H., D.L. Meadows, J. Randers and W.W. Behrens III. 1972. *The Limits to Growth: A Report for the Club of Rome's Project on the Predicament of Mankind.* Universe Books: New York. Retrieved from https://policycommons.net/artifacts/1529440/the-limits-to-growth/2219251/ February 14, 2024.

Michon, S. 2023. "Does it Matter How Much the United States Reduces its Carbon Dioxide Emissions if China Doesn't Do the Same?" *Climate.gov: Science & Information for a Climate-Smart Nation,* August 23. Retrieved from https://www.climate.gov/news-features/climate-qa/does-it-matter-how-much-united-states-reduces-its-carbon-dioxide-emissions January 2024.

Özokcu, S. and Ö. Özdemir. 2017. "Economic Growth, Energy, and Environmental Kuznets Curve." *Renewable & Sustainable Energy Reviews* 72: 639–647.

Panayotou, T. 1997. "Demystifying the Environmental Kuznets Curve: Turning a Black Box into a Policy Tool." *Environment and Development Economics* 2 (4): 465–484.

Porter, E. 2015. "The Economic Scene: Imagining a World Without Growth." *The New York Times,* December 1.

Ritchie, H. 2019. "How Do CO_2 Emissions Compare When we Adjust for Trade?" *OurWorldInData.org.* Retrieved from https://ourworldindata.org/consumption-based-co2 March 2024.

Roberts, S.H., B.D. Foran, C.J. Axon and A.V. Stamp. 2021. "Is the Service Industry Really Low-Carbon? Energy, Jobs and Realistic Country GHG Emissions Reductions." *Applied Energy* 292: 116878.

Sagoff, M. 1997. "Do We Consume Too Much?" *The Atlantic,* June.

Schmookler, A.B. 1991. "The Insatiable Society: Materialistic Values and Human Needs." *The Futurist* (July–August).

Shafik, N. and S. Bandyopaday. 1992. *Economic Growth and Environmental Quality: Time Series and Cross-Country Evidence.* Background paper for the 1992 World Development Report, The World Bank.

Shapiro, J. and R. Walker. 2018. "Why is Pollution from US Manufacturing Declining? The Roles of Environmental Regulation, Productivity and Trade." *American Economic Review* 108 (12): 3814–3854.

Suh, S. 2006. "Are Services Better for Climate Change?" *Environmental Science and Technology* 40 (21): 6555–6560.

Sustainable Europe Research Institute (SERI). 2009. *Overconsumption? Our Use of the World's Natural Resources.* Retrieved from https://friendsoftheearth.uk/sites/default/files/downloads/overconsumption.pdf 4 October 2024.

Waggoner, P.E. and J.H. Ausubel. 2002. "A Framework for Sustainability Science: A Renovated IPAT Identity." *Proceedings of the National Academies of Sciences* 99 (12): 7860–7865.

Wiedmann, T., M. Lenzen, L.T. Keysser and J.K. Steinberger. 2020. "Scientists' Warning on Affluence." *Nature Communications* 11 (1): 3107.

World Bank. 1992. *Development and the Environment.* The World Bank: Washington, DC. Downloaded from http://www.worldbank.org/en/publication/wdr/wdr-archive May 2018.

World Bank. 2023. Final Consumption Expenditure (% of GDP). *World Development Indicators Database.* Retrieved from https://data.worldbank.org/indicator/NE.CON.TOTL.ZS May 2023.

6 Human Population Growth

6.1 Recap and Introduction to the Chapter

In the previous chapter, we considered the topic of consumption and evaluated the degree to which "overconsumption" might be a reason for ecosystem service degradation. We saw that marketed consumption generates most of our incomes and that reducing consumption is roughly equivalent to reducing incomes.

We also evaluated the idea of overconsumption using the IPAT model, and found that a number of factors can mitigate straightforward links between market consumption and environmental degradation. Higher consumption and output in an environment of poor policies can certainly strain natural resources, but the higher incomes that come along with high levels of consumption may provide governments with resources to better protect environmental assets. As consumption and incomes per person rise, economies also tend to produce fewer goods and more services, which may be less environmentally damaging than producing physical goods.

Of course, it is also possible that higher consumption, production and incomes allow countries to simply offshore environmentally damaging activities like pollution to lower-income countries. Evidence on pollution offshoring is not completely clear, though some countries import almost as much embedded CO_2 as they emit in total via production. We should, of course, not count imported emissions twice - once at production and once when consumed. The realization that some high-income countries drastically increase their climate change footprints, if embedded imported CO_2 is counted, means we should also not dismiss the possibility that consumption via offshoring of CO_2 is a driver of climate change.

DOI: 10.4324/9781003308225-7

We now examine a second reason often cited for damage to environmental assets. Human population growth, which is a core component of the IPAT model, has been a source of concern since at least the 1700s. More than 200 years and several billion people later, the possibility of human overpopulation continues to make many people very uncomfortable. We therefore dig into why we might potentially be concerned, where "over" population might be an issue and why those areas may have high rates of population growth.

Spoiler alert: The contemporary evidence suggests that high population growth and fertility exist largely in low-income countries and especially in sub-Saharan Africa. High-income settings generally have very low birth rates and in many countries populations are falling. Excessive population growth is limited to some very specific settings with poor infrastructure, thin market development, limited old-age insurance, often low status of women and lack of contraceptives. Any population "problems" are therefore closely linked with people's life circumstances, the constraints they face and the many choices they must make.

6.2 Concerns about Population Growth

People worry and perhaps have always worried about overpopulation. It is often the subject of high emotion and drama and has captured our imaginations and caused concern for easily the last 50+ years. In his 1968 bestseller called *The Population Bomb* (and its co-authored 1991 successor *The Population Explosion*), Paul Ehrlich argues that only by aggressively controlling population can the world avoid hunger and major losses of environmental resources. Population also features strongly in *The Limits to Growth*, which was published in 1972, and in two subsequent updates of that work. Concerns do not seem to have abated during the ensuing years, and similar issues were raised in Stephen Emmott's 2013 book *10 Billion*, as well as by Lester Brown in *Plan B 4.0: Mobilizing to Save Civilization*, which was published in 2009.[1]

As of this writing, a Google search on the subject of "human overpopulation" yields 18.6 million hits and "overpopulation and starvation" offers up 1.15 million hits, including an article from 1985 titled "Starvation Result of World Overpopulation" (Deering, 1985).

The United Nations Population Division estimates that the human population of the Earth reached 8 billion people on or about November 15, 2022, the so-called "Day of 8 Billion" (UN, undated). This amount compares with about 1.6 billion in 1900, 2.5 billion in 1950 and 4.5 billion in 1980. Though population estimates are indeed estimates and the further back one goes the rougher the estimation can become, it seems likely that in 1700 the world human population was between 600 and 700 million. When Jesus of Nazareth lived, the population was probably about 256 million, which is roughly the 2023 population of Indonesia. It then took until the 1920s for the world population to reach 2 billion, but only about 50 more years to add the second 2 billion and, as shown in Table 6.1, by 2000 an additional 2 billion people were added. Under its medium projection scenario, the United Nations predicts that by 2100 the world population will be 10.9 billion. These numbers are large and for some potentially scary.

Table 6.1 World Population Estimates

Year	Population	Annual Growth Rate (%)
1950	2,557,628,654	1.46
1960	3,043,001,508	1.35
1970	3,712,697,742	2.09
1980	4,451,362,735	1.87
1990	5,288,955,934	1.56
2000	6,088,571,383	1.26
2010	6,866,332,358	1.13
2020	7,794,798,729	1.09
2024*	8,019,876,189	0.95

*US Census Bureau (2024).

Source: United States Census Bureau International Database. Available at https://www.census.gov/data-tools/demo/idb/#/table?COUNTRY_YEAR=2024&COUNTRY_YR_ANIM=2024&menu=tableViz

6.3 Measuring "Over" Population

The most straightforward and obvious measure of population is total population. At the world level, for some this can be a big, scary number, but even for particular countries, total population can be daunting. For example, in 2021, which as of this writing is the latest official United Nations statistics are available,[2] the US population was about 337 million, China had 1.43 billion people, the Russian population was 145 million and in Europe it was 745 million. India had a population of 1.41 billion people.

Think for a second about your first reactions to these figures. Does reading these numbers make you feel anxious? Do you have different reactions depending on the country mentioned?

Population *growth* can also be anxiety-producing. Did you know that at the time of this writing the planet is adding approximately 1.6 million people per week? 84 million people per year? By the time you finish reading this paragraph, according to the US Department of the Census population clock,[3] which ticks away as the population increases, the world will have added perhaps 80 people.

Challenge Yourself

Why do you think some people have strong reactions to overpopulation? Why might these reactions differ by country?

Not only are these large numbers, but they are also not very helpful for at least three important reasons. First, settings differ. Some countries have lots of resources – wealth, land, water etc. – to accommodate many people, whereas others don't. They also may differ in terms of ecological conditions, such that the environmental effects of more people could differ. Totals therefore don't tell us much about population problems.

Second, total population does not give us much insight, because it is not compared to anything. For example, is 1 billion people a lot? Hard to tell. Even population growth gives limited information without context. If China's population grows by 10 million people per year, is that significant? We don't know unless we have a bit of context. If the population were 200 million, then 10 million would represent an enormous 5% annual growth rate. If the base population were 1.3 billion, though, if China can effectively absorb 10 million people, 10 million would be a relatively modest 0.76% annual growth.

Third, total populations change very slowly, because they depend on baseline population levels. This dependence is easily seen in Eq. 6.1, which gives the definition of a population at any time t, where the subscript t indicates the current period and t-1 represents last year.

Equation 6.1

$$\text{Population}_t = \text{population}_{t-1} + \text{births}_t - \text{deaths}_t$$

Total population level is just the population we start with, plus births minus deaths. Even if population is growing at a very slow rate, a large past population will almost certainly imply a larger future population. For total population to fall, ignoring immigration and emigration, deaths_t in a country have to exceed births_t. As average life expectancy in the world is about 71 years, this is not something that is going to happen overnight![4] Bottom line: There is enormous *trend* in total population, effectively making it useless for decision-making.

Focusing our attention on the change in population is not much better, because *births and deaths are themselves functions of population$_{t-1}$*. For example, if we have a population of 100 million, we will certainly have more births and deaths – the difference depending on the birth and death *rates* – than in a population of 10 million. Measures that rely on changes in total population are therefore also highly trend-dependent and cannot be expected to change very quickly, even if birth and death rates change dramatically. You can think of populations as huge ocean-going ships that take a long time to change course.

Equation 6.2

$$\text{Population Growth Rate(\%)} = \frac{\text{population}_t - \text{population}_{t-1}}{\text{population}_{t-1}} * 100$$

A better measure of overpopulation is the population growth *rate*, which compares the change in population with the starting population level. The formula is given in Eq. 6.2. This measure adjusts for baseline population size, which is very important, and therefore provides a more objective measure of population change than those using totals. Importantly, this feature means we can compare across countries and over time within countries. If the population growth rate of Ethiopia is 2.4% per year and for Sweden it is 0.51%, these numbers are directly comparable, because each is based on its own starting population. This takes care of the problem that the population of Ethiopia is over 118 million and the Swedish population is about 11 million. The population growth rate of Ethiopia is therefore 4.7 times that of Sweden. Furthermore, if we know that 10 years ago the population growth rate of Ethiopia was 2.7% per year, we can say that the population growth rate has declined by 0.3 *percentage points* or 100* (2.40 - 2.70)/2.70 = 11 *percent*.

Though not as trend-dependent as total population or change in population, the population growth rate also depends on the baseline population, which is in both the top and the bottom of Eq. 6.2. Of course, population$_{t-1}$ is not something that can be affected. It therefore can take a long time to change population growth rates, even if people's fertility behaviors are changing very quickly. This is an important shortcoming of the measure.

All these numbers seem small. For example, are the 2021 annual population growth rates in Niger (3.7%) or Uganda (3.2%) "large?" To draw *normative*[5] conclusions about growth rates, we first need to think about the power of compounding. For example, with a 3.2% annual growth rate, after 10 years the cumulative growth will be about 37%, because each additional year of 3.2% growth applies to increasingly larger populations. In contrast to 3.2% annual growth, a 37% cumulative increase over 10 years seems large - basically 40%! This realization also means that if nothing changes, between 2021 and 2031 the Ugandan population would go from 46 million to about 63 million over a decade. Wow! A large change.

Equation 6.3

$$\text{Approximate Doubling Time} = \frac{72}{\text{Annual Growth Rate}}$$

A useful mathematical trick for judging whether a growth rate is large is the so-called Rule of 72. This method focuses on the time to double the initial value and has the advantage of being very easy to calculate. It also provides one possible measure of whether an annual growth rate is large. To find the approximate doubling time, simply divide 72 by the rate of increase. This formula, which is given as Eq. 6.3, takes into account compounding.

Using the Rule of 72 we find that Uganda has an approximate population doubling time of 72/3.2 = 22.5 years. That is a short timeline for doubling and therefore the rate is large. With a growth rate of only 0.92%, a country's population will take about 78 years to double.

Equation 6.1 makes clear that it is births and deaths that are changing now, and therefore these should be the focus of any population policies. Total births and deaths are dependent on population levels, but it is possible to convert to rates by measuring births and deaths per thousand people in the population. These are the so-called *crude birth and death rates*. The *Natural Rate of Increase*, where the word "natural" implies that it leaves out in- and out-migration, is just the difference between crude birth and death rates. The formula for the natural rate of increase is given in Eq. 6.4. Because neither of these components depends on population$_{t-1}$, this metric can change quickly as behavior changes. It is therefore a potentially very useful measure of ongoing population change.

Equation 6.4

$$\text{Natural Rate of Increase} = \frac{\text{Births}_t}{1000} - \frac{\text{Deaths}_t}{1000}$$

Equation 6.4 says that the natural rate of increase can fall by either reducing births or increasing deaths per 1000 population. As we will see in the next subsection, in areas where population may be an issue, high death rates do not seem to slow population growth rates or the natural rate of increase; indeed the opposite appears to be true. Policymakers, of course, do not want to increase death rates to affect total population. In fact, huge international development efforts, such as the United Nations 2030 Agenda for Sustainable Development, are intensively focused on **reducing** infectious disease and mortality rates. Population policy therefore, of course, focuses on reducing birth rates, which is the only metric that can potentially be affected by policy.

The most important measure of birth rate behavior is called the *Total Fertility Rate (TFR)*. The TFR is the average number of births that a woman of childbearing age can be expected to have during her lifetime. An average of 2.1 births per woman per lifetime is a replacement level of fertility, meaning that eventually the population growth rate will fall to zero; the population will stabilize and stop growing when births equal deaths.

The TFR can change very quickly as social norms change. For example, if human fertility behavior changes, this can show up in the TFR within a year. The TFR is therefore an especially useful and flexible way to assess whether a huge population ship is starting to turn away from its current course.[6]

6.4 Where is Population Growth Potentially an Issue?

Tables 6.2–6.9 offer data on population levels, population growth rates, crude birth and death rates, TFR and child mortality, which are the metrics that were just discussed. These data are from the United Nations (2022), which are some of the most authoritative data available at the international level. They are presented by official United Nations region and income group over time to assess changes.[7] There is also information on important related measures, such as infant mortality.

Before reading beyond the tables, please take the opportunity to analyze these data and decide for yourself where population level, growth, natural rate of increase and TFR are "problems." Only after drawing your own preliminary conclusions, go on to read my assessment of the data.

Table 6.2 Total Population (Thousands)

	1970	1990	2000	2010	2021
World	3,682,488	5,309,668	6,126,622	6,929,725	7,909,295
Income					
Low	198,589	323,227	425,993	558,333	718,255
Lower-Middle	1,188,054	1,900,946	2,305,031	2,708,711	3,396,756
Upper-Middle	1,275,035	1,880,875	2,112,612	2,294,244	2,519,864
High	1,019,422	1,202,698	1,280,673	1,365,643	1,245,566
Select Regions					
North America	231,029	280,633	313,724	344,129	375,279
Eastern Asia	978,113	1,368,592	1,496,284	1,575,320	1,663,697
Southern Asia*	741,603	1,189,261	1,451,933	1,702,991	1,989,452
Sub-Saharan Africa	282,743	491,498	642,172	840,390	1,137,939
Latin America and Caribbean	288,494	446,889	526,890	599,823	656,098
Europe	657,221	721,086	726,407	735,395	745,174

*In addition to the subcontinent, also includes Iran and Afghanistan.

Source: United Nations (2022).

Table 6.3 Annual Population Growth Rate in Percent (Average for Five-Year Period Plus 2021)

	1970-1975	1990-1995	2000-2005	2005-2010	2015-2020	2021
World	1.96	1.54	1.24	1.22	1.09	0.82
Income						
Low	2.46	2.83	2.73	2.69	2.56	2.62
Lower-Middle	2.35	2.03	1.67	1.56	1.36	1.13
Upper-Middle	2.32	1.36	0.85	0.80	0.68	0.23
High	0.93	0.67	0.59	0.69	0.47	0.16
Select Regions						
North America	0.95	1.05	0.92	0.93	0.78	0.34
Eastern Asia	2.16	1.14	0.53	0.50	0.40	-0.03
Southern Asia*	2.32	2.08	1.71	1.49	1.20	0.91
Sub-Saharan Africa	2.66	2.72	2.66	2.73	2.65	2.51
Latin America and Caribbean	2.43	1.73	1.36	1.24	0.94	0.59
Europe	0.60	0.19	0.07	0.17	0.12	-0.18

*In addition to the subcontinent, also includes Iran and Afghanistan.

Source: United Nations (2022).

Table 6.4 Annual Crude Birth Rate per 1000 Population (Average for Five-Year Period Plus 2021)

	1970–1975	1990–1995	2000–2005	2005–2010	2015–2020	2021
World	31.6	24.5	20.8	20.2	19.6	16.9
Income						
Low	46.5	43.9	40.8	38.9	34.5	34.7
Lower-Middle	38.9	30.9	26.6	25.1	21.6	20.5
Upper-Middle	33.4	21.2	15.5	15.1	14.1	10.6
High	17.3	13.7	12.1	12.2	10.8	9.8
Select Regions						
North America	15.6	15.4	13.8	13.6	11.8	10.9
Eastern Asia	30.6	18.1	11.8	11.8	11.5	7.6
Southern Asia*	39.7	31.1	25.8	23.7	19.6	18.1
Sub-Saharan Africa	47.1	43.4	41.2	39.7	35.5	34.6
Latin America and Caribbean	35.4	25.3	21.4	19.1	16.5	14.8
Europe	15.6	11.5	10.1	10.8	10.4	9.2

*In addition to the subcontinent, also includes Iran and Afghanistan.

Source: United Nations (2022).

Table 6.5 Annual Crude Death Rate per 1000 Population (Average for Five-Year Period Plus 2021)

	1970–1975	1990–1995	2000–2005	2005–2010	2015–2020	2021
World	12.0	9.1	8.4	8.0	7.5	8.8
Income						
Low	20.6	16.3	13.2	11.0	7.7	8.0
Lower-Middle	15.1	10.0	8.7	8.2	7.3	8.8
Upper-Middle	9.8	6.8	6.5	6.6	7.2	8.4
High	9.4	9.2	9.2	9.0	8.8	9.9
Select Regions						
North America	9.2	8.6	8.4	8.1	8.6	9.5
Eastern Asia	8.8	6.5	6.4	6.7	7.4	7.8
Southern Asia*	16.3	9.9	8.1	7.6	6.9	8.6
Sub-Saharan Africa	20.0	16.2	14.6	12.5	8.7	9.2
Latin America and Caribbean	9.6	6.6	5.8	5.8	6.3	13.0
Europe	10.2	11.2	11.7	11.3	11.0	8.4

*In addition to the subcontinent, also includes Iran and Afghanistan.

Source: United Nations (2022).

Table 6.6 Total Fertility Rate per Woman of Childbearing Age (Average for Five-Year Period Plus 2021)

	1970-1975	1990-1995	2000-2005	2005-2010	2015-2020	2021
World	4.48	3.04	2.62	2.56	2.47	2.32
Income						
Low	6.57	6.23	5.67	5.31	4.52	4.62
Lower-Middle	5.58	4.01	3.31	3.08	2.71	2.59
Upper-Middle	4.93	2.38	1.84	1.82	1.90	1.52
High	2.34	1.80	1.69	1.76	1.67	1.56
Select Regions						
North America	2.01	2.00	1.99	2.02	1.75	1.64
Eastern Asia	4.43	1.96	1.48	1.52	1.65	1.17
Southern Asia*	5.67	4.04	3.19	2.85	2.40	2.23
Sub-Saharan Africa	6.75	6.18	5.68	5.40	4.72	4.59
Latin America and Caribbean	5.03	3.01	2.52	2.27	2.04	1.86
Europe	2.17	1.57	1.43	1.55	1.61	1.48

*In addition to the subcontinent, also includes Iran and Afghanistan.

Source: United Nations (2022).

Table 6.7 Life Expectancy at Birth in Years (Average for Five-Year Period Plus 2021)

	1970-1975	1990-1995	2000-2005	2005-2010	2010-2015	2021
World	58.05	64.54	67.05	68.84	70.48	71.0
Income						
Low	43.74	49.11	53.07	56.93	60.27	62.5
Lower-Middle	51.52	59.94	62.86	64.61	66.30	66.4
Upper-Middle	60.62	68.46	71.32	72.65	73.83	75.3
High	70.38	74.35	76.15	77.53	78.80	80.3
Select Regions						
North America	71.39	75.82	77.38	78.36	79.16	77.7
Eastern Asia	62.56	70.56	74.02	75.58	76.58	79.1
Southern Asia*	49.37	59.67	64.01	65.82	67.74	67.6
Sub-Saharan Africa	44.66	48.90	50.32	53.83	57.18	59.7
Latin America and Caribbean	61.16	68.41	72.15	73.39	74.55	72.2
Europe	70.60	72.61	73.75	75.26	77.01	77.0

*In addition to the subcontinent, also includes Iran and Afghanistan.

Source: United Nations (2022).

I would suggest that you especially consider the following questions:

- **In what regions and types of countries would you say population growth is an issue?**
- **What defines population "progress?"**
- **Where has progress been made over time? How significant is that progress?**
- **Are there any areas where the population problem seems to have been "solved?"**

A lot can be inferred from Tables 6.2–6.9 and several details are left for you to consider in the questions for discussion and on your own. We can all draw somewhat different conclusions from these data, but probably we would all agree on a few conclusions. First, it is obvious that most of the world's population resides in middle-income countries. Most of these middle-income residents are in Asia.

Challenge Yourself

Why do you think particular regions and types of countries would have high or low total fertility and/or population growth rates?

Table 6.8 Infant Mortality Rate per 1000 Live Births (Average for Five-Year Period Plus 2021)

	1970–1975	1990–1995	2000–2005	2005–2010	2015–2020	2021
World	95	63	49	42	29	28
Income						
Low	141	113	84	70	48	44
Lower-Middle	126	79	60	52	37	33
Upper-Middle	79	42	29	23	12	10
High	25	12	8	7	5	4
Select Regions						
North America	18	9	7	7	6	5
Eastern Asia	66	37	23	16	9	6
Southern Asia*	141	83	61	52	36	30
Sub-Saharan Africa	133	111	88	75	51	49
Latin America and Caribbean	80	39	25	22	15	14
Europe	25	13	8	7	4	4

*In addition to the subcontinent, also includes Iran and Afghanistan.

Source: United Nations (2022).

Table 6.9 Under-Five-Year-Old Mortality Rate per 1000 Live Births (Average for Five-Year Period Plus 2021)

	1970-1975	1990-1995	2000-2005	2005-2010	2015-2020	2021
World	139	91	71	59	40	37
Income						
Low	235	186	135	109	71	64
Lower-Middle	187	113	85	71	49	44
Upper-Middle	108	53	37	29	15	12
High	31	15	10	9	5	5
Select Regions						
North America	21	10	8	8	7	6
Eastern Asia	91	46	28	18	11	7
Southern Asia*	206	116	82	68	44	37
Sub-Saharan Africa	226	184	142	117	78	72
Latin America and Caribbean	112	49	33	29	19	16
Europe	29	15	10	8	5	4

*In addition to the subcontinent, also includes Iran and Afghanistan.

Source: United Nations (2022).

Second, if we were to characterize regions of ongoing excessive population growth and fertility, we would probably all point to low-income countries and sub-Saharan Africa. These areas have annual population growth rates of over 2.50% (a high rate), versus the world average of 1.09%. With only a few exceptions, all countries with average annual population growth rates above 2.50% not due to in-migration are in sub-Saharan Africa. In this region, the average woman has almost 5 children during her lifetime versus 2.5 for the average country and crude birth rates are more than four times crude death rates. Furthermore, these metrics have not changed nearly as much over time as in other regions and income groups.

Third, high-income countries on average tend to have low fertility, population growth and natural rates of increase. In some high-income regions, in fact, TFR is well below the replacement fertility rate of 2.1 children per woman. This means that the population will continue to grow for some time as the big population ship changes course, but eventually it will decline. Looking deeper into the data in UN (2022), it seems there is no point worrying about population growth in Portugal (TFR of 1.36), Japan (TFR of 1.30), China (TFR of 1.16) or Iran (TFR of 1.69), because fertility is so far below the replacement rate. Indeed, more than 80 countries, as well as Europe, North America, South America and East Asia, have TFRs below 1.80 and 10 countries meet the UN definition of "very low fertility" with TFRs of less than 1.30.[8] As of 2021, the world was just above the replacement fertility rate at 2.32 children per woman of childbearing age (UN, 2022).

Challenge Yourself

What downside (if any) do you see from falling populations?

As of 2021, over 30 countries had deaths that exceeded births, including Germany, Italy, Japan, Portugal, Russia, Spain and Ukraine, plus East Asia, Europe and high-income countries. In these countries and regions, the big population ship has completely turned around in response to TFRs less than 2.1. Independent of migration, crude death rates above crude birth rates mean populations have begun to fall.

Fourth, in most areas of the world, there have been enormous declines in fertility and slowing population growth. Almost throughout the world, population growth has plummeted as the average number of births per woman nosedived. During the 1970–1975 period the average woman in the world had about 4.5 children during her lifetime, but by 2015 that figure was a mere 2.51 and in 2021, women averaged 2.32 children over their lifetimes. In the middle-income countries (which make up most of the world's population), over the last 40 years the TFR declined by about 60%, and much more in some countries. The Latin America average TFR in 2021 is only 37% of what it was in 1970–1975! Clearly, these are enormous changes.

With very few exceptions, it is only sub-Saharan African countries that still have the TFRs of 4–7 children per woman that were quite common in the 1950s and 1960s. The highest fertility in 2021 was Niger (6.8) and only 10 countries (all in sub-Saharan Africa) have TFRs above 5.0. Indeed, out of 30 countries with TFRs above 4.0 children per woman (which was the norm in the 1970s), only 1 is not in sub-Saharan Africa. As of 2021, high fertility is therefore a quite localized phenomenon.

Finally, the definitions in Eqs. 6.1, 6.2 and 6.4 suggest that areas with high crude death, infant mortality and under-five child mortality rates and low life expectancies should have smaller populations and low rates of population growth.[9] The data, however, suggest exactly the opposite. Countries and regions where mortality rates are highest tend to have the highest rates of population growth and natural rates of increase.

Challenge Yourself

Africa is a largely rural continent. How might such settlement patterns play into fertility decisions?

Areas with high infant and under-five-year-old child mortality also tend to have high TFRs. In sub-Saharan Africa, for example, we see infant and child mortality an order of magnitude (i.e., 10 times) greater than in high-income countries, with 6.4% of children under five and 4.4% of infants dying, versus less than 0.5% in high-income countries. Households may therefore be overcompensating for potential mortality through fertility. This speculation about humans' responses to their circumstances suggests there may be subtle responses that could affect fertility decisions.

As shown in Tables 6.8 and 6.9, since 1990 there has been major progress on infant and child mortality. In the last 25 years, infant and young child mortality have been cut by roughly 75%. This observation suggests – if indeed families are overcompensating with fertility for high child death rates – that TFRs could fall dramatically in sub-Saharan Africa and low-income countries as a whole as they did, for example, in Latin America and the Caribbean between 1970–1975 and 1990–1995.

You probably have already started thinking about potential explanations for high fertility and population growth in low-income country settings. Why exactly do low-income areas and especially sub-Saharan Africa have such high fertility rates compared to the rest of the world? The next subsection discusses day-to-day life in low-income countries and the subset of low-income countries found in sub-Saharan Africa, where population growth and fertility are highest. To organize our thinking, let us discuss two very different conceptual models of fertility behavior and consider why providing basic government services may affect fertility behavior.

6.5 Modeling Fertility Decisions Where Population Might be a Problem

To try to understand what factors affect fertility, why we have observed changes over time, and whether in the future we can expect improvements or continued problems, we need some sort of behavioral model. Human behavior is always very complicated, including in matters of the family. Models attempt to cut through such complexity and get to the heart of the issue so we can perhaps better understand it. In Chapter 3, we examined a model of price formation in markets often called the Supply-Demand model. In Chapter 5, we discussed a model of the key linkages between economies and the environment. Of course, these models did not deal with all aspects of the relationships, but instead focused in on what are believed to be the key aspects.

6.5.1 Malthusian Population Model

Thomas Malthus was an English clergyman and classical economist, who, in his 1798 essay titled *An Essay on the Principle of Population*, put forward perhaps the first – and still a very important – model of human fertility behavior. In his model, he presents two assumptions on which the model was built. These assumptions establish the essential context for looking at human behavior; model assumptions establish the boundaries of the debate.

> ### Box 6.1 Assumptions of Thomas Malthus' Model of Fertility (from Malthus, 1798)
>
> 1.   "First, that food is necessary to the existence of man."
> 2.   "That the passion between the sexes is necessary and will remain nearly in its present state."

Assuming that fertility is above replacement, from these two assumptions he infers that population grows exponentially (what he calls geometrically). For example, if we start with 2 couples, each of which has 4 children, after one generation we have a population of 4 parents + 8 children = 12. If the first generation dies and the remaining 4 couples have 4 children each, in the next generation we will have 4 couples*4 children per couple = 16 children + 8 parents = 24 people. If again the parents die and each of the 8 couples again has 4 children, the population is now 32 children + 16 parents = 48 people. If the parents again die and the 16 couples have 4 children each, the population is 32 parents + 16*4 = 98, etc. The population dynamics are shown in Fig. 6.1.

Malthus' first assumption is essentially that each person eats a fixed amount. Because population is growing exponentially, food demand therefore also grows exponentially. For example, if each person eats 2 hamburgers, 1/2 pound of rice and a salad each day, in the first generation we need 8 hamburgers, 2 pounds of rice and 4 salads to feed the population. By the sixth generation, however, we need 384 hamburgers, 96 pounds of rice and 192 salads to feed the population!

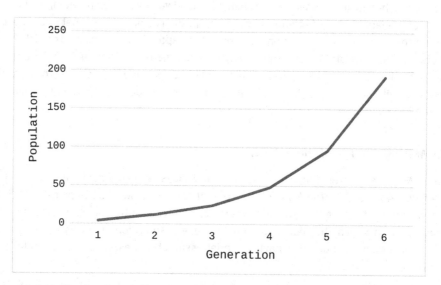

Figure 6.1 A Malthusian Population Model with No Grandparents and Four Children per Generation

Challenge Yourself

In what respects do you think the Malthusian model can apply more than 200 years after it was developed?

He then asserts that production increases "arithmetically." For example, each acre of land farmed by people produces four to five tons of grain. As land is fixed,

> The food must necessarily be distributed in smaller quantities, and consequently a day's labour will purchase a smaller quantity of provisions. An increase in the price of provisions would arise either from an increase of population faster than the means of subsistence, or from a different distribution of the money of the society.
>
> (p. 26)

Eventually humans bump up against a population limit that in the biological literature is often called a *carrying capacity*. Such limits act as a check on population or in the extreme there is a population collapse. Physical limits are therefore critical to the conclusion of the Malthusian model that there are limits to fertility.[10]

Of course, this model depends on human behavior and Malthus has in mind a fascinatingly myopic view of human behavior. In his *Essay*, Malthus sees human populations rising in response to more production, with people marrying more and having more children in good times, but reducing fertility when economic conditions are bad.

Children in a Malthusian model are therefore a consumption good like a television that people "buy" when times are good and cannot afford when times are bad. People always, though, "want" more children, because in Malthus' view people have uncontrollable lust. This view of fertility is presented in the *Essay* with what today seems like a very classist overlay. He concludes that support for the poor ends up doing no good, because they just end up having more children and finish where they started – in poverty. I leave you to read this seminal essay and draw your own conclusions.

6.5.2 A Modern Economic Model of Population

Modern economic models of fertility are based on very different assumptions than the model of Malthus. Their genesis is largely attributable to Nobel prizewinner Gary Becker, who in 1960 argued that decisions within the family can be viewed as rational choices. Rational does not mean people are super-optimizing computers. It just means they try to do the best they can, given their knowledge and the constraints imposed on them by their circumstances. That is, while husbands and wives may have different interests, those making decisions in households decide how many children to have and how much to invest in each child (e.g., how much to spend on education) to promote their own interests and the interests of their families. One way to express such a goal in a model is to say that mothers and/or fathers weigh the benefits of children against the costs and try to maximize the difference between total benefits and total costs. In such a setup, fertility is a sort of optimization problem.

Box 6.2 Assumptions of a Modern Economic Fertility Model

1. Children have costs.
2. Children have benefits.
3. Households are on average rational.
4. Men and women may not always agree.

Economic benefits need not be financial and can include the fun of having kids, social norm fulfillment (e.g., everybody else has kids, so we should too) and passing along genetic material into the future. Such non-monetary benefits are perhaps some of those Thomas Malthus had in mind. These are also likely what those living in richer countries might think of as the key benefits.

What about the costs of children? Certainly, there are direct costs, such as clothes, food etc., which can be expensive and differ from place to place. Probably the most important cost, though, is the time needed to raise children. Childcare is most often, but not always, primarily done by women, so it is the time cost to women that is especially important. The idea that Gary Becker proposed is that people will weigh, for example, the psychic importance of producing grandchildren for their own parents against the time and money costs. He said they will then choose how many children to have. Low opportunity costs/high benefits mean more kids and high costs/low benefits imply fewer children. Pretty straightforward.

These general categories of costs and benefits apply across a variety of contexts. To understand something about why low-income countries and particularly sub-Saharan Africa have high fertility and other areas are under the replacement rate, we need to think about people's life circumstances.

Before reading on, please think for a moment about livelihoods and life circumstances in high-income and low-income countries. For example, across the world, how do people typically earn their livings? What do they do with those resources? How and where do they get the basics of life, such as food, water and energy? Who takes care of such things when people get old?

For those who live in places like the US, Europe and Japan, these answers may seem obvious and we may naively assume they apply universally around the world. In high-income countries, people often work in jobs that pay wages and salaries. These resources are then invested or spent on consumables. If we have the opportunity to save some money, there are banks and other types of financial institutions available for that purpose. We may be able to borrow if we need to. People in high-income countries typically live in highly monetized economies.

Food is purchased in stores using earned income and water comes out of taps, usually from several places right inside homes (e.g., kitchen, bathrooms, laundry rooms, outdoor faucets, etc.). Municipalities typically provide this water, but in some places private companies supply water and are regulated by local governments. People use some of their incomes to pay for the pipes to be installed to move water to, in and away from their homes (i.e., sewage pipes), as well as to buy the actual water.

Energy is usually in the form of electricity and/or natural gas (e.g., methane), which is used for heating and cooking. People pay for electricity and gas to be provided at their homes by either municipalities or private companies that are contracted by governments to provide such services.

When people get old, they still need these and other necessities. People save for retirement during their lifetimes, but private companies, non-governmental organizations and governments also play critical roles by providing public pensions for the elderly.

For many people living in sub-Saharan Africa and much of the Global South, the context is completely different, and a number of these differences may contribute to larger families. Whereas in the US and other high-income countries energy is easily available – for a price – at home with the flip of a switch, people in the low-income world typically cook, heat and light with biomass fuels, such as wood and crop residues (e.g., corn stalks), in stoves located in or near their homes. About 40% of the world or almost 3 billion people cook with biomass fuels on a regular basis and in some countries in sub-Saharan Africa, the percentage can be virtually 100%.

In rural areas, people generally collect these fuels themselves in forests or from their own lands. For example, people go to the forest to cut wood or farmers plant trees around the edges of fields and cut branches. Collecting wood is, of course, time consuming and difficult. The branches or trees need to be brought home, dried and cut to the right size for cooking. Cooking takes longer than in richer countries because people must first make fires and smoke in kitchens can complicate cleanup. The whole process is much more difficult than turning on an electric stove!

Photo 6.1 "Rocket"-Style Improved Stove in Rwanda
Source: Author.

Photo 6.2 Fuelwood Collection in Rural Ethiopia
Source: Author.

And what about water? In low-income countries, where TFRs are high, water is often not piped into homes – even in cities. People must walk to fill water jugs and other containers every day and bring the water home. Sometimes they collect water from communal taps, which may only come on sporadically, but it is not uncommon for households to get water from rivers and streams.

I and colleagues found that in the Amhara region of Ethiopia, in 2005 about 60% of people got water from rivers and the rest from community taps. Nobody got water directly in their homes and half the people needed at least a 30-minute round-trip to collect water. Even as of 2016, Bogole (2020) found that worldwide about 144 million people got their water from "unimproved" surface water sources, including 35% of the Ethiopian population. In the West African country of Benin, in 2015 almost half of all users got their water from unimproved sources (Johnson et al., 2015).

Collecting water is onerous and creates strong incentives to conserve. After meals, dishes and pots must be washed with water carried to the home and there are no dishwashers, because there is no electricity. Cleanup after meals therefore takes a lot longer than it does in high-income countries.

Speaking of linkages between water and electricity, especially in rural areas of low-income countries washing machines basically do not exist. All laundry must be done by hand – perhaps while carefully conserving water – which is itself extremely time consuming and difficult. Just for the experience, those of us with washing machines should try doing a load of laundry by hand. To get the full experience, let's all be sure a pair of jeans is in the wash when we do it!

Particularly in rural areas of low-income countries, but sometimes in urban regions as well, financial institutions for saving and borrowing may not be available. This means that if households get in a pinch, they may not be able to borrow except from village moneylenders that can charge very high interest rates - often more than 100% per year.

They also may have nowhere to save money, so often people invest in animals, such as cattle, goats, chickens and pigs. Of course, raising animals is itself time consuming. They must be fed, watered and otherwise cared for, and in many parts of the Global South it is not uncommon to see children grazing cattle, water buffalo, horses and goats. Just as lack of energy and water systems creates labor demands, so too does not having access to financial institutions.

Finally, and of critical importance, in settings where population growth remains an issue, people often earn their livings as subsistence farmers. They therefore generally mainly eat what they produce. Agricultural technologies in such settings can be very rudimentary. Tractors, improved seeds, fertilizers and pesticides are often not available and whole fields may be plowed, planted, weeded and harvested 100% by hand. In some cases, animal traction for plowing may not even be available. In such settings, agricultural productivity can be low and farming may require lots of labor to produce the crops on which the survival of the family depends. For example, average output per hectare of land in East Africa is about 1/5 that of higher-income countries and East Asia, and 1/3 that of Latin America.

These huge labor demands in high-stakes, risky environments can create important pressures for large families. How is a three-person family in a rural area in a country like Ethiopia, Rwanda or Niger supposed to get done what it needs to do to get ahead? They cannot and an observant visitor to villages in such settings will quickly notice that larger families are often relatively prosperous and small households are in trouble. This is generally because basically all productive tasks there require so much labor.

In terms of our so-called modern model of fertility, in low-income countries households get critical production benefits from children that simply do not exist in high-income countries. Benefits are high and we would expect - not surprisingly once we think it through - bigger families.

A non-production benefit that is often important in lower-income countries, but is typically not as critical in high-income settings, is old-age insurance. We will all die and if we are lucky we will get old before that happens. We may therefore eventually not be able to work and earn a living. In high-income settings, though often far from perfect, public pension systems such as Social Security in the US are typically available. As we already discussed, financial systems also stand ready to help us save and invest money for "retirement."

Such institutions are very limited or non-existent in low-income settings and old-age insurance must be handled within households. What happens to old people who cannot work who have no children? They may be in big trouble, that's what happens. Families therefore substitute - for better or worse - for the pension and financial systems that support older people in higher-income countries. Our modern economic model predicts that pensions and robust saving mechanisms will reduce the demand for children.

What about the cost side of the equation? As was already mentioned, the #1 cost of children around the world is the time it takes mom and perhaps others to raise them. The opportunity cost of women's time is therefore critical to the cost of child rearing. If what is given up by taking time to raise kids (e.g., wages from work) is very valuable, opportunity costs will be high. This is roughly, but of course imperfectly, the case in high-income countries where educational and work opportunities are more available than in many low-income countries. Our model, all else equal, therefore predicts small family sizes if a lot of high-value opportunities other than motherhood are available to women. The appendix to this chapter presents these tradeoffs and the results in a graphical model similar to the supply and demand model presented in Chapter 3.

Countries where women's rights are protected by law and social custom tend to have lower average fertility than places without such protections. Examples include legal rights to own property, to inherit and to be free from discrimination in the workplace and in education. All these rights, which are enforced by law and governments, increase opportunities for women and boost the opportunity cost of child raising. Many of the countries with the highest fertility rates also have some of the worst records on women's rights. A number of countries in sub-Saharan Africa, for example, do not allow women to own land or inherit property.

Jobs for everyone can be very limited, particularly in rural areas of low-income countries. Like financial markets, which are poorly developed in many low-income countries, weak labor markets reduce opportunities for everybody, including women. Labor market development can therefore potentially also have very important fertility consequences.

This is perhaps an appropriate place to mention the importance of technologies used for family planning. Family planning is central to gender equality and women's empowerment, but the United Nations estimates that approximately 225 million women who want to avoid pregnancy are not using safe and effective family planning. The UN cites a variety of reasons, including lack of access to information, services and support from their partners or communities. Most of the women with unmet contraceptive needs live in the 69 poorest countries.[11]

6.6 Chapter Conclusions

People sometimes worry about overpopulation, but in terms of population growth and high fertility, population "problems" exist largely in low-income countries and especially in sub-Saharan Africa. Indeed, high-income countries typically have very low total fertility rates and even negative rates of natural increase. High population growth rates are therefore peculiar to certain settings that have circumstances like poor infrastructure, limited market development, limited old-age insurance, often low status of women and lack of contraceptives. The population "problem" is therefore intimately connected with people's life circumstances and the broader set of choices they make.

Issues for Discussion

1. Do you feel anxious and uncomfortable about high population levels? If so, exactly why?
2. In low-income countries many markets function poorly. For example, labor markets in which people get paid employment often don't exist. Markets where people can save and

borrow work badly and in most of Africa there is no insurance to speak of. How might these and other market problems affect fertility?

3. The highest fertility rates are in the poorest parts of the world. Do you think it is irresponsible for poor people to have lots of children? Why or why not?

4. The Malthusian model in Fig. 6.1 had about a 50-fold increase in population over 6 generations. How would this change if each couple averaged 2.5 children? 5.5 children? Are these increases still exponential?

   Hint: You might want to use a spreadsheet program, such as Microsoft Excel, with careful specification of each generation, to simulate these possibilities.

5. If a country has high population growth, high incomes and takes care of its environment, is rapid population growth OK?

6. In countries where the total fertility rate is less than 2.1, total populations may still be growing. Please explain clearly how this situation is possible.

7. Do you think negative population growth in some countries is "sustainable" in the long run?

8. What if anything might make countries with falling populations like China, Spain and Japan move toward growing populations?

Notes

1 Available at http://www.earth-policy.org/images/uploads/book_files/pb4book.pdf

2 All population data are from UN (2022), which as of this writing are the latest official UN population data.

3 https://www.census.gov/popclock/ or for a more dramatic representation see https://www.worldometers.info/world-population/

4 To see this result clearly, just subtract Population$_{t-1}$ from both sides of Eq. 6.1 and think about what would make the left side of the equality negative.

5 Meaning having to do with norms or evaluated with reference to reasonableness.

6 Demographers use statistical techniques to infer lifetime behavior from current behavior. The United Nations Population Division each year publishes World Population Prospects. The document as well as data and calculation methodologies are available at https://population.un.org/wpp/

7 All data are available at https://population.un.org/wpp/. You can analyze by country as well as groups of countries.

8 Cayman Islands, China, Italy, Japan, Malta, Republic of Korea, Singapore, Spain, Taiwan and Ukraine.

9 The conclusion about population growth comes from substituting Eq. 6.1 into Eq. 6.2 and evaluating.

10 The *Limits to Growth* model presented in Meadows et al. (1972) and discussed in Chapter 1 is sometimes called a "Malthusian" model, because of its emphasis on physical limits.

11 See https://www.unfpa.org/family-planning for more information.

12 It is perhaps worth noting that if the marginal benefit of an additional child always exceeds the marginal cost, the model predicts that the household will want to have as many children as it can. If instead the marginal cost always is greater than the marginal benefits, the optimal fertility level is predicted to be zero.

Further Reading and References

Becker, G. 1980. *A Treatise on the Family*. Harvard University Press: Cambridge, MA.

Bogole, G.G. 2020. "Hotspots of Unimproved Sources of Drinking Water in Ethiopia: Mapping and Spatial Analysis of Ethiopia Demographic and Health Survey Data 2016." *BMC Public Health* 20: 878.

Deering, F.J. 1985. "Starvation Result of World Overpopulation." *The Oklahoman*, January 13.

Ehrlich, P. 1968. *The Population Bomb.* Ballantine Books: New York.

Ehrlich, P. and J. Holdren. 1971. "Impact of Population Growth." *Science* 171 (3977): 1212–1217.

Johnson, R.C., G. Boni, Y. Barogui, G.E. Sopoh, M. Houndonougbo, E. Anagonou, D. Agossadou, G. Diez and M. Boko. 2015. "Assessment of Water, Sanitation, and Hygiene Practices and Associated Factors in a Buruli Ulcer Endemic District in Benin (West Africa)." *BMC Public Health* 15 (1): 801.

Malthus, T. 1798. *An Essay on the Principle of Population: An Essay on the Principle of Population, as it Affects the Future Improvement of Society with Remarks on the Speculations of Mr. Godwin, M. Condorcet, and Other Writers.* J. Johnson: London. Reprinted by the Electronic Scholarly Publishing Project, 1998. Retrieved from http://www.esp.org/books/malthus/population/malthus.pdf March 2024.

Meadows, D.H., D.L. Meadows, J. Randers and W.W. Behrens III. 1972. *The Limits to Growth: A Report for the Club of Rome's Project on the Predicament of Mankind.* Universe Books: New York. Retrieved from https://policycommons.net/artifacts/1529440/the-limits-to-growth/2219251/ 14 February 2024.

United Nations (UN). 2022. *World Population Prospects 2022.* Department of Economic and Social Affairs, Population Division. Retrieved from https://population.un.org/wpp/ March 2024.

United Nations (UN). Undated. *Day of 8 Billion.* Retrieved from https://www.un.org/en/dayof8billion March 2024.

US Census Bureau. 2024. "Census Bureau Projects U.S. and World Populations on New Year's Day." January 1. Retrieved from https://www.commerce.gov/news/blog/2024/01/census-bureau-projects-us-and-world-populations-new-years-day#:~:text=The%20projected%20world%20population%20on,the%20U.S.%20and%20world%20populations March 2024.

APPENDIX
Graphical Analysis of Optimal Fertility Choice

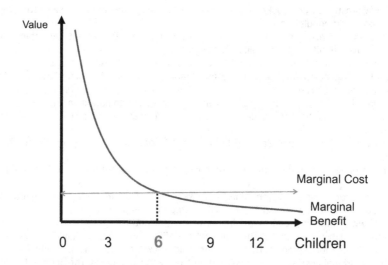

Figure 6.A.1 Value-Maximizing Fertility

The figure above presents a graphical cost-benefit model of fertility that presumes households choose their fertility to maximize the total net benefits of children. The above slopes may or may not be correct, but can be verified using empirical testing. For example, using actual data it would be possible to estimate the slope of the marginal cost function.

As written, though, the curves indicate that the marginal cost of each additional child increases slightly as family size increases, but the benefit of an additional child decreases. That is, like the ecosystem service presented in Fig. 3.1, getting the first child is absolutely critical (e.g., to pass along genetic material into the future, help during old age, graze animals etc.) and additional children are less critical. I would urge you to play around with the slopes of these curves, but be sure you are able to tell plausible stories about any changes![12]

Let us now examine why six children maximizes household well-being if the marginal benefit and marginal cost curves include all costs and benefits. Suppose instead of six children that the household chooses three children; let's evaluate whether in this model three children would be preferred. With three children the marginal benefit exceeds the marginal cost, which suggests that children 1, 2 and 3 add to household well-being. The fourth and fifth children also, though, have marginal benefits greater than marginal costs. The household can therefore increase its well-being further by adding those fourth and fifth children and stopping at six, because beyond six the costs are greater than the benefits.

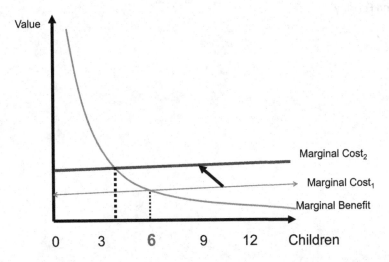

Figure 6.A.2 Higher Marginal Cost (e.g., Increasing Cost of Women's Time) Reduces Value-Maximizing Fertility

What happens if the marginal cost of children increases **for all levels of fertility**? This could for example happen if wages for women or educational opportunities increase. We see in the graph above that the optimal family size will decline to about four children. Decreasing the marginal benefits of children **for all levels of fertility** would reduce the family size further. We would model this change in the landscape (e.g., because of rural electrification or improved water supply) as a shift in the marginal benefit function to the left. The model then predicts that the optimal level of children would be lower. In the example below, the family would choose to have two children.

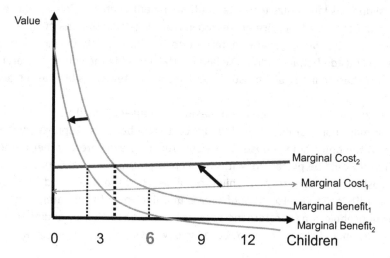

Figure 6.A.3 Lower Marginal Benefit (e.g., Rural Electrification) Further Reduces Value-Maximizing Fertility

7 Sustainable Economic Development

What You Will Learn in this Chapter

- What is sustainability and how is it related to savings and investment?
- About environmentally focused perspectives on sustainable development
- Human-focused views of sustainability
- Consumption over time, discounting and present value
- About adjusted net national income
- Sustainability metrics focused on capital stocks and adjusted net savings

---

7.1 Recap and Introduction to the Chapter

In Chapter 5, we discussed perspectives on overconsumption, including the national accounts definition of consumption and the relationship between marketed consumption and ecosystem services. We also attempted to pin down the potential environmental effects directly attributable to overconsumption per se, but instead found that overconsumption is an extremely slippery subject that is fraught with judgements about others' choices, lots of confounders and offsetting factors. Surprisingly, there are even difficulties figuring out if rich countries offshore environmental problems, such as pollution, to lower-income countries.

In the last chapter, we talked about human population growth, including the where, when and how of "over" population. In the contemporary context, high fertility is largely confined to sub-Saharan Africa and a few parts of Asia. The rather niche context, as well as the observation that so many high-income countries have low fertility and even falling populations, suggest that leaning into population as a cause of today's environmental problems may be a bit of a red herring (i.e., something that leads us astray). In fact, even a quick look at the data makes clear that the places with high fertility tend to be rural, with low incomes, weak infrastructure (e.g., water and energy), high child mortality, limited government support for the elderly and sometimes low status of women.

It is time to bring a bit more economic structure, à la Chapters 3 and 4, especially to our treatment of consumption. We therefore now turn our attention to the topic of *sustainability*,

DOI: 10.4324/9781003308225-8

which catapults some of the issues presented in Chapters 5 and 6 into a more dynamic context. In other words, we now try to incorporate the future.

7.2 What is Sustainable Economic Development?

Sustainable economic development, or "sustainability" for short, is all about what we who are living today leave for the future. In a background paper prepared for the World Conference on Environment and Development, which was held in 1992, the World Commission on Environment and Development to the United Nations[1] defined sustainability as, "Development that meets the needs of the present without compromising the ability of future generations to meet their own needs." This definition has largely stood the test of time and even after over 30 years forms the basis for most discussions of sustainability.

What does it mean to do right by the future? There are many details to fully answer this question, which we will get to in a minute, but at a fundamental level we need to leave the future assets, including natural resources, so future people have the tools they need to be well-off. These assets are resources that can be used as capital, allowing future humans to be at least as well-off in terms of goods, services (including ecosystem services) and other aspects of human well-being as we are today. Sustainable development therefore requires *investments*, which means we need to *save* in the present to replace and perhaps build up our stock of resources, such as machinery and tools (i.e., *physical capital*), education (i.e., *human capital*), natural resources (i.e., *natural capital*) and maybe even social structures (i.e., *social capital*). If we allow these resources to degrade over time, the situation would assuredly not be sustainable.[2]

We tend to think of "savings" as what is left over from our incomes after we purchase what we want to buy (i.e., consume), which is correct for analyzing personal finances. We want to consider societies more generally, though, and include both marketed and non-marketed services. Let us therefore think broadly in terms of benefits that are available to human societies in each period, with people deciding how much of these benefits to consume and invest for the future. Financial incomes are certainly part of this set of benefits, but so are non-marketed ecosystem services, such as climate stability, aesthetic values and recreation. With this way of thinking, "over" consumption then becomes a problem if humans do not save, which means we have little or no resources to invest. If we invest little, we pass fewer assets on to subsequent generations, which compromises the ability of future generations to meet their needs. This view of sustainability is one that would be immediately recognizable to most economists.

7.3 Environment-Focused Sustainability

So, what types of assets might we want to preserve or increase in order to benefit future humans and perhaps other species? Early work on this topic looked at the issue from materials and energy perspectives (e.g., Kneese et al., 1970). This "materials balance" approach, which emphasizes that energy cannot be created or destroyed, views savings as minimizing materials "loss" in the form of pollution or what early environmental economists called *residuals*. Sustainability is therefore a materials-focused process of "residuals management," to not waste too much matter in the form of smoke coming out of chimneys or solid waste expelled into freshwater and marine (i.e., saltwater) ecosystems.

Other ecologically focused notions of sustainability emphasize human appropriation of primary productivity. Primary productivity is the process by which organisms turn energy into biomass. The main source of productivity on the planet is plants, which convert solar energy, CO_2 and water into plant material. When humans take too much production for themselves or degrade ecosystems, it leaves less for nature to fuel its own activities and reduces natural "savings" for the future (e.g., see Vitousek et al., 1986; Folke et al., 1997). Examples of human appropriation of primary productivity include deforestation, farming, desertification of land, overfishing, the building of cities and pollution – anything that removes or damages productive organisms. Often, when focusing on appropriation of primary productivity as a measure of sustainability, there is some emphasis on human/nature carrying capacity, often leading to policy conclusions that humans need to operate within natural limits (e.g., IPBES, 2018).

The Environmental Performance Index (EPI), which is updated annually by Yale University and Columbia University,[3] offers what Wolf et al. (2022) call a "data-driven summary of the state of sustainability around the world." The authors use 40 environmental performance indicators to rank 180 countries on climate change performance (38% of the index), environmental health (20% weight) and ecosystem vitality (42% weight). Indicators include air pollution metrics, biodiversity protection, greenhouse gas (GHG) emissions, forest cover and much more. As noted in Wolf et al. (2022), "The EPI offers a scorecard that highlights leaders and laggards in environmental performance and provides practical guidance for countries that aspire to move toward a sustainable future."

Any guesses about the top ten countries? Before reading on, think about which countries you suspect might be the most environmentally sustainable.

By this measure, all of the top ten countries are high-income countries in Europe and we do not get a country from another continent until Australia (#17), followed by Cyprus (#22) and Japan (#25) in Asia. According to the EPI, The Seychelles is the most sustainable country in Africa (#32), followed by Botswana (#35), and the US is tops in North America (#43). Several Central American and Caribbean countries are in the top 50, starting with The Bahamas (#28), but South America lags, with Chile the top performer at #65. Many low-income countries are toward the bottom of the EPI, mainly because of low or falling ecosystem vitality and bad air quality. Three of the bottom five are in South Asia and the worst country in the western hemisphere is also the poorest – Haiti. Liberia is at the bottom of sub-Saharan Africa (#174), but most of the region is below #150. Importantly, while all of the high-income countries in the EPI top 25 have improved their scores over the past decade, the environmental quality in many countries with 2022 EPI rankings above #100 lost ground over the last ten years.

Before reading on, please ponder these findings. Are there any surprises? Do they seem correct to you or is something perhaps amiss?

Important contemporary sustainability perspectives that focus on the need for humanity to move back within biophysical limits come from the nature-centric planetary boundaries literature discussed in Chapter 1 (e.g., see Rockström et al., 2009; Steffen et al., 2015; O'Neill et al., 2018).

Table 7.1 The Top Ten Countries in the 2022 Environmental Performance Index

Rank	Country	Particular Environmental Performance Strengths
1	Denmark	Marine protected areas, GHG mitigation, wastewater treatment
2	United Kingdom	Marine protected areas, sanitation, drinking water
3	Finland	Marine protected areas, grassland loss, $PM_{2.5}$ emissions
4	Malta	Protecting rare national and global biomes, tree cover change
5	Sweden	Marine protected areas, grassland loss, wetland loss
6	Luxembourg	Protecting rare national and global biomes, SO_2 growth
7	Slovenia	Protecting rare national and global biomes, wetland loss
8	Austria	Protecting rare national and global biomes, grassland loss
9	Switzerland	Sanitation, drinking water, SO_2 growth, NOx growth
10	Iceland	Species habitat protection, $PM_{2.5}$ emissions, sanitation

Source: Wolf et al. (2022), available at https://epi.yale.edu/

Such approaches have sometimes ignored human costs associated with moving economic systems within biophysical limits, but recent research has been much more subtle and has recognized the economic aspirations of the Global South. It has also brought forward work on environmental policy instruments, much of which we will discuss in Section III of this book. Sterner et al. (2019), for example, argue for strict precautionary policies to quickly address the most urgent problems, such as species extinction and nitrogen/phosphorus emissions, while using other policies to steer economies back within safe operating spaces. They conclude that:

> Formulating policies that adequately address all boundaries is daunting, but the urgency is such that we cannot let complexity be an excuse for inaction... policy design needs to deal with a multitude of geographical levels, interconnected boundaries, and spatial, ecological and sociopolitical complexities.
>
> (Sterner et al., 2019, p. 20).

A very important recent nature-centric approach that attempts to incorporate planetary boundaries is *The Economics of Biodiversity: The Dasgupta Review*, by the Cambridge University economist Sir Partha Dasgupta. This work, which spans 600+ pages, lays down its key challenge in chapter 4. Relying on the work of Wackernagel and Beyers (2019) and Managi and Kumar (2018), Dasgupta (2021) estimates what it would take in terms of improved environmentally beneficial technical change to reach a sustainable level of pressure on the environment (what he calls "impact" à la the IPAT model discussed in Chapter 5) by 2030, which is the year the Sustainable Development Goals are to be achieved.

Using these sources, Dasgupta estimates that, even without any economic growth, technological improvements to reduce human impacts on natural resources, which have been increasing by about 3.5% per year, would have to increase to 5.5% per year to achieve balance by 2030. He also notes that global economic output increased by an average of 3.4% per year over the last 50 years. Dasgupta says that because growth puts additional demands on the biosphere, if this trend were to continue through 2030, efficiency would have to increase by 10% rather than 5.5% per year, which means it would need to double approximately every 7.2 years. As noted in The Review, this is basically impossible (Dasgupta,

2021, pp. 122-123), implying that humans need to also reduce production that degrades natural resources. Period.

An interesting related line of work focuses on trying to convert ecological effects into loss of land area. Savings and sustainability are therefore measured in land area not "used up" and left unaffected by humans. This approach to sustainability is called the Ecological Footprint and was developed in the 1990s by Matthias Wackernagel and William Rees (Wackernagel and Rees, 1996).

The Ecological Footprint conception of sustainability is influential and especially at the planetary level helps people visualize sustainability and unsustainability.[4] For example, according to the Global Footprint Network, there are a total of 12 billion hectares of biologically productive land and water on the planet, which can support significant photosynthetic activity. Humans in 2018 (most recent data as of 2023) drew on an average of 2.77 hectares per person to support their lifestyles, when the estimated biocapacity was only 1.58 hectares per person. This difference represents a 75% (8-billion-hectare) overshoot.

Not surprisingly, per capita ecological footprint differs dramatically across countries. The US is estimated to have an ecological footprint of 5.13 hectares per person, while Indians require 0.76 hectares each and Botswanans each need 1.5 hectares. In general, the higher the average income, the greater the footprint, with even countries that the EPI considers environmentally friendly, such as Denmark, having high per capita footprints.[5]

There are, of course, measurement issues in estimating the footprints of human production and consumption, including important details related to the effectiveness of reducing soil erosion and using tree planting as carbon "sinks." Furthermore, reducing all problems to land- and water-area problems, while easy to understand (which has value!), may just be too... reductionist. For a short but pointed critique of the Ecological Footprint approach, please see van Kooten and Bulte (2000), who note that "[The Ecological Footprint] is much like replacing measures of humidity, temperature and air pressure as indicators of weather with a single measure, altitude, since each of the former are (perhaps imperfectly) correlated with the latter" (p. 387).

7.4 Human-Focused Sustainability

In 1987, the economist Edward Barbier proposed that there are three aspects – typically referred to as pillars – of sustainability. The environmental pillar seeks to maintain and enhance the integrity of the biotic (i.e., living) and abiotic worlds, including biological diversity, environmental resilience and primary productivity. The environmental pillar is the focus of a number of the sustainability perspectives discussed so far in this chapter.

The economic pillar highlights the market economy aspects of sustainability. Marketed production, perhaps measured by GDP, is part of this pillar, though Edward Barbier is careful to limit this aspect of sustainability to increasing "useful" goods and services. Barbier does not, however, define useful, leaving that part to our imaginations. A second aspect of economic sustainability is meeting basic needs and reducing poverty, and the third is to improve equity by, for example, equalizing health outcomes and incomes across people, communities and countries.

The third sustainability pillar is social sustainability. This pillar includes cultural diversity, which focuses on protecting the cultural heritage of groups across the world. The social

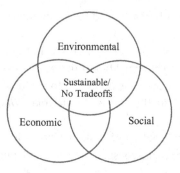

Figure 7.1 Sustainability Pillars
Source: Adapted from Purvis et al. (2019), based on Barbier (1987).

justice part of social sustainability includes fairness and equity in societies and the fostering of institutions, laws and policies that support those outcomes. Social sustainability includes societal stability and public participation in decision-making.

Barbier focused mainly on the process of development in lower-income countries and argued that while these three pillars operate in a human-environmental system, no pillar is most important. The driving idea was therefore that "any attempt to reduce environmental degradation will be counter-productive if there is failure to respect the needs and encourage the participation of those social groups which are most affected by any change" (Barbier, 1987, p. 103). Thus, taken from this perspective, it is not possible to have sustainability at the expense of the human inhabitants, but instead it involves a process of tradeoffs to sustain and enhance human welfare into the future.

The ideal combination of strategies focuses on the overlapping portion of Fig. 7.1 from Purvis et al. (2019). They call such approaches that enhance all three pillars the "sustainable" parts of this system, but sometimes they are instead called *win-win*, because they involve no tradeoffs. Barbier (1987) was clear, however, that, when necessary, tradeoffs must be made, including degrading natural resources for high-value, human-focused economic and social outcomes.

In the international policy arena, the contemporary, human-focused vision of sustainability is encapsulated in the Sustainable Development Goals (SDGs), which we discussed in Chapter 1. The 17 SDGs, which were adopted in 2015 by all 191 UN member countries, take a broad view of sustainability, which is very much in line with - and expands on - the original "three pillars" approach.

Challenge Yourself

Barbier (1987) and the UN Sustainable Development Goals include economic, social and environmental dimensions. Is this appropriate or should only environmental quality be the basis for sustainability?

Figure 7.2 The UN Sustainable Development Goals

Source: https://www.un.org/sustainabledevelopment/news/communications-material/

These goals are integral to the development strategy of the world through 2030 and are referenced in all sorts of international agreements. The SDGs have been part-and-parcel of environmental agreements since 2015, including subsidiary agreements related to the Framework Convention on Climate Change, such as decisions taken at the Sharm El-Sheikh Climate Change Conference held in Egypt in November 2022.

Though Barbier and Burgess (2017) argue it is possible, the SDGs perhaps cannot be perfectly categorized under environmental, economic and social sustainability pillars. Nevertheless, seven goals (#6, #7, #11, #12, #13, #14, #15) focus rather explicitly on environmental sustainability. Another four (#1, #2, #8, #9) are mainly in the arena of economic sustainability and the last six (#3, #4, #5, #10, #16, #17) are largely social goals.

How to operationalize the SDGs? As we discussed in Chapter 1, along with the 17 SDGs, the UN has adopted 169 quantitative and qualitative targets, along with numerous indicators of success.[6] Progress toward meeting these targets is monitored and reported every year by the Secretary General of the United Nations, and there is also a gender-based progress report and a voluminous statistical annex. Sachs et al. (2019) emphasize the transformational nature of the SDGs and identify six areas where transformations are necessary.[7] They also point to the institutions, including government agencies, that need to take the lead on making those transformations, which are mapped into the SDGs. The bottom line here is that the SDGs are the contemporary, policy-focused roadmap to international sustainable development.

7.5 Sustainability and Benefits to Humans over Time

How can we aggregate all the information needed to empirically assess sustainability? One way to measure whether humans are saving enough for the future is to think about the flow of consumption, including ecosystem services, over a broad stretch of time, such as 200 years. Here, the logic goes, if consumption is rising over time, how can we not be sustainable?

And there is little doubt that by many measures the human condition, often driven by increases in marketed consumption, has improved enormously and steadily since industrialization. These are changes that Pinker (2018) and others attribute to acceptance and operationalization of 1700s Enlightenment thinking, including those aspects related to scientific discovery. Average human health, women's rights and lifespan have improved dramatically over time. Total world output increased in real terms 90-fold between 1700 and 1998! And countries that had long been food-insecure and even subject to famine now export food (e.g., see Pinker, 2018; Maddison, 2001). Furthermore, in many countries, including most higher-income countries, forested area has grown dramatically and urban air and water quality are much better than in the past. So does that mean we are heading toward sustainability?

For humans to have more ecosystem services and human-produced consumption in the future, we need to save and invest. Suppose we were to aggregate all goods and services, including ecosystem services, into a set of total benefits that we think of as both marketed and non-marketed consumption. Let us analyze the problem of sustainable consumption across five periods. Suppose that consumption at any point in time (let's call time "t") is a function (remember functions from Chapter 3?) of past investments. In other words, to be better off at any time t than in t - 1, t - 2, t - 3 etc., we have to have saved and invested resources - i.e., consumed less than we would have otherwise - during those previous periods.

So, suppose we invest nothing in period 1 (i.e., consume everything), nothing in period 2 and play major catchup with lots of investments and low consumption in periods 3 and 4 (I think most of us can relate to this approach!). Is that sustainable? Is this consumption-investment time path better or worse than doing a lot of investments now and nothing in periods 2, 3 and 4?

It is a bit hard to tell with no numbers, but suppose we have numbers and are choosing between the consumption time paths in Box 7.1.

Box 7.1 Which Consumption/Investment Combination is Most Sustainable?

You (or all of humanity!) inherit natural resources, skills, equipment and other capital that can be used to improve human welfare. Assume that this amount, which you have available to you now in your youth, is 100 units, and you can choose to consume or save some, none or all of it. Consumption increases your welfare now; saving reduces your welfare now compared with if you consumed those resources.

Any savings are used for productive capital investments, which yield a return over time. Suppose there are five periods to your life: Youth, Young Adult, Adult, Middle Age and Old Age. Each period lasts ten years. Below are three consumption profiles that take into account the productivity of capital. Which one is the best?

	Consumption		
	Profile 1	Profile 2	Profile 3
Baby (t = 0)	0	0	0
Youth (t = 1)	20	50	30
Young Adult (t = 2)	30	30	30
Adult (t = 3)	40	30	30
Middle Age (t = 4)	30	20	30
Old Age (t = 5)	30	20	30

Which profile in Box 7.1 is best? Does it matter? What would such a choice depend on? Or suppose we add a Profile 4, which is the same as Profile 3, except we add 10 to old-age consumption. Presumably, Profile 4 would be better than Profile 3, but is it better than Profile 2?

Trying to compare consumption in period t = 0 (i.e., now) with consumption in period t = 4 is a lot like comparing mangoes and potatoes. These foods have totally different roles for humans, just as we will be totally different people (maybe literally!) in the future than we are today. To decide which consumption/investment profile is best, we need to have a way to evaluate consumption at different points in time. As part of this process, we need to take into account that consumption may have different meanings at different points in time.

Suppose that you live in Portugal and are traveling to South Africa. In Portugal, your money is the Euro and in South Africa they use the Rand. To spend money in South Africa, you therefore need to *exchange* your money for South African Rand. There has been some variation over, say, the last five years, but basically you would get 15-20 Rand for every Euro. This is the Euro-Rand *exchange rate*.

The *discount rate* is like the exchange rate, except it converts present money or consumption to the same in the future. Similar to how we compare exchange rates by asking which currency gets more of the other one, the first question to ask is whether present consumption is worth more than, less than or the same as future consumption.[8] As we discussed in Chapter 3, people are different, so there will be no universal answer that applies perfectly to everyone, so let us instead focus on an average person in a village, city, province or country.

In general, on average and viewed from a societal perspective, we should expect that consumption in the future would be worth less than consumption today. We suspect this is the case, because societies around the world reward savings by paying *interest*. That is, if you put your money in a bank or other savings institution, you almost always will get back more money in the future than you put in. The higher the interest rate and the longer you leave your money in the bank, the more you will get back in the future.

Let's call "PV" the *present value*, otherwise known as the amount invested now. "r" is the per-period (e.g., yearly) interest rate and "t" is the number of periods into the future when you will claim your money, plus interest. Suppose you leave your money in for one period. How much will you get back? You will receive your investment (PV), plus interest (r) on your PV. That is, you will get back PV(1 + r). Note that implicit in this equation is an exponent equal to 1. That is, you will get back $PV(1 + r)^1$ one period in the future. Suppose you left your money in the bank for two periods. Then you would get back your $PV(1 + r)^1$ once again multiplied by (1 + r)^1 or PV $(1 + r)^1 (1 + r)^1 = PV(1 + r)^2$. What about three periods? $PV*(1 + r)^1*(1 + r)^1*(1 + r)^1 = PV(1 + r)^3$, etc.[9] This general relationship is given in Eq. 7.1.

Equation 7.1

Future Value $= PV*(1 + r)^t$

While very valuable for figuring out how much one might get back in the future, Eq. 7.1 sort of tells us the opposite of what we want to figure out, because the example data in Box 7.1 give us the future values. We are not interested in the *future* value of something invested in the present. We want to know the *present value* of flows in the future. This is no problem, though, because we can just divide both sides of Eq. 7.1 by $(1 + r)^t$ and the equality will still hold. Of course, that means on the right side of the equation we will end up with PV, because $(1 + r)^t/(1 + r)^t = 1$, and the left side will have the Future Value, which we now abbreviate as "FV," divided by $(1 + r)^t$. The general equation for computing the present value of something in the future is given in Eq. 7.2.

Equation 7.2

$$PV = \frac{FV}{(1+r)^t}$$

So now we are equipped with the tools we need to evaluate which consumption profile in Box 7.1 is the best. We only need to know "r," which is the rate at which we are *discounting* future consumption to the present. Because our problem in Box 7.1 is more of a social analysis issue than a purely financial problem, our *discount rate* will be related to, but probably not exactly the same as, any financial interest rate. More to come on that in a minute.

Suppose the per-period discount rate is 25%, meaning that, for reasons we will soon discuss, each period in the future (in this case, stage of life) is worth 25% less than the previous one. With this discount rate, which consumption profile in Box 7.1 is best?

Before reading on, and especially before taking a look at Box 7.2, see if you can answer this question yourself.

From Box 7.2 we see that Profile 2 offers the highest total discounted consumption. We therefore can say that of the three possibilities that were offered, Profile 2 is the optimal or social welfare–maximizing consumption profile. Because getting to this point required saving for the future and investing in assets, we can also call Profile 2 the most "sustainable" consumption/investment profile.

Box 7.2 Worked Example: Which Consumption Combination is Most Sustainable in Terms of Maximizing Total Present Value?

In order to answer the question posed in Box 7.1, we need to make the consumption profiles fully comparable across time. We do this by computing the present value (PV) of each consumption amount. Assume that the per-period (stage of life) discount rate is 25%. As an example, the PV of undiscounted consumption in Period 2 as an adult under Profile 1 is $\frac{40}{(1+0.25)^2}$ = 26. **Please confirm all my calculations, which are rounded to have no decimal points.**

	Consumption Profile 1		Consumption Profile 2		Consumption Profile 3	
	Undiscounted	PV	Undiscounted	PV	Undiscounted	PV
Youth (t = 0)	20	20	50	50	30	30
Young Adult (t = 1)	30	24	30	24	30	24
Adult (t = 2)	40	26	30	19	30	19
Middle Age (t = 3)	30	15	20	10	30	15
Old Age (t = 4)	30	12	20	8	30	12
Total		97		112		101

Now that all values are fully comparable in terms of time, we can just add up the PVs to find out which consumption profile has the highest total PV. Adding up the numbers we see that Profile 1 has a total PV of 97, Profile 2's PV = 112 and for Profile 3, PV = 101. So, based on total PV, Profile 2 is the best – or we can say most sustainable – in the sense that savings and investment over time yield the highest total discounted consumption.

Of course, this conclusion depends on the discount rate used for the analysis. Try other discount rates. Can you find discount rates that make Profile 1 the most sustainable one? Can you do the same for Profile 3? Please try it, but – spoiler alert – the answer is yes, the choice of discount rate can critically determine which profile is viewed as the best, so we have to be pretty sure that the discount rate we are using is reasonable.

Given that discount rates are so important, where do they come from? What is the right choice? How do we decide? Fundamentally, discount rates come from people's behaviors and attitudes toward the future; in other words, the future is worth more, less or the same compared with the present, if people *act* as if it is worth more, less or the same.

The subject of choosing the appropriate social discount rate is too big to fully address here,[10] but from a societal view, the correct discount rate really depends on three factors. First, it depends on human impatience. If we humans are highly impatient, meaning that we view a bit of extra consumption in the present as lots *more* valuable than the future, it implies that our discount rate r in Eq. 7.2 would be larger and the present value of any future benefit would be smaller than if people were less impatient.

Second, we need to recognize that issues of sustainability are societal problems that span years or even generations. In our choice of discount rate, which captures how much we value future ecosystem services and human-produced consumption, we therefore must consider how much humans will have of those potential benefits in the future. After all, if it were clear that future generations will be 4x better off than us, why should we make big sacrifices to boost these future rich people's consumption levels? All else equal, in our decision-making we would *discount* their consumption *more* (with a higher level of r), giving a lower present value of future consumption.

Finally, we must take into account how much – on the margin (remember marginal analysis from Chapter 3?) – additional consumption translates into more human well-being. As we discussed in Chapters 3 and 4, higher *total* consumption should imply that *additional* consumption would be valued less than previous additions. Higher total consumption therefore means that marginal benefits, while still positive, should be less than if total consumption were lower.

In sum, there are three drivers of the social discount rate: impatience, future consumption levels and sensitivity of marginal human well-being to changes in consumption levels. But how might we measure how *additional* human well-being decreases due to a particular increase in consumption? After all, human well-being is in different units than consumption (if either have any units at all!).

It is actually not unusual to measure the responsiveness of an outcome (e.g., human welfare) to a change in an independent variable (e.g., consumption). Indeed, market demand functions themselves have different units; price is in monetary units and quantity is in a myriad of other potential units. A standard way to compare changes in different units is in terms of percentages. Once we calculate the percentage changes of the two variables, we can then compare them to see which is bigger. For example, we can form the two percentage changes into a ratio, which is typically called an *elasticity*.[11] If Y is the dependent variable, X is the independent variable and the Greek letter delta (Δ) means "change," the general form of an elasticity equation is given in Eq. 7.3.

Equation 7.3

$$\text{Elasticity} = \frac{\%\Delta Y}{\%\Delta X}$$

If the top is bigger than the bottom, the ratio > 1, which means that the dependent variable responds strongly to changes in the independent variable (i.e., it is elastic). If the numerator is smaller than the denominator, though, it means the dependent variable responds only weakly or perhaps not at all (i.e., the numerator is zero) to changes in the independent variable. Elasticities can be either positive or negative depending on the relationship between the variables. In our case, marginal benefit is our dependent variable and consumption is the independent variable. Because marginal benefits go down (though remain positive) as consumption rises, the elasticity is negative.

Let's now pull these three pieces together into a social discount rate equation. The theory we need to accomplish this goal is actually quite old and is due to Frank Ramsey of Cambridge University in England (Ramsey, 1928). The so-called "Ramsey equation" given as Eq. 7.4 forms the basis of most contemporary estimates of social discount rates, where r = discount rate, i = impatience[12] (likely >0), e = elasticity of marginal human well-being as consumption increases (likely < 0) and g is the growth rate of consumption.[13]

Equation 7.4

$$r = i - eg$$

This equation says that anything that increases impatience, growth in consumption or causes human welfare to increase slower as consumption rises (i.e., e is more negative) will drive up the discount rate and make future consumption less attractive. There are lots of estimates of these three components of discount rates, with much of the action happening around the parameter i. High-end estimates of social discount rates would be around 5% per year, meaning each year a dollar, euro, rupee etc. worth of potential benefits would be worth 5% less. Low estimates would be in the 1%-3% per year range (e.g., see Greenstone et al., 2013).

Challenge Yourself

Do you think that increasing average consumption in low-income countries is consistent with sustainability?

Using these discount rate estimates, we can evaluate sustainability using projections into the future of marketed consumption and ecosystem services. Typically, a set of multiple projections of the future based on what we know today is called *scenario analysis*. This type of analysis is useful, because the future is uncertain and it allows researchers to consider a few possibilities, perhaps informed by beliefs about the probabilities of each possible future occurring.[14]

In terms of method, once we decide on the discount rate, the calculations would more or less be similar to those used in Box 7.2. The question arises, though, where would one get information on consumption and investment. The answer is that much of the needed information is available from the national accounts that we discussed in Chapter 5, augmented by information on environmental investments and natural resource degradation. If we have these data, we can then compute the present values of potential development trajectories like those in Box 7.2 or perhaps just evaluate the trend in income, consumption and investment over time.

Let us start with gross national income (GNI), which measures the gross income people get due from marketed production.[15] We then subtract off any deterioration of human-made capital (typically called depreciation), as well as estimates of natural resource depletion, including the loss in value of fossil fuels, minerals and forests, which could represent losses in future provisioning ecosystem services. This gives us a measure of national income that is "net of" (i.e., minus) the wearing down of our human-made and environmental capital. This measure is called "adjusted net national income" (ANNI) and the basic relationship is given in Eq. 7.5.

Equation 7.5

Adjusted Net National Income (ANNI) = Gross Natinal Income −
Depreciation of Fixed Capital − Energy,
Mineral and Forest Resource Depletion

Figure 7.3 shows the 50-year trend in ANNI in current $US. The results indicate that, using ANNI as a measure of sustainability, the world has produced much more national income than it has lost due to capital depreciation and natural resource depletion. Though ANNI during the last decade seems to have increased at a much slower rate than in the past, ANNI has increased over 3x since 1990. This observation implies that because of our production we are investing much more than we are losing due to depreciation and natural resource

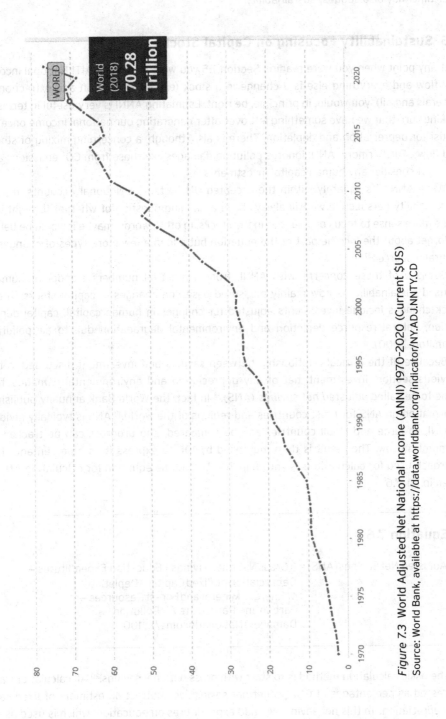

Figure 7.3 World Adjusted Net National Income (ANNI) 1970–2020 (Current $US)
Source: World Bank, available at https://data.worldbank.org/indicator/NY.ADJ.NNTY.CD

depletion and that the difference has been increasing over time. That ANNI has been rising over time may also suggest sustainability.

7.6 Sustainability Focusing on Capital Stocks

If at any point when you were reading Section 7.5 you were thinking WAIT! National income is a flow and everything else is a change in a stock (e.g., deforestation and extraction of minerals and oil), you would, in principle, be right. Estimating ANNI is very useful in terms of making sure that we have something left over after generating our national income once we adjust for depreciation and depletion. There is also, though, a concerning mixing of stocks and flows. Furthermore, ANNI ignores pollution damages like those from CO_2 emissions and does not consider any human capital investments.[16]

While ANNI fits squarely within the accepted UN System of National Accounts, from a sustainability perspective we will always have a nagging question of whether it might just make more sense to focus on changes in the *stocks*. In other words, maybe it would be better to forget about the income part of the equation but add in a few more types of changes in natural resources.[17]

Because of these concerns with ANNI, from a practical numbers standpoint human-focused sustainability is now mainly measured based on changes in capital stocks. These stock changes include investments adjusted for changes in human capital, capital depreciation, natural resource depletion and environmental degradation due to air pollution (Hamilton, 2000).

Because of the critical relationship between savings and investment discussed in the previous section, investment, net of any depreciation and environmental damages, has come to be called *adjusted net savings* (ANS).[18] In fact, the World Bank annually publishes information on ANS for most countries and regions of the world.[19] ANS is typically divided by GNI, so large and small countries can be compared, and progress can be tracked as economies grow. The result is then multiplied by 100 to express it as a percentage. The approach used for calculations is shown in Fig. 7.4 and the equation for calculating ANS is given in Eq. 7.6.

Equation 7.6

$$\text{Adjusted Net Savings (ANS)} = \{(\text{Gross National Savings} + \text{Education Expenditures} - \text{Depreciation of Fixed Capital} - \text{Depletion of Energy, Mineral and Forest Resources} - \text{Carbon and Particulate Air Pollution Damages}) \text{National Income}/\} * 100$$

The basic calculation method is to start with gross national savings.[20] All calculations are expressed as percentages of GNI. To get net saving, we subtract an estimate of fixed capital depreciation. To this net savings we add expenditures on education, which is used as an

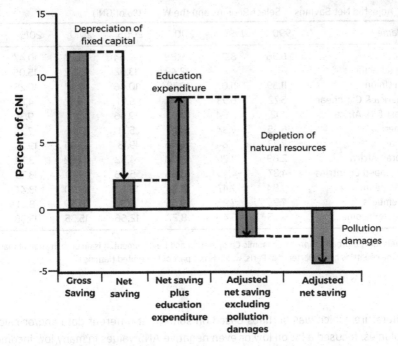

Figure 7.4 How to Calculate Adjusted Net Savings
Source: Harris, J.M. and B. Roach (2013).

admittedly imperfect proxy for saving/investment in human skills. Depletion of (marketable) natural resources like forests (net of any forest additions), energy and minerals is subtracted off next to get adjusted net savings excluding pollution damages. Finally, we need to subtract estimates of air pollution damages from CO_2 and particulate emissions (using methods discussed in Chapter 4), giving our ANS estimate. Table 7.2 provides World Bank estimates of ANS as percentages of GNI for the world and broken down by official United Nations geographic and income categories.[21]

The core of the ANS logic is that it is the total stock of assets available to the future that matters. If one asset is drawn down and the resources are invested in other assets that are even more valuable, total assets increase and the shift increases ANS, presumably increases ANNI in the future and improves sustainability.

As shown in Table 7.2, based on the criterion of ANS, as was the case for ANNI, for the last 30+ years the world has generally been sustainable. In 2021, the world added about 10% of GNI to its total stock of capital (environmental and human-made) and this rate of accumulation has been pretty steady for the last 25 years. The most sustainable region is South Asia, with 2021 ANS of 16.68% of GNI (about 1 out of every 6 Rupees, Takas etc. is saved), and the least sustainable region is Latin America (3.91%), followed by North America (4.57%). That ANS estimates are positive means that all regions added to their 2021 stocks of capital.

It is notable that there is no clear relationship between income category and ANS; indeed, low- and middle-income countries and the European Union have very similar ANS estimates.

Table 7.2 Adjusted Net Savings in Select Regions and the World (% of GNI)

Region Name	1990	1995	2000	2005	2010	2015	2021
World	6.96	8.28	9.88	9.64	9.27	10.47	9.58
East Asia & Pacific			12.49	13.52	15.02	15.05	13.19
European Union	11.39	11.07	10.78	10.66	8.37	10.25	11.59
Latin America & Caribbean	5.72	6.93	6.14	6.91	7.32	4.13	3.91
Middle East & N. Africa	2.12	7.54	11.01	19.66	18.11	9.31	7.95
North America	6.35	7.34	8.70	5.71	3.14	7.36	4.57
South Asia	7.33	11.84	12.46	19.18	21.94	19.21	16.68
Sub-Saharan Africa	2.89	3.25	5.08	4.52	1.82	3.72	7.08
Least Developed Countries	-4.37	-4.49	3.66	5.56	9.70	13.73	17.63
Low- & Middle-Income	0.84	7.47	8.94	12.57	15.20	13.46	12.80
OECD* Members	7.95	7.95	9.65	8.39	6.09	8.61	7.53
Upper Middle-Income	-1.55	6.40	8.27	12.66	15.35	13.38	12.77

*OECD stands for the Organisation for Economic Co-operation and Development. It is an organization of mainly higher-income countries headquartered in Paris, which is not part of the United Nations.

Source: World Bank as in note 19.

Earlier literature, which was perhaps based on somewhat different data and/or calculation methodologies, focused a lot on low or even negative ANS values in many low-income countries and especially in sub-Saharan Africa, which is the poorest region of the world (e.g., see Hamilton and Clemens, 1999; Hamilton, 2000). These calculations led to the conclusion that the poorest region on the planet was also the least sustainable (Arrow et al., 2004), which set up a serious clash between intuitive notions of sustainability and ANS. After all, how can countries that have GNI per person of $1000 be the least sustainable? How are such countries supposed to "save" more when average incomes are so low? As shown in Table 7.2, based on the ANS metric, sub-Saharan Africa and least developed countries as a whole have over time become more sustainable.[22]

Box 7.3 Sovereign Wealth Funds Help Assure ANS-Style Sustainability in Minerals and Energy

You might find it counter-intuitive, but countries with large mineral and fossil fuel reserves typically have *worse* economic performance than resource-poor countries (Badeeb et al., 2017). The main reason for this so-called natural resource "curse" is that big influxes of money can distort market economies and disrupt or corrupt existing social institutions (van der Ploeg, 2011), reducing economic growth. These countries may also have more land degradation and water pollution due to the mining and pumping of non-renewable resources.

To try to ensure that extraction of marketable resources promotes ANS-style sustainability, some countries have created country or state/province-level wealth funds that park oil and mineral earnings and dole out financial gains (sometimes only the interest on funds' financial investments) over time.

These *sovereign wealth funds* exist in most oil- and gas-producing countries (e.g., the US, Qatar, Saudi Arabia) and some mineral-producing countries (e.g., Chile and Botswana). The largest fund is the Norway Global Pension Fund – Global, which was established in 1990 and holds investments of about $1.2 trillion or $240,000 per resident (James et al., 2022).

Chile is the largest copper exporter in the world. To stabilize public spending as copper prices fluctuated, in 1985 it established what has become the Economic and Social Stabilization Fund. As of 2023, this fund held about $6 billion in assets. Botswana has a similar fund called the Pula Fund to manage revenues from diamond sales.

For more information:

- Norway: https://www.nbim.no/en/the-fund/about-the-fund.
- Chile: https://www.hacienda.cl/english/work-areas/international-finance/sovereign-wealth-funds/economic-and-social-stabilization-fund.
- Botswana: https://www.bankofbotswana.bw/content/pula-fund.

Before reading on, what do you think about ANS as a measure of sustainability? Do you find it convincing? Why or why not? Is there anything important that you think is left out of ANS?

You may be trying to reconcile this positive assessment of global sustainability with some of the material on environmental stability presented in Chapters 1 and 2, things you may have read on your own, and Sections 7.1–7.3 of this chapter. For example, according to the EPI, three of the least sustainable countries are in South Asia, but based on the ANS, South Asia is the most sustainable region in the world. What is up with that?

In Section 7.1, we learned from Dasgupta (2021) that it is basically impossible for efficiency in natural resource use to increase fast enough to achieve environmental sustainability by 2030, the year the SDGs are to be achieved. On the other hand, ANS estimates tell us we are doing just fine. In Chapter 2, we learned that climate change is progressing much faster than originally expected by experts, the Earth is experiencing continued extreme weather shocks and species extinctions are at about 1000 times the historical level. ANS, on the other hand, purports to take climate change damages into account and says all is well.

So why are there differences in these conclusions? To answer this question, we need to look at the assumptions that underpin ANS calculations. First, note that each component of the ANS equation in Eq. 7.6 is either added or subtracted. This means that if Bangladesh invests $1000 in education and depletes forests by $800, all else equal, sustainability is improved. That is, the ANS methodology presumes that *the components of Eq. 7.6 are perfectly substitutable*. Allowing for or assuming perfect substitutability of investments is often referred to as "weak" sustainability. So-called "strong" sustainability requires preserving specific natural resources for future generations (e.g., see Pearce and Barbier, 2000, p. 21).

Second, although ANS calculations are "normalized" by dividing by national income, elements in the numerator in Eq. 7.6 are also functions of national income. It is therefore possible that as income increases, some ANS components systematically grow or shrink faster than others, tilting the composition of ANS as income growth progresses. We therefore can end up in a world with lots of capital, education and forests, but few minerals and a highly unstable climate. From Chapter 3 and the previous subsection in this chapter, we know that marginal benefits increase as availability declines. It is therefore possible that the asset tradeoffs allowed by the ANS calculation methodology may not truly make future generations better off.[23]

Finally, ANS focuses on measurable capital. As we discussed in Chapter 4 and will delve into in the next section of this book titled *The Challenge of Protecting Natural Resources*, many of our arguably most cherished and at-risk ecosystem services are not well-measured. For example, while education expenditures and mineral depletion can be calculated, loss of biodiversity values may go pretty much unmeasured and the costs of losing climate stability are difficult to estimate.

Please take a moment to consider ecosystem services that are left out of ANS that could be relevant for the three pillars of sustainability.

7.7 Chapter Conclusions

This chapter moved some of the key material previously covered in this book into the dynamic realm. We saw that sustainable economic development is a difficult goal to pin down, kind of like avoiding overconsumption and human overpopulation. Sustainability can be measured a lot of different ways, partly because people have different views on whether only people should be considered in such calculations. But the 1987 Nobel prizewinner Robert Solow perhaps said it best when he quipped that "It is very hard to be against sustainability. In fact, the less you know about it, the better it sounds" (Solow, 1991).

Environment-focused sustainability measures are very much in the spirit of nature-centric models, which treat humans as just another species. Most of those measures have little good to say about our sustainability behavior, though according to the Environmental Performance Index high-income countries, especially in Europe, tend to be top performers, with low-income countries lagging.

The main human-focused sustainability metrics, such as adjusted net national income and adjusted net savings, suggest that during the past 30 years all major regions have significantly added to total capital stocks after subtracting natural resource depletion and capital depreciation. These measures include relatively limited forms of capital and assume perfect substitutability between human-made capital and natural resources which, within the contemporary world of pushed planetary boundaries, may not be appropriate.

At the international policy level, the 17 Sustainable Development Goals represent an important navigational guide. The SDGs, which build on the Millennium Development Goals and Barbier (1987), expire in 2030 and include economic, social and environmental metrics. Interestingly, the UN does not attempt to aggregate the individual goals into an overall sustainability metric as do other measures; countries and the world as a whole are therefore either on track to meet specific goals or they are not.

With the barrage of bad environmental news coming at us every day, some of us may be uncomfortable with any suggestion that economic development can be sustainable. As Dasgupta (2021) and the planetary boundaries literature try to warn us, humans are pushing past several important and interrelated environmental limits, including global warming, nitrogen emissions and biodiversity loss. They therefore urge us to do a much better job of considering the effects of human activities on the biosphere.

Dasgupta, Sterner et al. (2019) and many others also argue for a better understanding of how the characteristics of ecosystem services, human incentives and people's behaviors interact with natural resources. Indeed, we can perhaps shed important light on the prospects for sustainability if we allow ourselves to think beyond consumption and population as the principal drivers of natural resource degradation.

Let us give this approach a try. With our understanding of environmental challenges, economic fundamentals and sustainability firmly in hand, we now turn our attention to *The Challenge of Protecting Natural Resources*, which is the subject of the second of the three sections in this book. We start with trying to answer the fundamental and critical question: "Why are environmental assets so difficult to manage?"

Issues for Discussion

1. Using data discussed in the chapter, compare measures of adjusted net savings (ANS) with the Environmental Performance Index (EPI). Are the countries that rank high (low) on the EPI the same ones that have high (low) levels of ANS? What are some examples?
2. Please consult the World Bank ANS database at https://data.worldbank.org/indicator/ NY.ADJ.SVNG.GN.ZS. Check ten sub-Saharan African countries. Are they sustainable? Can you find some that have negative ANS values? Which ones? Where do they lag in terms of the components of savings?
3. Please consult the World Bank ANS database mentioned in Issues for Discussion 2. Choose four regions or income categories in the world. Analyze the five components of ANS (gross savings, depreciation, natural resource depletion, human capital accumulation, CO_2 and particulate pollution damages). What parts of the calculation drive changes in estimated ANS values? Have any parts become less important over time?
4. The economics literature puts the social discount rate at anywhere between 1% and 4% per year. This means that investing today to receive $1000 in benefits (e.g., climate change or food production) 100 years in the future is worth somewhere between $19.80 at 4% (i.e., $1000/1.04^{100}$) and $369 at 1% (i.e., $1000/1.01^{100}$). High discount rates make it difficult to justify long-term public investments. What do you think of this issue? Is there a problem, and if so, what, if anything, should be done?
5. Rather than using a human-focused sustainability measure like ANS, would it be better to use a more environment-focused indicator, such as the Ecological Footprint? Please explain your answer in light of footprints published by The Ecological Footprint Network at https://www.footprintnetwork.org/our-work/ecological-footprint/.

Practice Problems

1. During the last ten years, birdwatching has become increasingly popular around the world, including in China. Assume that ten years ago the average person in China was willing to pay 100 yuan for a day of birdwatching. Since that time, willingness to pay

grew by an average of 5% per year. What is the willingness to pay today? **Hint: use the method from Eq. 7.1.**

2. The human population in a region was 1000 people five years ago and is now 1700. Over the same time period total income rose from 17 to 33. Did income per capita increase during the five-year period? By how much? What is your estimate of the elasticity of income with respect to population? How about population with respect to income? **Hint: use the method from Eq. 7.3.**

3. Recalculate the present values in Box 7.2 using different discount rates. What is the cutoff discount rate to make Profile 1 the most sustainable? What about Profile 3? **Hint: Use the method in Box 7.2 as a guide. Hint #2: If you have access to Microsoft Excel or similar, use it to make the calculations.**

4. Accra, which is the capital of Ghana, is considering three projects to improve human health and plans to choose the one project with the highest total present value. The first project is to retire and replace old buses with new buses, which will reduce air pollution. The second project enhances key urban green space so residents can spend time in nature. Finally, the city government is considering investing in upstream protections for the Densu River, which supplies over half the drinking water of the city. Environmental professionals, including economists, have estimated the costs and benefits of each project (in millions of Ghanaian cedi) over the next five years. These values are in the table below, with positive values indicating benefits and negative values meaning costs. The discount rate used by Accra is 7% per year.

Which project should the city undertake? How would your answer change if the discount rate was 3% as was used by Greenstone et al. (2013)? **Hint: Use the method in Box 7.2 as a guide. Hint #2: If you have access to Microsoft Excel or similar, use it to make the calculations.**

	Values in Millions of Ghanaian Cedi for Each Year		
	Old Bus Replacement	Urban Greenspace Enhancement	Densu River Protections
Year 0 (i.e. now)	-100	-50	-35
Year 1 (after 1 year)	33	22	12
Year 2	30	14	13
Year 3	27	16	14
Year 4	27	9	15
Year 5	27	7	16

5. Using the data in Practice Problem 4, after the initial investment in Year 0, which project's benefits grew fastest between Year 1 and Year 5? Compare pairs of projects. What is the *percentage point* difference in growth comparing each project? How do these percentage point differences compare with the initial benefit values of each project? **Hints: For reference, see Box 7.2 for how to calculate present value. Please note that the values in the table are per year, so you will need to calculate the total or cumulative values by adding up the annual values.**

6. The table below presents estimates of the inflation-adjusted investments and natural resource depletion in a country during the period 2024-2030. Calculate adjusted net savings (ANS) as presented in Fig. 7.5 and Eq. 7.6. Positive values indicate gains and negative numbers mean losses. Do the data suggest that the country is becoming more sustainable over time? Please briefly explain your method, your understanding of the drivers of the answer and any caveats you would like to mention.

Year	Inflation-Adjusted Values in Millions					
	Gross Savings	Education Expenditures	Capital Depreciation	Energy, Minerals and Forest Depletion	Air Pollution Damages	Gross National Income
2023	$300	$210	-$21	-$210	-$150	$3000
2024	$243	$170	-$17	-$193	-$139	$2700
2025	$195	$137	-$14	-$165	-$120	$2295
2026	$203	$143	-$14	-$160	-$115	$2157
2027	$233	$168	-$16	-$170	-$119	$2244
2028	$257	$189	-$18	-$185	-$125	$2401
2029	$285	$210	-$20	-$185	-$135	$2617
2030	$317	$222	-$22	-$204	-$148	$2878

7. In Practice Problem 6, suppose national income remained constant at the 2023 value. How would that affect ANS? Please redo your calculations and discuss your findings.

Notes

1 This commission is sometimes called the Brundtland Commission report after the chairperson Gro Harlem Brundtland.

2 One of the first to make this link between savings, capital formation and future human welfare was the 1987 Nobel Prize in Economic Sciences winner Robert Solow (Solow, 1956).

3 See https://epi.yale.edu/ for details.

4 According to Google Scholar, Wackernagel and Rees (1996) has been cited by other authors an astonishing almost 11,000 times.

5 See https://www.footprintnetwork.org/our-work/ecological-footprint/ for more details and for links to country-level estimates, data and methods. The Ecological Footprint Network also publishes country-level analysis. Perhaps not surprisingly, from this perspective many land-poor countries, such as Singapore, Bermuda, Israel and Cyprus, fare the worst with respect to their actual land areas, with footprints 19-105 times their actual country areas.

6 For more information see https://unstats.un.org/sdgs/iaeg-sdgs/, https://unstats.un.org/sdgs/files/report/2022/E_2022_55_Statistical_Annex_I_and_II.pdf and https://unstats.un.org/sdgs/files/report/2022/secretary-general-sdg-report-2022--EN.pdf

7 Quoting directly from the article, these transformations are the following: 1) education, gender and inequality; 2) health, well-being and demography; 3) energy decarbonization and sustainable industry; 4) sustainable food, land, water and oceans; 5) sustainable cities and communities; and 6) digital revolution for sustainable development.

8 Let us ignore comparisons with past consumption because the past is... past.

9 Note that when multiplying equations with exponents you add the exponents.

10 Chapter 10 in Boardman et al. (2018) offers an awesome, if a bit technical, treatment of discount rates.

11 More elastic = "stretchier." That is, the dependent variable changes more in percentage terms than the independent variable.

12 Impatience is often called the "pure rate of discount" and can be estimated using behavioral experiments.
13 Often Greek letters rho, delta and nu are used instead of r, i and e. There are several technical issues that we are sidestepping here, but if you would like to know more and especially if you would like to understand the circumstances in which a discount rate is a declining function of time (i.e., the discount rate is a function rather than just one number), please see Arrow et al. (2014).
14 Chapter 11 in Boardman et al. (2018) offers a detailed discussion of how to incorporate uncertainty into such analysis.
15 Recall from Chapter 5 that national income is roughly equal to production due to the circular flow of the market economy. "Gross" here refers to before subtracting or adding relevant adjustments and is the opposite of net.
16 Human skills are often called *human capital*. Interestingly, human capital makes up the vast majority of capital investment in virtually all countries around the world. A typical country – regardless of income level – will have 50%–75% of its total capital and annual capital investment in the form of education!
17 With some mathematics added for rationale and emphasis, this point is the core message of Dasgupta (2009).
18 It was originally referred to as *genuine investment*, but the terminology has changed over time.
19 Available at http://data.worldbank.org/indicator/NY.ADJ.SVNG.GN.ZS
20 It is "gross," because we have not yet added or subtracted anything. Remember the national accounts from Chapter 5? An overview of the calculation method is provided by the United Nations at https://www.un.org/esa/sustdev/natlinfo/indicators/methodology_sheets/econ_development/adjusted_net_saving.pdf
21 A detailed overview of the calculation method is available from the Environment and Natural Resources Global Practice at the World Bank (World Bank, 2023a).
22 Barbier (2019), which examines ANS and natural capital depletion going back to 1970, emphasizes that low-income countries are relying on high levels of natural capital depletion to drive positive ANS values.
23 See Sterner and Persson (2008) for more ideas on this issue.

Further Reading and References

Arrow, K., P. Dasgupta, L. Goulder, G. Daily, P. Ehrlich, G. Heal, S. Levin, K.-G. Mäler, S. Schneider, D. Starrett and B. Walker. 2004. "Are we Consuming Too Much?" *Journal of Economic Perspectives* 18 (3): 147-172.
Arrow, K.J., M.L. Cropper, C. Gollier, B. Groom, G.M. Heal, R.G. Newell, W.D. Nordhaus, R.S. Pindyck, W.A. Pizer, P.R. Portney et al. 2014. "Should Governments Use a Declining Discount Rate in Project Analysis?" *Review of Environmental Economics and Policy* 8 (2): 145-163.
Badeeb, R.A., H.H. Lean and J. Clark. 2017. "The Evolution of the Natural Resource Curse Thesis: A Critical Literature Survey." *Resources Policy* 51: 123-134.
Barbier, E.B. 1987. The Concept of Sustainable Economic Development." *Environmental Conservation* 14 (2): 101-110.
Barbier, E.B. 2019. "The Concept of Natural Capital." *Oxford Review of Economic Policy* 35 (1): 14-36.
Barbier, E.B. and J.C. Burgess. 2017. "The Sustainable Development Goals and the Systems Approach to Sustainability." *Economics: The Open-Access, Open-Assessment E-Journal* 11 (2017-28): 1-22.
Barbier, E.B. and J.C. Burgess. 2021. *Economics of the SDGs: Putting the Sustainable Development Goals into Practice*. Palgrave Macmillan: London and New York.
Barbier, E.B. and J.C. Burgess. 2023. "Natural Capital, Institutional Quality and SDG Progress in Emerging Market and Developing Economies." *Sustainability* 15 (4): 3055.
Boardman, A.E., D.H. Greenberg, A.R. Vining and D.L. Weimer. 2018. *Cost-Benefit Analysis: Concepts and Practice* (5th ed.). Cambridge University Press: Cambridge, UK.
Dasgupta, P. 2009. "The Welfare Economic Theory of Green National Accounts." *Environmental and Resource Economics* 42 (1): 3-38.
Dasgupta, P. 2021. *The Economics of Biodiversity: The Dasgupta Review*. HM Treasury: London.

Folke, C., Å. Jansson, J. Larsson and R. Costanza. 1997. "Ecosystem Appropriation by Cities." *Ambio* 26 (3): 167-172.

Greenstone, M., E. Kopits and A. Wolverton. 2013. "Developing a Social Cost of Carbon for US Regulatory Analysis: A Methodology and Interpretation." *Review of Environmental Economics and Policy* 7 (1): 23-46.

Hamilton, K. 2000. *Genuine Saving as a Sustainability Indicator*. World Bank Environment Department Paper 77. The World Bank.

Hamilton, K. and M. Clemens. 1999. "Genuine Savings Rates in Developing Countries." *The World Bank Economic Review* 13 (2): 333-356.

Harris, J.M. and B. Roach. 2013. "National Income and Environmental Accounting." In *Environmental and Resource Economics: A Contemporary Approach* (3rd ed.). M.E. Sharpe: Armonk, NY.

IPBES (Intergovernmental Science-Policy Platform on Biodiversity and Ecosystem Services). 2018. "Summary for Policymakers of the IPBES Assessment Report on Land Degradation and Restoration." IPBES: Bonn.

James, A., T. Retting, J.F. Shogren, B. Watson and S. Wills. 2022. "Sovereign Wealth Funds in Theory and Practice." *Annual Review of Resource Economics* 14: 621-646.

Kneese, A.V., R.U. Ayers and R.C. D'Arge. 1970. *Economics and the Environment: A Materials Balance Approach*. Resources for the Future: Washington, DC.

Maddison, A. 2001. *The World Economy: A Millennial Perspective*. Development Studies Centre of the Organisation for Economic Co-operation and Development. OECD: Paris.

Managi, S. and P. Kumar. 2018. *Inclusive Wealth Report 2018: Measuring Progress towards Sustainability*. Routledge: New York.

O'Neill, D.W., A.L. Fanning, W.F. Lamb and J.K. Steinberger. 2018. "A Good Life for All within Planetary Boundaries." *Nature Sustainability* 1: 88-95.

Pearce, D. and E.B. Barbier. 2000. *Blueprint for a Sustainable Economy*. Earthscan: London.

Pinker, S. 2018. *Enlightenment Now: The Case for Reason, Science, Humanism, and Progress*. Viking: New York.

Purvis, B., M. Yong and D. Robinson. 2019. "Three Pillars of Sustainability: In Search of Conceptual Origins." *Sustainability Science* 14 (3): 681-695.

Ramsey, F.P. 1928. "A Mathematical Theory of Saving." *Economic Journal* 38 (4): 543-549.

Rockström, J., W. Steffen, K. Noone, Å. Persson, F.S. Chapin III, E.F. Lambin, T.M. Lenton, M. Scheffer, C. Folke, H.J. Schellnhuber et al. 2009. "A Safe Operating Space for Humanity." *Nature* 461 (24): 472-475.

Sachs, J.D., G. Schmidt-Traub, M. Mazzucato, D. Messner, N. Nakicenovic and J. Rockström. 2019. "Six Transformations to Achieve the Sustainable Development Goals." *Nature Sustainability* 2: 805-814.

Solow, R. 1956. "A Contribution to the Theory of Economic Growth." *The Quarterly Journal of Economics* 70 (1): 65-94.

Solow, R. 1991. "Sustainability: An Economist's Perspective." Text of lecture given June 14, 1991 at the Woods Hole Oceanographic Institute, Woods Hole, MA. Reprinted in R.N. Stavins, ed. *Economics and the Environment: Selected Readings* (7th ed.). Edward Elgar: Cheltenham.

Steffen, W., K. Richardson, J. Rockström, S.E. Cornell, I. Fetzer, E.M. Bennett, R. Biggs, S.R. Carpenter, W. de Vries, C.A. de Wit et al. 2015. "Planetary Boundaries: Guiding Human Development on a Changing Planet." *Science* 347 (6223): 736.

Sterner, T., E.B. Barbier, I. Bateman, I. van den Bijgaart, A.-S. Crépin, O. Edenhofer, C. Fischer, W. Habla, J. Hassler, O. Johansson-Stenman et al. 2019. "Policy Design for the Anthropocene." *Nature Sustainability* 2 (1): 14-21.

Sterner, T. and U.M. Persson. 2008. "An Even Sterner Review: Introducing Relative Prices into the Discounting Debate." *Review of Environmental Economics and Policy* 2 (1): 61-76.

van der Ploeg, F. 2011. "Natural Resources: Curse or Blessing?" *Journal of Economic Literature* 49 (2): 366-420.

van Kooten, G.C. and E.H. Bulte. 2000. "The Ecological Footprint: Useful Science or Politics?" *Ecological Economics* 32 (3): 385–389.

Vitousek, P., P.R. Ehrlich, A.H. Ehrlich and P.A. Matson. 1986. "Human Appropriation of the Products of Photosynthesis." *BioScience* 36 (6): 368–373.

Wackernagel, M. and B. Beyers. 2019. *Ecological Footprint: Managing our Biocapacity Budget.* New Society: Gabriola Island, BC.

Wackernagel, M. and W. Rees. 1996. *Our Ecological Footprint: Reducing Human Impact on the Earth.* New Society: Gabriola Island, BC.

Wolf, M.J., J.W. Emerson, D.C. Esty, A. de Sherbinin, Z.A. Wendling et al. 2022. *2022 Environmental Performance Index.* Yale Center for Environmental Law & Policy: New Haven, CT. Retrieved from https://epi.yale.edu September 2022.

World Bank. 2023a. *Estimating the World Bank's Adjusted Net Saving: Methods and Data.* Retrieved from https://datacatalogfiles.worldbank.org/ddh-published/0037653/DR0045503/ANS%20Methodol ogy%20-%20April%202023.pdf October 6, 2024.

World Bank. 2023b. *DataBank: Adjusted Net Savings.* Retrieved from https://databank.worldbank.org/ source/adjusted-net-savings/preview/ July 27, 2023.

SECTION II

The Challenge of Protecting Natural Resources

8 Why are Environmental Assets so Difficult to Manage?

8.1 Recap and Introduction to the Chapter

We should not rule out consumption and, perhaps in some places, human population growth as the crux of why we have the important environmental problems we discussed in Chapters 1 and 2. Chapters 3 and 4 also pointed to a few potentially intriguing economic issues that deserve further attention. For example, in contrast to marketed goods, such as Mercedes-Benz cars and candy bars, ecosystem services give us enjoyment, but often require limited human effort. Environmental assets therefore offer us real magic because we can get benefits with nature doing some or all of the work for us!

That we undermine natural capital is particularly frustrating because some of the ecosystem services we are losing are extremely valuable – much more like a Mercedes-Benz than a candy bar. Services like healthful air, pollination from bees and stable climate are perhaps more desirable and valuable than almost any service we can imagine and yet they are often degraded. Not always, but often, such ecosystem services are non-excludable, which has implications. For example, businesses have a tough time turning a profit on some of the most important ecosystem services and without businesses we have no markets for those services.

DOI: 10.4324/9781003308225-10

You might be saying, "Right on!" Perhaps the last thing you want to see is businesses controlling nature. Unfortunately, though, without businesses and markets we also have no prices. You may again be saying, "Hurray! Who wants prices on such ecosystem services?" but consider that this means we have no objective *measure* of ecosystem service scarcity and value that is comparable with marketed services like timber and fish. An even more important problem than lack of valuation metrics, though, is that without markets and prices, effective prices are zero – those ecosystem services are free. People therefore get the erroneous message that top-shelf ecosystem services like a stable climate and good habitat services for other species, which are unpriced, are worthless. This is exactly the wrong message. How can we counteract it? The chapters in this section get us started on answering this question.

8.2 Conflicting Ecosystem Services

In Chapter 4, we learned that most valuable environmental assets – fertile land, a clean river, clean air – typically produce a number of diverse ecosystem services. Let's think about this in the context of a specific example. Think about a stretch of clean river or a lake near your home – hopefully there is one! What does that body of water do for people?

Before reading on, please think about this question. Please make a list of the top three ecosystem services it provides.

Do people fish in it? Swim? Canoe, kayak, raft, motorboat or sail? Extract drinking water? Irrigate crops? Give people a place to dump sewage? Provide a nice place for people to camp or just sit and enjoy the view? Offer a nice place for hiking, off-road vehicle driving, mountain biking or off-road motorcycling? Serene water sounds? Offer habitat for non-human species, some or all of which are appreciated by humans? Produce drugs from some of the riparian (i.e., shoreline) species?

All these ecosystem services can come from rivers and lakes, which means that those assets are quite productive![1] Notice, though, that several ecosystem services conflict with each other, which means to get more of a particular ecosystem service we must accept less of others. For example, swimming and fishing do not go together very well. Camping and off-road vehicles coexist uneasily, and some habitat ecosystem services are degraded by basically all other services. Irrigation, which uses over 80% of the water in the western part of the US, conflicts with just about everything, because it can significantly reduce water flows. We only need to look to the Colorado River in the American West, which provides many, many ecosystem services, most of which are significantly impacted by irrigation.[2] Like habitat, serenity values, which are part of cultural ecosystem services, are extremely easily degraded, even by something as quiet as mountain biking.

Challenge Yourself

Can you name any environmental asset that produces only one ecosystem service?

Ecosystem services that degrade environmental assets to benefit humans – e.g., fishing, irrigation and off-road four-wheeling – not surprisingly have the most conflicts with other ecosystem services. In contrast, nature photography, biodiversity preservation, hiking and most forms of recreation conflict less with other ecosystem services.

Ecosystem services that degrade environmental assets often have the property that something is harvested. That harvesting or "extraction" is occurring, means that the extracted ecosystem service is excludable (e.g., I can extract or harvest fish and keep them for myself) and therefore marketable. This is, of course, a key reason why some think of the marketed economy and GDP as incompatible with environmental protection. How can they be compatible when it always seems we are taking out too much stuff from the environment and messing it up?

I would like to mention the ecosystem service that almost by its nature degrades environmental assets and conflicts with most other ecosystem services. Water bodies are often used to dispose of liquid wastes produced by humans. That is, they act as a *sink* for – a place to dispose of – wastes, such as human sewage, that we inevitably produce. Typically, this function of water bodies is not even considered an ecosystem service. After all, ecosystem services are good things, right, but what is good about water pollution? Certainly, there is little to recommend human sewage, but it must go somewhere, mustn't it?

What if magically the natural environment disappeared and we had no place to put all our solid, liquid and gas wastes (i.e., our various types of "garbage")? That would be inconvenient for us, wouldn't it? Maybe we would just have to keep it in our houses! I would therefore suggest that a broad-minded and potentially useful approach to thinking about ecosystem services might consider the sink function of the environment as a provisioning ecosystem service that benefits humans.

Challenge Yourself

Is pollution immoral?

The problem is, of course, that by its nature pollution messes up everything else. Imagine if you can for a moment that pollution did not degrade all other ecosystem services, it just got rid of our garbage. Pollution would then just be a positive ecosystem service like the beautiful view one enjoys looking across a large, pristine lake. Thinking of the disposal function of the environment in this way in fact makes the term pollution sound a bit overly pejorative!

I am reminded of the classic children's book titled *Everyone Poops* by Taro Gomi and Amanda Stinchecum, which tries to make young children comfortable with going to the toilet. It's alright to go, the book urges! What if there were nowhere to go? Would Gomi and Stinchecum still say it was OK? On a similar wavelength is *Where's the Poop?* by Julie Markes and Susan Hartung. According to the promotional blurb this book "...shows children that all creatures have a place to poop: tigers in the jungle, kangaroos in the outback, and monkeys in the rain forest... your child will see[3] that he or she has a place to poop, too." Where exactly would that place be if the natural environment did not exist or we so prioritized other conflicting ecosystem services like freshwater ecosystem health that we could not use any of this sink function?

8.3 Ecosystem Service Costs

Some of the ecosystem services provided by water bodies are marketed or have major links to things that are traded. Irrigation water is a key input into agricultural crops, which are traded. Fish are bought and sold.[4] Furthermore, pollution with relatively few exceptions can be reduced using technologies. For example, tertiary sewage treatment reduces solids, bacteria and nitrogen and phosphorus that are responsible for algae blooms. Sewage treatment is generally provided by cities and towns and if only lower-level, primary treatment (i.e., settling ponds) is used rather than more sophisticated tertiary treatment, this saves municipalities money. The *assimilative capacity* of the natural environment – its ability to process wastes for us if we don't overload natural systems too much – therefore often saves cash-strapped municipalities real money. This is a different kind of market link.

When we reduce ecosystem services with tight links to markets, it costs real money. If we reduce irrigation extractions, as has been done from the Klamath River in southern Oregon and many other rivers in the American West, it costs farmers money. Either they need to get the water from somewhere else, shift to lower-value crops to save water, pray for rain or just leave their fields fallow. A shift in ecosystem service emphasis from irrigation to in-stream flows therefore comes out of farmers' pockets.

Similarly, if fishers are required to reduce their fish catches to allow stocks to recover, it costs them money. When factories accustomed to disposing of untreated or poorly treated wastes in water bodies are required to take steps to protect water quality, that also costs real money. They may have to find alternative and more costly disposal sites, install treatment equipment or perhaps even go out of business. All these costly steps are paid by the private sector to protect mainly non-marketed ecosystem services like habitat, recreation etc. that benefit society as a whole.

Sometimes marketable ecosystem services conflict with each other, creating monetary costs. For example, pollution impacts commercial fishing and excludable recreation (e.g., rafting services) and those losses are measurable in terms of money. Other costs are related to non-marketed services like swimming, drinking water quality, riparian hiking and habitat values. If pollution goes wild and trashes those ecosystem services, what then are the monetary costs? The answers are not quite as obvious, though in Chapter 4 we learned some methods to estimate them.

In the introduction to this chapter, I emphasized that nature is as close to a real magician as most of us are ever going to see in our lifetimes. Nature is like a magic wand that creates something great for people (and other species too!) without much, if any, effort. We see, though, that from the purely human perspective, lack of effort does not mean ecosystem services are "free." In fact, **when the way environmental assets are managed, costs are incurred**. Ecosystem services are given up by humans – opportunity costs are incurred – to

get more of other services, simply because they come from the same natural asset and conflict with each other.

8.4 Criteria for Managing Environmental Assets

So, just like in the allocation of all other valuable resources, decisions must be made. This
leads us to the question of how we know when management is "good" or, even more ambi-
tiously, the "best." From the purely human perspective, this definition depends on the basic
values of those making the decisions and the *social norms*[5] that follow from them. We will
discuss the role of collective action norms in Chapter 13, which focuses on social institutions
for better managing ecosystem services.

Given the basic values and norms, though, at any point in time the "best" course of action
would depend on the value people get from any shift in ecosystem service emphasis. As
discussed in Chapter 3, again from a purely human standpoint, the value of any shift is likely
to depend on how much of a particular ecosystem service we have vis-à-vis all the other
ones. In other words, the value we humans get from a particular ecosystem management
system depends on what we start with.

Box 8.1 Air Pollution Chokes Chinese Cities

Many Chinese cities experience very high levels of air pollution, particularly in winter
when temperature inversions occur and there may be little wind. In December 2015,
for example, the *New York Times* reported that "Red Alerts" were announced in
Beijing that closed its 3200 schools and forced half the cars off the road. A town near
Shanghai had measurements of $PM_{2.5}$, the fine particles that are most dangerous to
human health, of 265-268 micrograms per cubic meter. This is more than 10 times
the healthful standard set by the World Health Organization of 25 micrograms per
cubic meter.

Sources: Piao and Boehler (2015) and Wong (2015).

For example, if we have very low levels of small particles ($PM_{2.5}$), an important air pollutant, in the atmosphere, increasing air sink ecosystem services by a bit probably will not
matter much (i.e., we have lots of the ecosystem service respiratory health).

In polluted areas like Beijing, China in the winter, though, increasing $PM_{2.5}$ to increase
economic output may be viewed as a very costly and bad idea. Under such circumstances,
simply because we already have so much pollution, humans would give up a lot if we allowed
additional pollution.

This concept that the value we place on changes in ecosystem services generally depends
on how much we start with is very important for evaluating natural resource management
strategies. I reproduce Fig. 3.1, which illustrated this feature, here as Fig. 8.1.

Let's consider the implications of Fig. 8.1 for natural resource management. There are
many ecosystem management approaches that could be chosen, ranging from a 100% focus

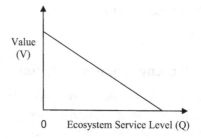

Figure 8.1 Marginal Benefit from Ecosystem Services

on market exploitation to pure ecosystem conservation. Suppose an environmental asset is close to pristine and therefore has very high levels of non-excludable ecosystem services. Though it is a bit hard to imagine in 2024, under such circumstances something akin to a frontier management approach that emphasizes marketable ecosystem services could be justified for some time if all other ecosystem services were truly abundant and people are extremely poor.[6]

Box 8.2 CAMPFIRE Program Attempts to Balance Human and Wildlife Needs in Zimbabwe

The Community Areas Management Programme for Indigenous Resources (CAMPFIRE) was started in 1989 in the wildlife-rich southern African country of Zimbabwe. The program attempts to better balance human and wildlife needs by giving local communities a stake in wildlife management. As of 2001, the program had generated $20 million in total revenues from activities such as trophy hunting, photography, fishing and livestock grazing and over half goes to local communities. CAMPFIRE also works to mitigate conflicts between wildlife (mostly elephants) and people. For further details, see www.campfirezimbabwe.org.

An alternative approach - still driven by the marginal valuation of humans who control environmental assets - would be to emphasize the non-marketed, non-excludable ecosystem services. Such a conservationist management approach might be justifiable because non-excludable ecosystem services are available to everyone, implying that the most people possible benefit. Also, non-excludable ecosystem services are not extracted and therefore do not degrade ecosystems. Of course, such a management strategy is likely to be most palatable in areas where such services are very highly valued, and people don't feel the need to overly exploit ecosystems. As was discussed in Chapter 5, this situation typically means that humans have a reasonable or even good standard of living, including enough to eat, good houses and the chance for people to have the things needed to enjoy their lives.

Between these two extremes would be an impartial weighing of the conflicting ecosystem service values - excludable or non-excludable - that flow from environmental assets. This approach is called management based on *environmental cost-benefit analysis*, where the goal is neither to extract whatever can be marketed from the environment nor to conserve environmental assets, but to choose a mix that maximizes the total value of ecosystem services.

Cost-benefit analysis-based decision-making considers not only the benefits of lifting up a particular set of ecosystem services, but also the opportunity costs. This decision-making criterion leads to the question of the shape of the marginal cost function. Marginal benefit functions should generally slope downward according to the law of demand, but there is no similar "law of supply" that specifies the form of marginal opportunity cost functions as ecosystem services increase. Figure 8.2 adds in a typical, upward-sloping cost function shape, which indicates that as natural resource management emphasizes one ecosystem service, on the margin opportunity costs increase.

What is the best level of the ecosystem service in this model? Suppose our criterion for choice is to maximize the total net benefits from the ecosystem service. *Net* in this context means that something is subtracted from something else. In this case, total opportunity costs are subtracted from total benefits.

Before reading on, please think about how one might infer maximum total net benefits from Fig. 8.2.

Total benefits at a particular level of ecosystem service are just the sum of all the marginal benefits or, if units of the service can be made very small, just the area under the marginal benefit function. For example, at ecosystem service level B, the total benefit would just be the area under the MB curve at quantity B. Notice that with a linear MB function, the total benefit is made up of a rectangle $V_1 * B$ and a triangle of $(V_2 - V_1) * B * 0.5$. Again, if the units are very small, total costs are just the area under the MC function. If MC were linear, that would mean that at ecosystem service level B, the total cost would be the triangle $(V_3 - 0) * B * 0.5$.

So where is the difference between total benefits and total cost the largest? This is where a bit of marginal thinking is especially helpful. Suppose we consider ecosystem service level B as a candidate for the optimum. At B, as well as everywhere between 0 and B, the marginal benefits exceed marginal costs. Are these levels of the ecosystem services contributing to human welfare?

The answer is yes if MB includes all benefits and MC includes all opportunity costs. But have we added up as many marginal net benefits as possible? The answer is no, because between quantity B and A, MB > MC. Though marginal net benefits are smaller than those up

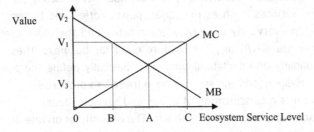

Figure 8.2 Environmental Cost-Benefit Analysis and Optimal Levels of Ecosystem Services

to level B, they are still positive and would therefore add to total net benefits. At what eco-system service level have we added up as many net benefits as possible? Level A, which is where the marginal benefits = marginal costs, maximizes total net benefits.

Equation 8.1

Optimal Ecosystem Service Level is Where MB = MC

Going beyond point A means that MC > MB. This net loss would need to be subtracted off from all the marginal net benefits added up through level A, implying that points such as level C cannot be optima. Using a cost-benefit analysis ecosystem service management lens therefore means choosing the mix of ecosystem services from natural resources such that the marginal benefits equal marginal costs. This result is presented in Eq. 8.1.

8.5 Common Pool Ecosystem Services

Cost-benefit analysis can help us characterize, identify and perhaps measure the best set of ecosystem services from environmental assets. Some of these ecosystem services may conflict with each other, which makes management of environmental assets difficult. Equally important, though, is that many ecosystem services themselves often have inconvenient properties. Perhaps by looking into the properties of ecosystem services we can better understand why we so often fall short on environmental management and do not reach the balanced set of ecosystem services we characterized as quantity A in Fig. 8.2?

We have talked about the non-excludability of many ecosystem services, which makes the private sector uninterested in providing them. There is a second dimension to goods and services, though, that is also important. Some goods and services deplete or degrade with use, whereas others do not. For example, the coffee I am drinking as I write this sentence is not only excludable, but I also deplete it as I use it. You and I can therefore not simultaneously consume the same sip of coffee. It is impossible. This depletability (often called consumption rivalry because we are "rivals" for the service) implies that I very well may care if you were to drink my coffee or sneak a bite of my sandwich, because you would be depriving me of it. Furthermore, coffee and sandwiches are excludable, so if I am selfish and I don't want to share, I don't have to do it. Whether or not to share is totally my choice.

This combination of depletability/rivalry and excludability is ideally suited to commerce. *Private goods and services*, such as massages, paper, coffee and laptops, don't instantly become available to everybody once they are created and therefore everyone needs to buy their own; depletability/rivalry is therefore good for business. These two properties together – excludability and depletability/rivalry – basically define the private sector and most exclusively private goods and services have these two properties.

Figure 8.3 presents a classification of goods and services depending on the degree to which they are excludable and deplete with use. The opposite of private goods and services are *pure public goods and services*. Services like photosynthesis in our gardens, ocean sailing

	Excludable	Difficult to Exclude
Depletable / Rival	**Private Goods and Services** (e.g., timber, fish, mushrooms, massages, legal service, iPhone, paper, coffee, laptop, tablet etc.)	**Common Pool Goods and Services** (e.g., climate stability, clean water from forests, respiratory health from clean air, biodiversity)
Not Depletable / Non-Rival	**Club Goods and Services** (e.g., music, movies, apps, uncrowded park recreation, swimming in large pools, travel service from uncrowded roads, education in schools)	**Pure Public Goods and Services** (e.g., streetlights, garden photosynthesis, ocean sailing, view of a large mountain)

Figure 8.3 Characteristics and Classifications of Goods and Services

and good views all have the properties that they do not deplete (i.e., we can all enjoy the services at the same time), and it is difficult or impossible to exclude people from enjoying the benefits. The private sector will not get involved with such services, because they are not only non-excludable, but they are freely shared. They are often not a management challenge, though, because they are freely available. Once they are created, they are available to everyone.

Club goods and services are a hybrid of public and private goods and services. They are excludable, but they do not deplete like a cup of coffee or a meal in a restaurant. Examples of club goods and services include a variety of intellectual and creative services, such as music, apps on your phone and movies. All these services are excludable – usually through technologies that are specially designed to exclude – but in principle can be shared without reducing the experience of individual users. They certainly don't deplete or degrade with use. Other examples of club goods include educational services from schools and transport services provided by roads. These services are excludable due to the existence of doors, on-ramps and gates, but unless the roads and schools get very crowded, they can accommodate more people and not affect the experience of current users. Uncrowded park services are natural resource club services.

In Chapter 3, we discussed how some ecosystem services can be "privatized." Many of these privatizable ecosystem services are pure private goods, such as oil, minerals, timber, fish etc. that are harvested from the environment. Others, though, are *club goods and services*, such as private ecotourism sites. In these realms, government and privately provided services often coexist. The same is true for educational services in schools (e.g., public schools, state universities), various intellectual and creative endeavors (e.g., music and theater) and recreation facilities (e.g., clubs, swimming pools, tennis courts etc.). Club goods and services have the interesting feature that they are often simultaneously provided by both the public and private sectors.

The last main class of goods and services – and the most relevant for understanding why environmental assets are difficult to manage – is *common pool goods and services*. This group of goods and services depletes or degrades with use and is difficult to exclude. Imagine something that is valuable (i.e., people want it, and it is costly to provide) that depletes or degrades, but it is also very difficult to exclude people from degrading those services. If you think about it, such characteristics make for a very tough management challenge!

Most ecosystem services we worry about have these features. Clean air provides many ecosystem services, including respiratory health for humans and other species. These services are degraded by using the sink ecosystem services of the air. Though we have air pollution laws that aim to limit the use of the sink function, think about how difficult it is to stop seemingly wasteful air pollution!

Box 8.3 Moss Reveals Air Pollution Hotspot in Portland, Oregon

"Portland's air is dirtier than we thought. The discovery that urban moss could be used as air quality monitors throughout the city was a tremendous scientific step by researchers at the U.S. Forest Service. That moss revealed 'hot spots' throughout Portland with concentrations of toxic heavy metals like cadmium, arsenic, nickel and lead. So far, two cadmium hotspots have been linked to colored glass makers Bullseye Glass and Uroboros Glass. The nickel hotspot has been attributed to Precision Castparts. Since the toxic hotspot bombshell hit a month ago, inadequate pollution regulations have been exposed, neighbors have expressed their anger and state air quality officials have resigned from their posts."

Source: Schick (2016).

Indeed, despite regulation and generally heavy emphasis on everything environmental in Portland, Oregon, in March 2016 it was revealed that glass makers and other manufacturers located in heavily populated areas had for years been emitting toxic air pollution that may have been causing serious health problems. Nasty heavy metal emissions raised concentrations to almost 50 times the state limit (Forbes, 2016) in a place where people are generally environmentally conscious. It is just too difficult to police everything all the time and, as we will discuss in Chapter 15, hundreds of pollutants are either impossible or very costly to measure.

Excessive car and truck idling is another example of how hard it is to "exclude" air polluters from degrading our clean air. Do you ever see people sitting in their idling cars with the windows open on a perfect day? What a waste of respiratory health and climate stability ecosystem services! The other day I saw someone park their car, set the alarm using the key fob and leave the engine idling while going into a store!

Such behaviors indicate that non-pollution ecosystem services are not considered when making choices. It is also tough to stop - i.e., it is *difficult to exclude* people from enjoying sink ecosystem services, because monitoring and enforcement costs are so high. Would you be willing to ask someone idling their engine for no good reason to stop their wasteful polluting? If you are, you are braver than me.

If it is difficult for individuals to monitor and enforce, collective action institutions can be very helpful. Many local governments across the US, including Houston, Texas, have passed

Photo 8.1 No-Idle Zone
Source: Author.

anti-idling laws and the State of New York prohibits idling by diesel trucks and buses for more than five minutes. But even then, how likely do you think it would be for someone to be fined for excessive idling?

Most environmental problems are related to common pool ecosystem services. Habitat loss leading to the extinctions and loss of biodiversity discussed in Chapter 1 is an important example. People may clear or degrade lands, such as forests, that are used for habitat and often these land use changes are difficult or impossible to reverse. Virtually all net deforestation at the national level (i.e., loss of forests after accounting for any forest area expansion) is in the Global South in or near tropical regions (Ritchie, 2021; Pan et al., 2011). Forestland is cleared for a variety of purposes, but one of the most important is for agriculture. The Food and Agriculture Organization of the United Nations estimates that clearing land for cropland and grazing is the #1 cause of deforestation in much of Asia and Africa (FAO, 2022). It is also a critical threat to natural habitat.

Loss of natural habitat, especially from agricultural expansion to produce excludable and depletable ecosystem services, such as crops and meat, is a key reason for the human-caused extinctions that were highlighted in the planetary boundaries literature. Of course, throughout the Earth's long history extinctions have occurred, but since the coming of humans and especially since the industrial era, extinctions have increased dramatically. The Harvard University biologist E.O. Wilson has estimated that the contemporary extinction rate is now about 1000 times faster than before humans. For example, in the US during the period 1895-2006, 57 freshwater fish species were driven to extinction, which is about 10% of the previous total. This extinction rate is just under 900 times the rate before industrialization (Wilson, 2016). Because extinctions – largely due to our use of land for narrow, excludable human purposes – have sped up so dramatically since the 1800s, the author Elizabeth Kolbert warns of a potentially catastrophic species collapse that she refers to as *The 6th Extinction* (Kolbert, 2014).

As late as the second half of the 20th century, the common pool ecosystem services provided by the natural world seemed too abundant for humans focused on private ecosystem services to ever seriously degrade them.[7] That perspective turned out to be amazingly wrong. An important example of a big common pool ecosystem service that has been seriously degraded by our bias toward private ecosystem services is climate stability – the opposite of climate change. Much later than 1950, it seemed ridiculous to most people that humans could be changing the *climate*. How could such a thing be possible? The climate is just too big! Well, somewhere along the way, we got very big as well and this idea, which is still advanced by a shrinking pool of climate change deniers, turned out to be wrong.

The point here is that the stable climate, biodiversity, respiratory health from clean air, swimmable oceans, fishable rivers etc. are all common pool ecosystem services. They degrade due to competing ecosystem services, such as pollution, that are difficult to stop. It is possible to limit self-serving behaviors that degrade common pool ecosystem services, but it can be challenging. Crafting public policy instruments, such as those we will discuss in Section III of this book, requires careful attention to the incentives of those who unnecessarily damage natural resources. Modeling such human incentives and predicting our responses therefore become especially important.

8.6 Chapter Conclusions

Environmental assets, such as forests, lakes, the air and oceans, typically simultaneously produce many ecosystem services. This is one expression of the magic of nature, but unfortunately often ecosystem services conflict with each other. Respiratory health and unacidified soils coexist very well, but both those ecosystem services are seriously degraded – as is climate stability – by fossil fuel combustion that releases air pollution into the atmosphere. Most any significant environmental asset we might point to produces conflicting ecosystem services in which the levels of some services are inversely related to levels of other services; generally so-called "sink" ecosystem services – also known as pollution – conflict with everything else.

Left to our own devices, we humans may emphasize the ecosystem services that benefit us the most, to the detriment of other services. This is because it is (opportunity) costly to maintain ecosystem services. That is, because levels of some ecosystem services are inversely related to each other, often humans must give something up to get more of a particular ecosystem service. From that perspective, it is hardly a surprise that, faced with management decisions, often our first reaction is to emphasize the marketable ecosystem services that benefit us most.

There are many ways to view "best" management of the conflicting ecosystem services flowing from environmental assets, but whatever the definition used, best or even good management can be a challenge. Many of the most important ecosystem services are "common pool." Difficulty in excluding people from overusing, combined with degradability of ecosystem services, implies a potentially tough management challenge. Especially if we are talking about unobservable behaviors, we are forced to recognize that human incentives – and willingness to work together for the common good – are probably especially important dimensions of any solution to the problem of environmental degradation. In the coming two chapters, we therefore consider potential models of human behavior when there are incentives to emphasize private interests vis-à-vis the environment.

Issues for Discussion

1. Can you think of any environmental asset that does not produce multiple ecosystem services?
2. Identify an environmental asset in your neighborhood that simultaneously provides at least four ecosystem services. What are the main conflicts between those services?
3. Are there environmental assets that do not produce ecosystem services that conflict with each other?
4. Do you think that intellectual club goods should be freely available because they are not depletable/rivalrous? Why or why not?
5. Do you think an objective weighing of the costs and benefits, as suggested by environmental cost-benefit analysis, is an appropriate decision criterion?
6. Are there important environmental assets that don't produce any common pool ecosystem services? Which ones?
7. Why do you think people waste gasoline and other resources that cost money and produce pollution? Isn't such behavior irrational?
8. How might we encourage people to consider the common pool nature of some ecosystem services when they make day-to-day decisions?

Notes

1 If you think about it, human-made capital tends to be rather specialized. For example, trucks haul stuff, bulldozers move dirt and computers – one of the most flexible types of human-made capital – analyze information.
2 For more information on how US water policy and law prioritize irrigation, see Reisner (1993) and Sabo et al. (2010).
3 Using the environment to dispose of human waste without protection and treatment can spread serious infectious diseases, such as cholera and typhoid fever.
4 They are also species within broader aquatic ecosystems and therefore themselves produce multiple ecosystem services.
5 Merriam-Webster defines social norms as "standards of proper or acceptable behavior."
6 Such was the approach in the western US in the 1800s, which emphasized marketable ecosystem services like agricultural soil fertility, hunted animals and minerals. Of course, not all humans endorsed this view. In particular, many native people preferred other approaches. There is also abundant evidence that the frontier approach went on for long enough in the American West that it was extremely costly for non-marketed ecosystem services. For example, wild American plains bison, of which there were 25–30 million in the 16th century, were down to only 100 by the late 1880s (Taylor, 2007). There

are also a variety of negative incentive effects associated with the *open access* that underpins any frontier management. The topic of incentives will be tackled in the following chapter.

7   One of my favorite quotes from the BBC television series *Downton Abbey*, which is set in 1920s England, is by the dowager countess played by the late Maggie Smith. In one of the episodes, she notes that one of the great things about nature is that "...there is so much of it."

Further Reading and References

Food and Agriculture Organization of the United Nations (FAO). 2022. *State of the World's Forests*. Rome.

Forbes, D. 2016. "State Finds Alarmingly High Arsenic, Cadmium Levels Near Two SE Portland Schools." *The Portland Mercury*, March 3.

Government Accountability Office. 2014. *Federal Rulemaking: Agencies Included Key Elements of Cost-Benefit Analysis, but Regulations' Significance Could be More Transparent*. GAO Report GA-14-714. A Report for Congressional Requesters.

Kolbert, E. 2014. *The 6th Extinction: An Unnatural History*. Henry Holt and Co Publishers: New York.

Pan, Y., R.A. Birdsley, J. Fang, R. Houghton, P.E. Kauppi, W.A. Kurz, O.L. Phillips, A. Shvidenko, S.L. Lewis, J.G. Canadell et al. 2011. "A Large and Persistent Carbon Sink in the World's Forests." *Science* 333 (6045): 988–993.

Piao, V. and P. Boehler. 2015. "In China, Diners Pay for Clean Air with Their Entrée." *The New York Times*, December 15.

Reisner, M. 1993. *Cadillac Desert: The American West and Its Disappearing Water*. Penguin Books: New York.

Ritchie, H. 2021. *Deforestation and Forest Loss*. OurWorldInData.org. Retrieved from https://ourworldind ata.org/deforestation March 2024.

Sabo, J.L., T. Sinha, L.C. Bowling, G.H. Schoups, W.W. Wallender, M.E. Campana, K. Cherkauer, P. Fuller, W. Graf, J. Hopmans et al. 2010. "Reclaiming Freshwater Sustainability in the Cadillac Desert." *Proceedings of the National Academy of Sciences* 107 (50): 21263–21269.

Schick, T. 2016. *7 Things You Need to Know about Portland's Toxic Air Situation*. March 15. Oregon Public Broadcasting. Retrieved from https://www.opb.org/news/series/portland-oregon-air-pollution-glass/7-things-you-need-to-know-about-portlands-toxic-air-situation/ May 2021.

Taylor, M.S. 2007. "Buffalo Hunt: International Trade and the Virtual Extinction of the North American Bison." Working Paper #12969. National Bureau of Economic Research: Cambridge, MA.

Wilson, E.O. 2016. "The Global Solution to Extinction." *The New York Times*, March 12.

Wong, E. 2015. "Beijing Issues a Second 'Red Alert' on Pollution." *The New York Times*, December 17.

9 Bad Incentives, Market Failures and Property Rights Problems

9.1 Recap and Introduction to the Chapter

Most environmental assets produce multiple ecosystem services for humans. For example, a tropical forest sequesters carbon, filters water that humans drink, provides habitat for animals that people care about, offers recreation, is a place for grazing domesticated farm animals and produces timber and non-timber forest products, such as mushrooms. These ecosystem services are all produced at the same time, plus many more. This aspect of the magic of nature is notable, because human-made assets rarely produce more than one or two services. Unfortunately, valuable ecosystem services can conflict with each other.

Ecosystem services are "free" in that nature provides them, but this does not mean they are costless. When one service is emphasized to the detriment of conflicting services, ecosystem services are given up to get that particular service and opportunity costs are incurred. Pollution - the waste assimilation ecosystem service of the environment - conflicts with virtually all other ecosystem services. Reducing pollution is (opportunity) costly because scarce pollution control methods must be used to reduce pollution.

Most ecosystem services we are especially worried about losing, such as climate stability, biodiversity services and capture fisheries, are common pool. This observation means they can be depleted, usually by degrading environmental assets themselves, but it is difficult

DOI: 10.4324/9781003308225-11

to keep people from depleting those resources. Non-excludability is an especially terrible complement to depletability because it implies that not only can an ecosystem service be depleted, but it is difficult to stop people from doing it.

You probably suspect that without intervention, this situation could be a recipe for disaster. Such a prognosis depends on the reactions of humans individually and through institutions, such as businesses, communities and governments. We now begin our examination of the ways people might interact with environmental assets and their ecosystem services. We therefore now consider human incentives and discuss models of human behavior vis-à-vis natural resources.

9.2 What Have You Done to the World Today?

Challenge Yourself

In the last four hours can you name two things – big or small – that you knowingly did that degraded the environment, but which you could have avoided?

One way to begin to understand some of the incentive problems associated with management of environmental assets is to consider the issue from a personal perspective.

For example, what have you knowingly done today that you could have avoided, that reduced the quality of environmental assets and degraded common pool ecosystem services?

Here is my true confession. Today I put a newspaper in the garbage can rather than the recycling bin. I could have fished it out and moved it to our recycling container, but I would have had to reach down into the garbage can and I did not want to do that. The newspaper will therefore go to the landfill rather than be recycled, degrading our land, because I did not want to risk getting my hands dirty ☹. But why do I even have a physical newspaper, which causes water and air pollution during its production and delivery to my house, rather than just reading it online, which seems to be what the newspapers prefer anyway? Answer: Because I like a physical paper.

It so happens that I did not drive my car today, but did you? Did you really need to drive or could you almost as easily have walked? Did you sit in an unnecessarily idling car? Did you leave your reusable coffee cup behind when you got coffee, because you did not want to have to wash it afterward or it was inconvenient to carry? Do you take long, hot showers? Do you take energy-intensive saunas just because you enjoy them? More significantly, perhaps, did you buy a gas-guzzler rather than an electric car because it was cheaper or had features you preferred? Does a big, heavy car just seem "safer?" Did you build a new house rather than buy one that was already there because you always wanted to be the first owner? If you work for a government agency or company, did your institution make choices that harmed the environment to save money? Yada yada yada...

Every single day we each face big and small choices that are basically non-observable to anyone but us. Via choices of commission or omission, we deliberately choose options that harm the environment. We do this so we can have lower costs, get better quality or deal with fewer hassles than if we took the more environmentally friendly approach. It's the way of the human world and often those actions are completely legal and socially acceptable; and even if they are not, your friends, neighbors and family will never know. It is all on us to make the choices.

Behaviors by individuals, firms, governments and other institutions that deliberately push costs or risks onto societies to benefit themselves are called *moral hazard*. Moral hazard is certainly not limited to environmental problems. Calling in sick to work when we are not sick would also be an example. But because so many human interactions with the environment are hidden, the scope for moral hazard is enormous. Neutralizing temptations associated with environmental moral hazard, especially when incentives for socially conscious behaviors are weak, is, in a nutshell, the challenge of protecting natural resources.

What is viewed as an unobservable infringement on the environment can depend very heavily on community norms and may change over time. For example, I was in Sri Lanka for some meetings, lectures and time on a beautiful Indian Ocean beach. I could not help but notice, though, that it seemed to be absolutely acceptable to drop trash on that beautiful beach. Though the beach was still fantastic, it was littered with beer cans, plastic bags, bottles and other trash. People just walked down the beach and stepped over the trash.

The key point is that at some point someone dropped that trash on the beach and likely did so with the approval of others. They probably littered guilt-free, because it was considered acceptable. In the US the norm used to be similar. Up through at least the 1970s, Americans routinely littered. Now in some locations it is difficult to imagine someone deliberately throwing trash on a beach. Norms have changed to the point that even if we are not observed, many of us would be embarrassed to litter.

But attributing moral hazard has limits. Clearly, we do not want to force an 87-year-old to ride her bike or walk everywhere. Driving everywhere could easily make sense for such an older person. If you really love half-and-half in your coffee, despite the need to raise polluting cattle to produce it, perhaps it is OK to have a bit of cream. Maybe a steak is alright every once-in-a-while despite the environmental effects? If you love hot tubs or traveling on greenhouse gas-producing airplanes, maybe that can be justified as well.

Challenge Yourself

What individual norms of behavior can you think of that changed in the recent past and improved environmental quality?

But how and when can we challenge people to do the best they can and, if possible, not trash the common pool ecosystem services we all share? Are there ways to think about incentives that perhaps offer us systematic approaches to solving environmental problems?

Some of these challenges, such as species extinctions and climate change, are basically at emergency levels, so understanding how to address environmental moral hazard can be critical. Shedding light on these two questions is the subject of the rest of this book and answering them is a major challenge of our time.

We have so far emphasized that environmental assets are complicated, producing multiple, conflicting, often common pool ecosystem services. These observations say little about the explicit incentives involved and how people will react to those incentives. Two very important analytical approaches to the moral hazard associated with common pool ecosystem services are to conceptualize it as market failures and as property rights problems. In the next chapter, we examine collective action complexities, which offers a third lens through which to view environmental moral hazard.

Market failures take markets as their starting point and ask a critical question: What can go wrong with markets that might affect human behavior and welfare? The market failures of special relevance to environmental issues are so-called *negative externalities* in which costs from markets spill over onto either other markets or non-marketed, common pool ecosystem services.

A second way of thinking about incentives to trash the environment for our own gain is in terms of property rights problems. The most important of these property rights issues is open access, where ecosystem services are "free" and therefore may be overused.

9.3 Market Failures

When the operation of markets is an important reason for environmental degradation, the notion of a market failure can be helpful. Businesses and those who buy their products often cause environmental damage, such as pollution. A market failure lens views environmental damage from the perspective of markets and asks the question: What is wrong with this market?

The market failure approach takes as its starting point a well-functioning market in which the price and quantity bought and sold fully take into account the costs and benefits of the product or service. This means that the market weights the costs and benefits of what is transacted, balancing them against each other. The result of such a process is what is typically called an *efficient* market. Such a market outcome maximizes total net benefits, which are defined as total benefits minus total costs, and it is in this sense that an efficient market is "best" or "optimal."

For example, suppose the market for wholesale coffee is in equilibrium, such as we saw in Fig. 3.4 in Chapter 3, adapted here as Fig. 9.1. We found in Chapter 3 that our model predicts equilibrium occurs at the quantity such that marginal benefit equals marginal cost, where "marginal" means extra or additional. That the coffee market is in equilibrium implies that it is at a resting point and there is no major tendency for the price or quantity of coffee transacted to change, in this case from quantity B and price P_B. To some degree, such equilibria are hypothetical situations, because markets are always changing, but suppose we took a snapshot at a particular point in time and found no major tendencies for change.

The price-quantity combination (B, P_B) is the equilibrium to which the market itself tends. If *all* marginal benefits and marginal costs are included in defining that equilibrium, it

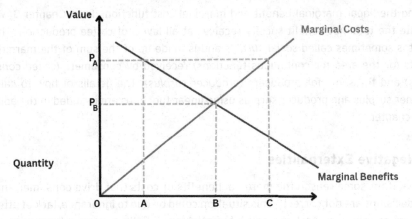

Figure 9.1 Market Equilibrium Price and Quantity

is also an optimum or efficient outcome. To see why, let us have a very simple decision rule for determining when any pound of coffee should be produced. Using a logic that is similar to the one we used for environmental cost-benefit analysis in the previous chapter, let's produce any pound of coffee when the (marginal) benefit of producing that pound exceeds the cost.

Now consider quantity A, which is less than B, and ask whether, given our decision rule, it makes sense to produce A pounds of coffee. Is it a good thing to produce this Ath pound of coffee? Yes, of course. At this quantity, the benefit of that pound of coffee greatly exceeds the cost, so we should produce it. Indeed, benefits exceed costs for all pounds of coffee before A as well (and indeed the difference between marginal benefits and costs is greater), which implies that it is also in society's interest to produce those pounds of coffee.

What about production of coffee after A pounds? Yes, but no more than quantity B, because after quantity B the marginal costs exceed the benefits. Producing up to the Bth pound therefore yields positive *marginal net benefits* (i.e., marginal benefits exceed costs), but beyond point B marginal net benefits are negative (marginal benefits < marginal costs).

It is a no-brainer that if we would like to *maximize* the total net benefit, we should produce any and all pounds of coffee such that marginal benefits are greater than marginal costs (i.e., we should add up all positive marginal net benefits). The point where *total net benefits* are maximized occurs at quantity B, and this quantity is therefore known as an efficient level of production. If we consider the possibility of very small changes in quantities of coffee, there is no need to add vertical distances between marginal costs and marginal benefits to get total net benefits. As quantities become infinitely small, the total net benefit of B pounds of coffee (not the Bth pound) is just the total area between the marginal benefit and marginal cost curves between 0 and quantity B.

If the marginal cost and marginal benefit functions are accurate, B is not only the market equilibrium, but it is also the efficient, best or optimal quantity in that it maximizes total net benefits; to the extent this model reflects reality, this is a potentially very powerful endorsement of what markets can do for humans.

Using the linear marginal benefit and marginal cost functions from Chapter 3, we can calculate the total net benefit society receives at all levels of coffee production. This net benefit is sometimes called *social surplus* and is made up of the sum of the marginal net benefits (or the area for continuous functions) received by consumers (called *consumer surplus*) and the same for producers (*producer surplus*). The details of how to calculate consumer surplus and producer surplus using linear functions are included in the appendix to this chapter.

9.4 Negative Externalities

But what if for some reason the marginal benefits or costs that drive consumer and producer decisions are not correct? This situation could be due to ignorance, lack of attention by buyers or sellers or, perhaps, self-interested, essentially antisocial behaviors. Let us take the case where real costs exceed the private market costs. We therefore amend Fig. 9.1 to show a distinction between true or *social* marginal costs and perceived or *private* costs.

An important example of where private and social marginal costs differ is when markets produce pollution. We previously thought of pollution as an ecosystem service (i.e., waste-absorbing service provided by nature) that conflicts with most other ecosystem services, but for the moment let's look at this problem wholly from the standpoint of a market. Suppose in our coffee example the coffee is produced on plantations that rely heavily on pesticides to control a variety of coffee pests and pathogens.[1]

One way to model pollution is that costs created by the polluting market are greater than the private costs on which producers and buyers make decisions. This situation is presented in Fig. 9.2, where at quantity A the difference between the value P_A (the social marginal cost of the Ath pound) and P_C (the private cost of the A_{th} pound) is the monetary value of the pollution costs the coffee market imposes on society, perhaps measured using valuation methods we learned about in Chapter 4. Of course, often the costs are inherently non-market (e.g., when the organophosphate diazinon kills birds, bees and insects).

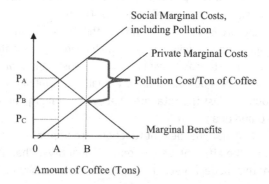

Figure 9.2 Negative Externality

Challenge Yourself

Can you identify the area of total net benefits in Fig. 9.2 by which quantity A is "better" than quantity B?

One of the many things we might note, after amending our marginal cost to reflect the reality of pollution, is that it turns out quantity B is not really the optimum at all. Quantity A is the efficient quantity once pollution costs are included as a true cost, because it perfectly balances off **all** costs and benefits. Quantity B is "too much" in that for all pounds of coffee between A and B, the social marginal costs exceed the benefits. If quantities of goods and services produced and consumed in equilibrium are less than or greater than the optimum, this is what is meant by a *market failure.*

The market "fails," because the quantity produced and consumed does not maximize net benefits. We also notice that in this case, this excessive quantity is supported by a market equilibrium price that is too low. In other words, the market price is P_B when if all costs were considered by the market, the equilibrium market price should be P_A.

Pollution in this framework is one example of a so-called *negative externality*, which is an important, but hardly the only, type of market failure.[2] Negative externalities occur when there are costs created by markets that are not incurred by sellers and buyers. Again, if we suppose that very small quantities can be analyzed, the total negative externality (e.g., the total pollution cost) is just the area between the social marginal cost and private marginal cost curves between 0 quantity and any positive quantity we want to analyze.

For example, we can consider the value of the externality at quantity B and see that it is greater than the value of the externality at quantity A. It is important to point out, though, that at quantity A, which is the optimum, the negative externality is not zero; positive pollution is therefore optimal in the sense that costs and benefits are appropriately balanced.

Box 9.1 Worked Example: Calculating the Inefficiency of Markets When there is an Environmental Externality

Suppose the marginal benefit and the private marginal cost functions in the coffee market are linear, like those in Chapter 3. Private marginal cost or supply = $50 + 3Q and marginal benefit or demand = $300 - 2Q. These are the same equations as in Practice Problem 3 of Chapter 3. Setting marginal cost = marginal benefit (i.e., assume the market is in equilibrium), we get $50 + 3Q = $300 - 2Q. Let us recall that if we add, subtract, multiply or divide both sides of an equality by the same value, the equality still holds. Subtracting $50 and adding 2Q to both sides give us 5Q = $250 or Q = 50. Plugging Q = 50 into either price equation tells us that the equilibrium price should be $200.

But suppose there are a number of negative environmental externalities due to plantation coffee-growing practices that reduce biodiversity and increase water runoff. Also part of this set of externalities is worker health concerns due to the spraying of pesticides and human exposure. Suppose the environmental externality is constant at $50 per ton of coffee produced.

To get the *efficient* or welfare-maximizing price and quantity of coffee, we need to include all costs, not only private costs. We therefore add the $50 pollution cost to the cost side of the equation. That is, the *social marginal cost* is really $50 + 3Q + **$50** or $100 + 3Q. Setting this true marginal cost function equal to the marginal benefit (which is not distorted by the externality), we now set $100 + 3Q = $300 - 2Q. Adding 2Q to both sides and subtracting $100, we find that 5Q = $200 and Q = 40. Plugging Q = 40 into the social marginal cost or marginal benefit equation and solving for P tells us P = $220. We therefore find that with the negative externality, equilibrium price is $20 below the efficient price ($220 - $200) and quantity is 10 tons too high (50 - 40).

We can actually calculate the value of the social welfare lost due to the negative externality by calculating the difference in *consumer surplus* and *producer surplus* generated at the efficient and inefficient quantities. The appendix to this chapter demonstrates how to make those calculations.

What does this market-centered, market failure-focused approach suggest about how to *internalize* (i.e., encourage buyers and sellers to fully consider) environmental externalities? The model suggests that to internalize the externality, we would need to increase unit costs to polluters by the monetary value of the pollution, which is the vertical distance between the social and private marginal cost functions. In Fig. 9.2, if the social and private marginal cost functions are parallel, this would mean that the market needs to "feel," perceive and make decisions based on costs that are higher than the private costs by the amount of P_A minus P_C; this is the monetary value of the pollution created by producing a pound of coffee.

Box 9.2 How Costly are Climate Change and CO_2 Emissions for People? The Social Cost of Carbon and the Clean Power Plan

When a US federal government regulation is expected to have a substantial effect on the economy (>$100 million) or raise other important issues, the agency proposing the regulation must be confident the benefits exceed costs. This requirement is from Executive Order 12866, which went into effect in 1993 during the presidency of Bill Clinton; despite the possibility that subsequent presidents may have wanted to roll back E.O. 12866, it still endures today.

Many regulations related to energy, transportation and the environment have climate change implications, and some of them, like the President Obama-era Clean Power Plan (CPP), which was an executive order to reduce CO_2 emissions into the atmosphere, are all about reducing carbon pollution. The CPP, which was abandoned by President Donald Trump in 2018, sought to reduce CO_2 emissions by 32% in 2030 compared with 2005. The US Environmental Protection Agency regulatory impact cost-benefit analysis* estimated that this would add an additional 16% onto the total 17% CO_2 reduction that would have occurred without the CPP.

Such regulations have costs. Electricity producers, which were the focus of the CPP, would have needed to burn coal more efficiently, increase the use of natural gas and increase the use of zero-emissions power generation like wind and solar power. All these steps are costly and the cost-benefit analysis estimated that annual costs of the CPP would have been somewhere between $1 billion and $8 billion (in $US 2011).

But what about the benefits, and especially the climate change benefits of the CPP? These were measured by a group of US government agencies using integrated climate and economy simulation models that tried to quantify the monetary value of the climate damages that would be avoided by the CPP. The per-ton estimate of those damages was called the *social cost of carbon*, the first estimates of which were issued in 2010** and subsequently updated until 2017, when President Donald Trump shut down the effort. The midline 2020 estimate in $US 2016 was $49/ton of CO_2, which meant that if a ton of CO_2 was emitted in 2020, it would cause $49 worth of damages in terms of increased air conditioning costs, lost skiing days, flooding damages, health costs, reduced crop production etc. The social cost of carbon was estimated to rise to $80 by 2050, because it was estimated that by then climate change would be much worse and the value of reducing CO_2 emissions by a ton would be much greater than in 2020.

* USEPA. 2015. "Regulatory Impact Analysis for the Clean Power Plan Final Rule." Available at https://archive.epa.gov/epa/sites/production/files/2015-08/documents/cpp-final-rule-ria.pdf
** Interagency Working Group on Social Cost of Carbon. 2010. "Technical Support Document: Social Cost of Carbon for Regulatory Impact Analysis under Executive Order 12866." Available at https://19january2017snapshot.epa.gov/sites/production/files/2016-12/documents/scc_tsd_2010.pdf

One potential mechanism for making private costs equal to social costs is taxation. Taxation's main function is to raise money for governments to run their operations, but in this case the purpose of the tax would primarily be to affect the behavior of coffee buyers and sellers.[3] If the pollution tax per pound of coffee were set at P_A minus P_C, in principle the market would fully internalize the pollution externality by reducing the quantity of coffee transacted in the market to quantity A.

Often, though, the marginal cost of pollution - i.e., the extra damages from an additional unit of pollution - increases with the amount of pollution. If additional production is related to more pollution, the correct tax rate would depend on how much is produced.

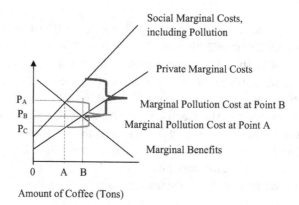

Figure 9.3 Externalities that Increase with Production

For example, when a river has very little organic suspended solids like food waste or sewage in it, the river is easily able to absorb a bit more organic garbage without eco-system services being terribly affected. That is, when there is only a little pollution in a river, it is less likely to make you sick or give you a rash than when it is heavily polluted; stated formally, the marginal cost of that pollution is small. Figure 9.3 takes this pos-sibility on board by making the marginal damages a positive function of the level of production.

Challenge Yourself

Figure 9.3 supposes that the monetized costs of pollution are known or not too expensive to determine. If this is not correct, what might we do?

Leaving aside the important difficulties associated with valuing the pollution externality from a pound of coffee, let us consider whether such an output or product tax is a good policy. We will analyze environmental taxes in some detail in Chapters 15 and 16, so for now let us just ask ourselves what behaviors such taxes will incentivize. As in the worked example in Box 9.2, suppose after some analysis it is estimated that the externality value is $50 per ton, but whether and how much coffee growers apply pesticides are essentially non-observable; to internalize the externality in Costa Rica (a major coffee producer), a tax is therefore levied on all coffee farmers.[4] Let us further assume that Costa Rican coffee growers are in it for the money and try to make as much profit as possible.

Box 9.3 Worked Example: Calculating the Effect of Environmental Product Taxes on Equilibrium Price and Quantity when Negative Externalities Increase with Production

Suppose we are analyzing the market for coffee in Costa Rica. The marginal benefit and the private marginal cost functions in the coffee market are again linear, and private marginal cost or supply = $20 + $6Q. Marginal benefit or demand = $80 - $4Q. Subtracting $20 and adding $4Q to both sides give us $10Q = $60 or Q = 6 tons. Plugging Q = 6 into either price equation tells us that equilibrium price should be $56. This is the market equilibrium without any environmental taxes.

The government of Costa Rica knows that coffee production reduces ecosystem services, and the marginal externality increases as production increases. It therefore decides to tax coffee producers $2.00 times the number of tons produced.

To include this policy in the model, we simply add the tax = $2.00 * Q to the private marginal cost. The marginal cost, including the tax, is therefore $20 + $6*Q **+ $2*Q =** $20 + $8*Q. Setting this marginal cost + tax equal to the marginal benefit and solving for Q, we get Q = 5 tons, which is less than without the tax. Equilibrium price is $60 per ton of coffee, which is higher than without the tax. A graph of the predicted effects of such a tax is given in Fig. 9.4.

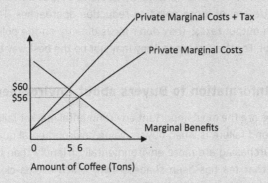

Figure 9.4 Pollution Externalities as a Function of Output

Suppose the marginal benefit and the private marginal cost functions in the coffee market are linear, like those in our first worked example in this chapter. Private marginal cost = $50 + 3Q and marginal benefit = $300 - 2Q, giving us an equilibrium quantity of 50 and price of $200. Now suppose that we add a tax equal to the $50/ton externality (here assumed constant at all levels of coffee production).

Adding this $50 tax to the cost makes the marginal cost $100 + 3Q as in the first worked example, which is, of course, the full social marginal cost. Setting this cost function equal to the marginal benefit, we now set $100 + 3Q = $300 - 2Q. Solving for the equilibrium quantity, we get Q = 40 and a price of $220 rather than $200 without a tax. The policy is therefore predicted to fully internalize the externality as long as we correctly choose the tax! Notice that the $50 pollution tax only increases the price by $20, which means that coffee sellers would absorb $30 of the tax, presumably because they do not want to lose too many customers.[5]

With this setup, how can coffee growers reduce their tax liabilities? The only way they can avoid paying taxes is to grow less coffee, which indeed will reduce pollution from pesticides. But is this the only or even the best way to reduce pesticide runoff from coffee production? Probably not. To reduce the environmental effects of pesticides, it also matters *which* pesticides are used and *how much* is applied.

As always, circumstances differ. Some coffee growers may live in environments that are very prone to insect infestations or in areas that more readily cause pesticides to decay. In both these cases, more pesticides should probably be used than when environments are highly sensitive or there are no bugs anyway. Choosing an environmental tax that better differentiates environmental effects and compliance costs will almost certainly be better than an output tax.

To be direct, the problem with output taxes as environmental policy instruments is that they are blunt. They presume that the only way to reduce the amount of the environmental threat is to reduce the amount of the product being bought and sold. Reducing output will reduce pollution, but as we saw in Chapter 5, it matters *how* and *where* production occurs. In some cases, it may be possible to use technologies specifically designed to reduce pollution before it is emitted into the environment. Unfortunately, because output taxes focus only on *output*, they do not incentivize any of those pollution reduction approaches. This, therefore, is our primary concern with output taxes: They don't focus directly on the pollution, which is what we actually care about. For these reasons, they may not be the best way to reduce pollution.

9.5 Inaccurate Information to Buyers about Environmental Damages

Negative externalities are the most important environmental market failure, but they are not the only ones. A second failure is when buyers value environmental quality and believe the products they are purchasing are more environmentally friendly than is true. For example, buyers may believe that tea has been shade-grown, which avoids clear-cutting of native forests and helps birds and other wildlife. They might also think the tea was grown organically, without chemical fertilizers and pesticides, but perhaps neither of those beliefs are justified; for any number of reasons, buyers' information about environmental protection practices may be incorrect.

The way we model such a situation is similar to the method used for a negative pollution externality, but instead of social marginal costs exceeding private marginal costs, the marginal benefits are higher at all quantities than is really the case. This situation is shown in Fig. 9.5.

We see in the figure that, as was the case for a negative externality, the private equilibrium quantity is higher than the optimal quantity that truly balances all costs and benefits. As the quantity of tea transacted is above the optimal level, there is therefore a market failure. The key difference between Fig. 9.2 and Fig. 9.5 is that the market failure in Fig. 9.5 comes only from buyers – buyers must care about the environment for there to be a market failure, whereas in Fig. 9.2 *anyone* who cares about environmental effects like pollution or

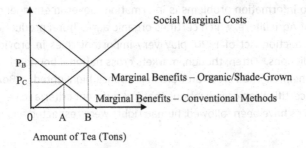

Figure 9.5 Inaccurate Information to Buyers about Environmental Effects

wildlife habitat in coffee-growing regions would be included in the social cost. We also notice that while the quantity effects of the information problem are the same as for negative externalities, the price distortion is exactly the opposite; with inaccurate information, the high quantity is supported by a price that is too high.

Box 9.4 The Forest Stewardship Council® Certifies Sustainable Forestry Management and Documents Forest Product Sources

Founded in 1993, the Forest Stewardship Council® (FSC®) is an international, nonprofit, non-governmental organization that works to promote responsible management of the world's forests. It works toward this goal by helping consumers distinguish sustainable from unsustainable wood products. As they note on their website, "We enable businesses and consumers to make informed choices about the forest products they buy, and create positive change by engaging the power of market dynamics."

FSC has certified the management of almost 200 million hectares (about 500 million acres) of forest area since its founding and issued more than 1500 certificates in 83 countries. The organization also certifies the origins and sustainability of forest products through so-called "chain of custody" certificates. They have issued over 33,000 of these certificates focusing on 122 countries, making it one of the largest forest and forest product certifiers in the world.

As stated on the organization's website:

"The Forest Stewardship Council (FSC) logo on a wood or wood based product is your assurance that it is made with, or contains, wood that comes from FSC certified forests or from post-consumer waste."

Source: https://uk.fsc.org/.

The solution to information problems is information. Government agencies, such as the US Department of Agriculture, which certifies organic agricultural production under the US Organic Food Production Act of 1990, play very important roles in providing information within their jurisdictions.[6] Often, though, markets cross national boundaries and third-party certifiers can be helpful. The Forest Stewardship Council highlighted in Box 9.4 is one of a number of forest certification organizations that use different criteria to certify that sustainable forest practices have been followed, human rights were respected and product qualities are accurate.

9.6 Property Rights and Open Access

Imagine that the president of your company, organization or university kindly sets up 30 long banquet tables in a public place. On those tables, she arranges to be placed boxes that contain many different types of pizza. She announces widely that this pizza is available for free as a thank-you gift to the community. That is all she says.

This gesture would be a generous one and potentially very much appreciated. What other reactions might we expect? First, the event would not only be appreciated, but it would likely be quite popular as well. What would that popularity mean for behaviors? Unless the society has especially strong norms against gluttony, probably many people would take a lot of pizza. Now suppose someone had pizza for dinner every night during that week and was therefore a bit tired of pizza. Probably, they would still take at least a few slices or maybe even more, because... they are free. This behavior occurs despite the reality that taking pizza when one won't enjoy it very much does not make much sense.

Moving this idea into the formalized structure of Chapter 3, this means that such people will be getting very low marginal benefits from pizza and, unless it is effectively unlimited, will be taking pizza from those who have not had pizza for a long time and therefore would value it a lot. There is therefore a *misallocation* of pizza from those who value pizza highly to those who are way down their marginal benefit curves due to high consumption. This is one problem with free or *open access*.

What else might happen? Well, those who are taking too much pizza may have a bit more complex logic than merely wanting a lot of free pizza. They might (reasonably) be thinking that the pizza will be exhausted very quickly and they will only get one shot at it. Perhaps they will therefore take a bit more than they would have otherwise, just in case they are not as sick of pizza as they thought or if it turns out to be especially delicious; people are likely to engage in strategies that include the reactions of others.

We do not want to caricature for the sake of the example, but it is certainly possible that arguments could break out simply because of the open access. Some may claim they are getting too little and others too much. Perhaps those who arrived first – maybe those who could move the easiest (i.e., the young or those without disabilities) – would get the majority of the pizza, which would not be fair.

Some may feel they have more of a claim to the pizza than others. For example, unless there is a requirement that everyone must actually work for the company or study at the university to get pizza, and a way to check, nonaffiliated people may come to get a slice or two. This allocation may not be what was intended, and the additional people could contribute to exhausting the pizza. The kindly president likely intended a relaxed pizza party, but with open access there is the potential for perverse incentives that do not contribute to a reasonable, rational or efficient allocation of pizza. In the end, the event could be anything but relaxed and possibly anything but a party.

Open access natural resources, which offer provisioning ecosystem services that can be directly extracted and used by people, are like unrestricted, free pizza. The private ecosystem services discussed in Chapter 8 potentially get used up quickly and typically this means the asset itself is degraded.

I and some colleagues analyzed carbon sequestration in forests across Nepal. Nepal has an interesting forest management system based on community management, but many forests are also open access. We compared forests that were managed by communities with government forests that were open access. We found that forests that were under active community management had about 1/3 more carbon than forests that were open access, presumably because they were no longer a free-for-all (Bluffstone et al., 2018).

Challenge Yourself

Are there resources – natural or otherwise – in your community that are open access? What are the results?

But for most natural resources, the issue goes deeper than mere extraction, because natural resource quality can often be improved by reducing stresses, allowing regeneration or through direct investments.[7] Forests are an interesting example of a renewable natural resource that can be improved via a variety of investments. Afforestation, which is the active planting of trees to increase forest cover, can improve forest resources. Furthermore, forests grow back if harvest pressure is reduced. They therefore regenerate if we leave them alone and allow them to grow. This is also a type of savings and investment – forgoing harvests today so forests can accumulate biomass through natural growth.

What does this recognition imply about open access? Would we expect people to invest in natural resources that are open access and are always under threat due to uncontrolled use? Such people would have to be especially uninterested in their own welfare to invest under such conditions! They would be investing effort, money etc. to improve the quality of natural resources when it is very likely that the fruits of those efforts will be enjoyed and even undermined by others.

We therefore at best expect *under*-investment in natural resources when access is free and open. More likely, without explicit policies, there will be no investment. This is sort of like the free rider problem most of us experience around cleaning up our kitchens when we have roommates who are messy. We may be willing to put in the effort to wash dishes, clean counters and otherwise keep our kitchen clean, but our willingness to do it is likely to depend on whether our roommates are also willing to keep the kitchen up. Why should we spend a lot of time cleaning the kitchen when our roommates will just mess it up again?

Open access is undoubtedly the enemy of natural resources, which is an old insight, dating back to the 1950s when H. Scott Gordon analyzed the economics of fishing (Gordon, 1954). He did not use the words open access, but he was the first to discuss problems with free access. In 1968, open access was front-and-center in the important and famous paper by Garrett Hardin titled *The Tragedy of the Commons* (Hardin, 1968). In this fascinating paper, Hardin applies the idea of open access (what he calls "the commons") to a variety of issues from overgrazing to population policy.

If you think about it, most of the natural resources we are most worried about have at least some elements of open access. For example, every time anyone burns unsustainably created fuels, carbon is emitted into the atmosphere and contributes to climate change. How is it possible to monitor even most such activities? In many cases, it is not possible.[8] The same can also be true for open ocean fish, wildlife and biodiversity, all of which are ecosystem services that are degraded by harvests. Often the environmental assets people are exploiting are so large that under many circumstances they cannot be controlled.

If we are unable to exclude users from harvesting privately valuable ecosystem services, we can end up with some version of an open access problem. So, what is the solution? When environmental problems are thought of in this way, the solution is to close open access, implying some sort of *property rights*. Property rights over ecosystem services are rights to the stream of benefits associated with a resource, combined with the ability to make decisions about those resources.

This is a very interesting and historically quite influential approach, particularly in the US and other highly market-oriented countries. A key challenge with viewing environmental problems as property rights issues, though, is that for many ecosystem services, establishing property rights can be very difficult. For any environmental asset we care about (e.g., clean air or even a small lake), there may be many, many stakeholders. Evaluating who has rights to the various ecosystem services (and therefore the right to deprive others of conflicting ecosystem services) may be next to impossible.

It is unlikely that an exclusive focus on property rights will save our natural world, because property rights are difficult to define for complex natural resources with a lot of users having many, difficult-to-measure, conflicting uses. Thinking in terms of securing property rights can, however, be useful when natural resources have basically one use (i.e., they are like pizza). For example, in analyzing rapid depletion of an underground aquifer or overharvesting of firewood from a village woodlot (i.e., a forest plantation created to produce firewood for cooking), examining property rights and the degree of open access can be very helpful. In Chapter 13, we will consider some innovative property-rights solutions for protecting natural resources and the ecosystem services they produce. In Chapters 17 and 18, we discuss cap-and-trade systems for pollution control and related "tradable rights" instruments, which can be thought of as property rights mechanisms.

9.7 Chapter Conclusions

There are some basic bad incentives associated with common pool ecosystem services and every day we all take shortcuts that harm environmental assets for our own benefit. How can we characterize and model such antisocial behaviors so we can figure out how to steer people away from them?

In the previous chapter, we viewed environmental problems in terms of the characteristics of ecosystem services. This framework is important, because it gives us reason to expect environmental moral hazard. One way to model that moral hazard is in terms of deviations from market optima. Negative externalities are one such type of market failure in which costs of market activities, such as environmental costs, are not fully incorporated into prices and therefore "too much" is produced. A critical example of a negative externality is pollution, which affects marketed and non-marketed services, imposing costs on people in the process.

Thought of in this way, the solution to some environmental problems may be to figure out ways to be sure markets incorporate non-marketed environmental costs, as well as normal market costs, into their decisions. The goal then is to be sure quantities of marketed products and services are at efficient levels. This approach is probably best thought of as a guide to policies, because focusing on reducing quantities of goods and services is often not the best way to address environmental issues.

A straightforward fix is available when information on the environmental quality of a product is overstated or otherwise incorrect. In this case, the costs of the product may be right, but it is the benefits that are not correctly incorporated into markets. Though such market failures only deal with actual buyers in the market and do not include those outside the market who are affected by its activities, improving information available to buyers may help them make decisions with regard to their own preferred levels of environmental stewardship. Government and non-governmental "third-party" certifiers can therefore be very helpful.

Lack of property rights is not a market failure per se, because it does not presume a failure of markets, but instead a failure of the legal structure supporting markets. For the complicated environmental assets yielding multiple ecosystem services to hundreds, thousands or even millions of people, delineating property rights can be very tough. In some cases though, especially when uses of natural resources are limited and reasonably well-defined, securing property rights can be important; open access is undoubtably the enemy of natural resources.

Issues for Discussion

1. Can you name a natural resource that is often open access? Are there any outcomes you notice that are attributable to the free access?
2. Is it possible to model the open access problem in a supply-demand framework? If yes, how would you do it and if not, why not?
3. The externalities view of environmental problems often leads to a conclusion that market prices are "distorted" from what they should be (e.g., too low) and the solution is to "get the prices right." Do you think proper pricing – pricing that includes all costs, including environmental costs – is a key part of the solution? Why or why not? What, if any, are the limits of such a policy?

4. With so many unobservable actions by millions or even billions of people affecting the environment, government monitoring has limits. What potential voluntary alternatives do you see for encouraging better behavior by people vis-à-vis natural resources?

5. Littering is often much higher in low-income countries than in higher-income settings. What do you think are the key reasons for such observed differences?

6. In 1992, then World Bank Chief Economist Lawrence Summers (later US Treasury Secretary and President of Harvard University) caused a firestorm by suggesting that because different countries at different levels of development place differing values on environmental protection (i.e., are willing to give up different amounts of other things), pollution should be disproportionately concentrated in places with low such values, which tend to be lower-income countries (https://www.nytimes.com/1992/02/07/business/furor-on-memo-at-world-bank.html). Do you see any value in this argument?

Practice Problems

1. The marginal benefit of eating quinoa is given by $30 - 0.3Q and the marginal cost of producing quinoa for sale in stores in the US is $20 + 3Q, where Q is in tons.

    a. What are the equilibrium price and quantity?
    b. Suppose quinoa, which is grown in mountainous areas, requires a lot of irrigation water that is currently free. The government starts to charge a price for water that equals $2.00 per ton of quinoa produced. How does this change the equilibrium?
    c. Quinoa, which for decades or more had been considered a specialty grain, is now found to have important health benefits that are valued at $6.00 for each ton of quinoa consumed by buyers. What now is the social welfare-maximizing output level?
    d. Assume 1b and 1c happen at the same time.

2. Roundup is a brand of glyphosate-based herbicide made by the Monsanto Company, which is owned by Bayer, Inc., that is widely used in the US to control weeds. In May 2019, a California jury awarded $2 billion to a couple who claimed their cancer (non-Hodgkins lymphoma) was caused by Roundup exposure. This news was widely publicized.

    Suppose the public reacts to this news by demanding less Roundup at all prices. The federal government also imposes a tax on Roundup to finance cancer research. Model both these developments graphically in our marginal benefit - marginal cost/supply-demand framework. How do these two developments - knowledge of the possible cancer-causing properties of Roundup and the tax on Roundup - affect the equilibrium price and quantity?

Notes

1   Coffee farmers may use a variety of pesticides, including endosulfan, chlorpyrifos, diazinon and several other potentially toxic chemicals (https://www.coffeehabitat.com/2006/12/pesticides_used_2/).

2 An important type of market failure we will not consider is monopoly power. Monopoly occurs when a market is controlled by one firm or just a few firms. Models predict monopolies reduce equilibrium quantity *below* the efficient level – i.e., produce too little – to increase price. For more information, please see any principles of microeconomics textbook.

3 Because the purpose of such taxes is to reduce pollution rather than raise revenue for governments, some authors prefer the term "fee" or "charge" (e.g., see Oates, 1988).

4 This assumption is not as extreme as it might at first seem. Coffee is grown in tropical and subtropical regions where many countries have low average incomes, relatively weak governments and limited environmental enforcement capacities. Pesticide applications may therefore truly be unobservable.

5 This topic of how responsive buyers and sellers are to changes in prices is called *price elasticity* and applies to anything that might change prices. For more details, readers should see any introductory microeconomics text.

6 See https://www.ams.usda.gov/grades-standards/organic-standards for additional information.

7 Mineral resources, such as metals, petroleum and coal, are perhaps the only resources where extraction alone is the key issue.

8 Of course, this conclusion depends on resources and technologies available for monitoring. Particularly in lower-income countries, government capacity and resources can be very limited, making many ecosystem services essentially open access. Even in high-income countries, emissions, such as those from transportation, are difficult to monitor.

9 Producer surplus can be thought of as "profit," but in reality it ignores any costs that do not change with output.

Further Reading and References

Bluffstone, R.A., E. Somanathan, P. Jha, H. Luintel, R. Bista, M. Toman, N. Paudel and B. Adhikari. 2018. "Does Collective Action Sequester Carbon? Evidence from the Nepal Community Forestry Program." *World Development* 101: 133–141.

Gordon, H.S. 1954. "The Economic Theory of a Common Property Resource: The Fishery." *Journal of Political Economy* 62 (2): 124–142.

Hardin, G. 1968. "The Tragedy of the Commons." *Science* 162 (3859): 1243–1248.

Oates, W. 1988. "Taxing Pollution: An Idea Whose Time Has Come?" *Resources* (Spring). Reprinted in *The RFF Reader in Environmental and Resource Management 1999*. Resources for the Future: Washington, DC.

APPENDIX
Calculating Producer and Consumer Surplus Using Linear Functions

Suppose we take as an example a revised version of Fig. 9.1, which is a linear model of a coffee market. Demand takes the form marginal benefit = $a_1 - b_1 Q$, where a_1 and $b_1 \geq 0$, and supply is represented by marginal cost = $a_2 + b_2 Q$, where a_2 and $b_2 \geq 0$.

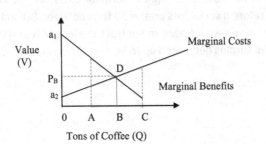

Coffee Market

Consumer surplus is the sum of the marginal net benefits (net of costs) to consumers received from marketed or non-marketed goods or services. For buyers of market goods like coffee or services like oil changes, the consumer surplus is just a measure of the degree to which buyers get an awesome deal. This metric is the marginal benefit taken from the demand function, minus expenditures on the good or service. Like coffee lovers, users of common pool ecosystem services increase their well-being when they enjoy services like nature hikes, clean air or litter-free beach experiences. If we know the marginal benefits (to add them up) or the marginal benefit function (to calculate the area), we can subtract any cost of getting the ecosystem services (though there is no market price), such as hiking boots to take a hike, which would give us the consumer surplus.

Producer surplus is the net benefit received by producers of marketed goods and services. It is the difference between the market price and the cost to produce a good or service. Of course, many ecosystem services are produced only by nature and in any case, it is not possible for firms to make profits from non-excludable ecosystem services.

In the equilibrium coffee market, the net benefits to consumers will be the triangle that shows the full area above the equilibrium price (P_B), but below the marginal benefit (demand) function. This is the area a_1, D, P_B. Producers also get surplus, which measures the amount of profit they get from selling the good. It is the area between the market price and the marginal cost function[9] or a_2, D, P_B.

Using the parameter values from Practice Problem 3 in Chapter 3, let a_1 = $300, b_1 = 2, a_2 = $50 and b_2 = 3. Assuming both functions are linear, we find that the equilibrium price (P_B) = $200 and quantity (B) is 50 tons. The formula for the area of a right triangle is area = (base * height)/2. For both surpluses the triangle base is the same, just the equilibrium quantity (50 - 0), but the height of the triangle will differ for consumers and producers. For consumers, the height is ($300 - $200) = $100 and the area of the triangle, which measures consumer surplus, is ($100 * 50)/ 2 = $2500. The height of the producer surplus triangle is ($200 - $50) = $150. Producer surplus at Q = 50 is therefore ($150 * 50)/ 2 = $3750. The total social surplus generated by the coffee market at the equilibrium is therefore $2500 + $3750 = $6250.

Any quantity other than the competitive equilibrium of 50 tons (B in the figure) will generate less social welfare. For example, if quantity A in the graph were Q = 30, if the market produces 30 tons rather than 50 tons, it means all the consumer and producer surplus the market would have received on the 20 tons between 30 and 50 would be given up. Furthermore, point C (something like Q = 80) represents a point of over-production. Each ton of coffee produced past 50 tons implies marginal costs are incurred that exceed the benefits. There is therefore a social *loss* on the 30 tons between 80 and 50. To calculate the social surplus at 80 tons, we would need to subtract the loss between 80 and 50 tons from the total positive social surplus generated up to 50 tons. This value is clearly less than simply producing 50 tons.

10 Collective Action

10.1 Recap and Introduction to the Chapter

Given the multiple and often conflicting ecosystem services produced by environmental assets and the common pool nature of the ecosystem services that are most at risk, humans face a difficult management challenge. Often it is our environmental moral hazard behaviors that create these challenges. It is therefore of critical importance to try to understand the key features of human activities that cause us to excessively degrade ecosystem services and environmental assets themselves.

There are a number of related ways to think about the problem, each suggesting different solutions. The first is as a property rights problem. In this model, the key issue is insufficient incentives for saving and investment within open access environments. The challenge is therefore to figure out how to give individuals, households, firms and other institutions personal stakes in natural resources and close free access to ecosystem services. This conceptual model suggests that governments should give attention to clarifying property rights to ecosystem services rather than directly regulating human behaviors per se. Two important examples of property rights-focused policies are cap-and-trade, which are covered in Chapters 17 and 18, and community natural resource management, which is discussed in Chapter 13.

The second approach looks at environmental problems as a negative externality and asks how market performance can be improved for the benefit of the environment. This

DOI: 10.4324/9781003308225-12

way of thinking about human-environment interactions quickly leads to solutions focused on "internalizing" externalities. Pricing of negative externalities, such as pollution, and "getting the prices right" through instruments like carbon taxes, water charges and development fees fall out quite naturally from this way of thinking about the problem.

Though quite useful, thinking of environmental problems only as externalities and issues of poorly defined property rights is limiting and perhaps insufficient to manage the broad range of environmental issues the world faces. Indeed, our #1 environmental challenge, climate change, probably cannot be fully addressed via either or both of these conceptual frameworks; a wider view is needed. We now add a broader conceptualization to our understanding of the challenge of protecting natural resources.

10.2 Collective Action is Important, but Often Difficult

So, what if establishing property rights over natural resources to better balance ecosystem service flows is not appropriate, and optimizing environmental spillovers (externalities) and focusing on solving information problems do not make sense? What then should be the policy focus? One possibility is to promote *collective action*.

Natural resource collective actions are activities that group members undertake together, on their own or via their governments, to improve natural resource quality. It involves cooperation, which is a synonym for collective action, and is a very flexible and comprehensive – if sometimes difficult – way to think about environmental problems. It is also consistent with the way many people (especially the young) conceptualize environmental issues. Often, indeed, when my students advocate for reduced consumption to limit stress on the environment, they do so from the standpoint of "working together," "changing people's minds" and advancing "new paradigms" of consumption and cooperation.

Box 10.1 Why is Collective Action Critical?

Paraphrased quote from DIVESTPSU* panel discussion on February 16, 2015:

Question: Why are you a part of DIVESTPSU?

Answer: I'm already a vegetarian, don't have a car and shower infrequently. What more can I do? I want the institutions that I am a part of to step up and address the climate problem on my behalf.

*DIVESTPSU is a student-led organization formed to encourage Portland State University to divest its foundation accounts of investments in fossil fuels.

One plea for collective action is given in Box 10.1. This paraphrased statement was made by a student at a #DivestPSU panel discussion held at Portland State University in 2015. She says that she is doing all she can to save the planet, but it is not enough, and she wants her university to take steps that support her private efforts. In other words, she wants the university to help the community coordinate toward the common purpose of reducing its carbon footprint.

There are many types of collective action other than those related to natural resources. These include efforts on behalf of vulnerable groups (e.g., the mentally ill, children, the poor and the elderly), religious organizations (e.g., church groups), public safety (e.g., voluntary or professional fire departments, coastal rescue, police etc.), public libraries and public art. A variety of institutions have emerged to facilitate our collective action, but the most important by far are government institutions. In fact, a potentially useful way to think about government agencies – federal, state/province and local – is as the most important organizations for financing non-excludable goods and services that must be provided via collective action.

Challenge Yourself

Can you name five non-governmental environmental collective action institutions? What about five governmental institutions?

Natural resource collective action is the flip side of acting for one's own benefit to the detriment of the environment. Collective action is costly to group members in a variety of ways, some of which were mentioned in previous chapters. Collective action may involve actual effort, such as neighborhood cleanup campaigns, political collective action to protect parks from conversion to other uses, invasive species removal in public forests or stream restoration. It could also involve paying money to finance collective action, including through taxes. Collective action may also be in terms of *not* doing something in order to enhance the quality of a natural resource. For example, natural resource collective action may involve driving only when necessary, not idling engines unnecessarily or keeping the heating thermostat at a moderate level. It could also mean complying with norms or formal rules to avoid littering, using single-use plastics, dumping into rivers untreated ship ballast water that could potentially be carrying invasive species, urinating publicly, grazing animals or harvesting timber, fuelwood or fodder from common forests.

Elinor Ostrom spent much of her career at Indiana University in the US state of Indiana and won the Nobel Prize in Economic Sciences in 2009. Ostrom helped us understand that when people agree on what should be done, groups can cooperate without external intervention. Until Ostrom wrote her path-breaking book titled *Governing the Commons*, which was published in 1990, conventional wisdom almost assumed à la the "Tragedy of the Commons" (Hardin, 1968) that group cooperation was essentially impossible. Ostrom showed us that under reasonable conditions, cooperation for mutual gain is possible. We come to specific examples in Chapter 13.

If we all worked together on behalf of the environment, ecosystem services that are currently at risk would be in much better shape. Unfortunately, a number of factors make pulling together in such a fashion difficult. First, as discussed in Chapter 3, people are different. Some are very passionate about environmental protection, some don't care and others may even be hostile to the notion of environmental protection. This may come from "preferences" that are particular to each person or could be related to self-interest. For example, if someone is an avid kayaker, they may be especially passionate about river water quality. Parents who have children with severe asthma may advocate particularly strongly for air pollution controls. Members of religious groups who believe other species were created to be exploited by humans may view habitat protection and biodiversity conservation as a waste of resources or even counterproductive.[1]

Some may have vested interests – for example, financial interests – that cause them to advocate against protection of common pool ecosystem services that could affect those interests. Non-excludable ecosystem services cannot be commercialized and therefore those who have financial interests in their status share those interests with others.[2] For example, open space (Lutzenhiser and Netusil, 2001), street trees (Netusil et al., 2010) and clean air (Harrison and Rubinfeld, 1978) increase housing prices, which is a very important financial benefit. Apple tree growers in Washington State benefit financially from pollination services from bees that are sensitive to pesticide applications by home gardeners and farmers. As we discussed in Chapter 3, that home gardeners and farmers benefit from *degrading* pollination services – that something must be given up (garden and farm yields) – is indeed part of what tells us that pollination services are scarce and valuable. In other words, pollination is not "free."

However, often financial benefits from non-excludable ecosystem services are so diffuse that they are swamped by concentrated interests. For example, in the early 1990s, almost immediately after the Intergovernmental Panel on Climate Change was established to provide information on climate change, the largest oil, gas and coal companies formed the Global Climate Coalition (GCC). This organization's mission was to spread skepticism about the existence of climate change and oppose policies to reduce greenhouse gas emissions. Throughout the 1990s, the GCC implemented a multimillion-dollar advertising campaign to discredit climate science and advocate against international agreements, such as the Kyoto Protocol. The GCC was disbanded in 2002, but similar work continues today.[3]

In his seminal 1965 book titled *The Logic of Collective Action: Public Goods and the Theory of Groups*, Harvard professor Mancur Olson argued that group size matters. If interests are concentrated, he noted, those who are members of such groups have greater incentives to pursue their interests than larger, more diffuse groups. Each member of a very large group (e.g., one concerned about climate change) is likely to gain less and therefore have less stake in success than more concentrated opposition groups (e.g., some fossil fuel companies).

In sum, people disagree about the focus and direction collective action should take. Some have strong personal preferences. Some have important incentives to promote particular ecosystem services. Others want to undermine services that conflict with services in which they are invested. Plus, "special interests" may have an advantage over those with interests that are more diffuse. These basic features of collective action make protecting common pool ecosystem services difficult.

But even without different viewpoints and interests, there are challenges. Let's abstract from reality for a moment and consider a form of collective action in which we may actually have personal experience. For example, have you ever had to work on a group project, either at school or at work? Suppose your professor or supervisor forms you and others into a group to complete a specific task. For example, assume your group needs to complete a research project and write a report by a specific date. In contrast to most environmental challenges, there is no disagreement about the goal; everyone agrees that a high-quality report must be written.

We all have had experiences working in such groups. Before moving on to the next paragraph, reflect for a moment on your experiences. What were the advantages of working in groups? What about the disadvantages? From your experience, what were the potential pitfalls?

Most of us have participated in groups that have gone well, but when I ask these questions to my students, normally I get something akin to a collective groan and many people say they hate group work. Typically, students point to "free riding," which is a process of enjoying gains from a group without sufficiently contributing; normally, those who may self-identify as free riders keep pretty silent. Students also, though, acknowledge the possibility for accomplishing complicated tasks more efficiently and at higher levels of quality than if they worked alone. Many, for example, point to tapping into the skills of others that they may not have (e.g., statistical analysis or technical writing). They also say that with many people, jobs go more quickly if everyone pitches in.

Challenge Yourself

Why have group projects in which you have been involved gone well? Why have others gone badly?

Herein lies the major rub with group work: Will everyone pitch in? If not, students and many others report that those who pick up the slack can end up with **more** work than if they were not part of a group. Not only do these leaders end up covering for free riders, but they also waste time and effort figuring out that they need to do more than their share of work; at least at the beginning, leaders must participate in meetings to assess others' performance and determine they have to do the work themselves. All these efforts are costly in terms of time and frustration, and, if free riding is bad enough and nobody steps up to cover for the slackers, groups may completely fall apart. Adding to the complexity, the "slackers" may not even agree they are not doing their shares. Instead, they may view the "leaders" as overly anxious busybodies who have "taken over" the project. Oy vey!

Mancur Olson addressed such issues in his 1965 book and noted that:

> ... it does not follow that perfect consensus, both about the desire for the collective good and the most efficient means of getting it, will always bring about the achievement of the group goal. In a large, latent group [i.e., one where one member is relatively insignificant] there will be no tendency for the group to organize to achieve its goals through the voluntary, rational action of the members of the group, even if there is perfect consensus.
>
> (Olson, 1965, pp. 59-60)

10.3 Modelling Collective Action

What might we make of such inferences? One possibility is to simply despair that important collective action will ever occur, because people are simply too selfish (except those of us who do all the work!). Was Olson right that groups cannot organize to achieve their goals?

To properly evaluate human behavior, we need to have some sort of vision or "model" of human cooperation. We are no strangers to models, having in several chapters in this book used models. For example, in the supply-demand model people buy based on the enjoyment they get, and the more they have consumed in the past, the less contemporary enjoyment they receive. People also do not pay more than what something is worth to them (i.e., they don't waste money) and businesses try to maximize profits.

Just as supply-demand is a useful model of market behaviors, the fundamental model of collective action is called the prisoner's dilemma. This model is part of a class of models often called "games" and the mathematics used to analyze them is "game theory." The prisoner's dilemma model was developed in 1950 by Princeton University mathematician Albert Tucker. It was originally phrased in terms of two prisoners facing potentially long prison sentences (Runciman and Sen, 1965), hence its name.[4] Since its development, the model has been used to analyze a variety of collective action problems.

Quoting Runciman and Sen (1965, p. 554), here is the basic setup:

> Two persons are thought to be jointly guilty of a serious crime, but the evidence is not adequate to convict them at a trial. The district attorney tells the prisoners that he will take them separately and ask them whether they would like to confess, though of course they need not. If both of them confess, they will be prosecuted, but he will recommend a lighter sentence than is usual for such a crime, say 6 years of imprisonment rather than 10 years. If neither confesses, the attorney will put them up only for a minor charge of illegally possessing a weapon, of which there is conclusive evidence, and they can expect to get 2 years each. If, however, one confesses and the other does not, the one who confesses receives lenient treatment for providing evidence to the state and gets only 1 year, and the one who does not receives the full punishment of 10 years.

Figure 10.1 presents the game, based on the numbers (called "payoffs") in Runciman and Sen (1965). Given this setup, what can we expect to happen? As was true for the supply-demand model, we need to make an assumption or two about how these prisoners behave. A common-sense behavioral assumption would be that each prisoner tries to minimize their punishment.

		Prisoner 2 (Second number in pair gives years in jail)	
		Not Confess	Confess
Prisoner 1 (First number in pair gives years in jail)	Not Confess	2, 2	10, 1
	Confess	1, 10	6, 6

Figure 10.1 Prisoner's Dilemma Game 1

Because punishment is tied to the response of the other accomplice, though, and because the two are separated, each prisoner needs to predict what the other criminal will do. There are a variety of assumptions that could be made, all of which would affect the results, but a standard assumption would be that each prisoner will suppose the other person tries to minimize their sentence, given their expectation of the other prisoner's best strategy. This way of thinking about strategic behavior is known as a *Nash strategy* and the equilibrium (i.e., the expected result) is called a *Nash equilibrium* after the 1994 winner of the Nobel Prize in Economic Sciences John F. Nash.[5]

Challenge Yourself

What are the best and worst outcomes for the prisoner "community" in Fig. 10.1?

What is the Nash equilibrium of this game? Suppose Prisoner 1 does not confess, thinking that Prisoner 2 would do the same. Given the assumptions about Prisoner 2's behavior (tries to minimize sentence and assumes Prisoner 1 will do the same), this is not a good strategy. If Prisoner 2 believed Prisoner 1 would not confess, they could get an even lower sentence (1 year rather than 2) by confessing and Prisoner 1 would end up with 10 years. Not confessing is therefore clearly not the best way to go (i.e., it is not "Nash"). Prisoner 1 therefore confesses and given that it is in the interests of Prisoner 1 to confess, Prisoner 2 should also confess (otherwise they will get the 10-year sentence). The Nash equilibrium of the prisoner's dilemma game is therefore that both prisoners confess and each gets 6 years.

What is notable about the Nash equilibrium of the prisoner's dilemma game is that it can never be the best outcome. It does not have to be the worst as in the Runciman and Sen (1965) setup (i.e., if the lower right-hand quadrant of the figure was 4,4, the result would be the same), but it can never be the best. The best outcome is always where neither confesses, but this outcome is not Nash given the payoffs and the assumptions about behavior. Furthermore, even if the prisoners could communicate, given the model assumptions the results are predicted to be the same. The prisoner's dilemma game therefore offers a dim view of the potential for collective action.

Let us move this model into the natural resource management world. An easy example would be to recast Fig. 10.1 as focusing on pollution damages by two beer producers located on a river. The beer producers get their water from the river, but they also dispose of water pollution into the river. These two competitors need to decide how much to pollute, given that (as discussed in Chapter 8) pollution reduction is costly, but both firms are negatively affected by water pollution. The numbers now represent the value of pollution damages (e.g., loss of sales due to reduced beer quality in millions of dollars) for each beer company. The potential decisions are to pollute little (i.e., don't confess) or pollute a lot (confess). Given we have just changed the labels and left the numbers intact, the Nash equilibrium is for both beer companies to emit a lot of pollution.

Suppose instead of prison sentences or pollution damages, we are talking about open ocean fishing by commercial fishers. Fish are a renewable resource capable of reproduction,

	Representative "Other" Fisher (Second number in pair gives tons of fish caught)		
Fisher 1 (First number in pair gives tons of fish caught)		Allow Fish Stocks to Recover Now	Try to Harvest a Lot Now
	Allow Fish Stocks to Recover Now	15, 15	0, 25
	Try to Harvest a Lot Now	25, 0	5, 5

Figure 10.2 Prisoner's Dilemma Game 2: Fishing Outcomes Depending on Current Behaviors

but reproduction is reduced by over-fishing. That is, if a lot of fish are harvested now, in the next period there will not be as many breeding fish and future harvests will suffer.

The fishers therefore might choose between something like "Allow Fish Stocks to Recover Now" or "Try to Harvest a Lot Now." Suppose that we are analyzing one fisher's outcomes vis-à-vis all the other fishers, but we use one as "representative" of the group. The setup might look something like Fig. 10.2, where the numbers are now "good" (i.e., fish landings) rather than "bad" as in Fig. 10.1. I leave it to you to confirm that both fishers will "Try to Harvest a Lot Now." The fishing and the pollution control examples are just game theory representations of the open access problem discussed in the previous chapter and in Hardin (1968).

Specific behavioral assumptions underpin the so-called tragedy of the commons. People must act strictly in their own interests and they must assume that others do the same thing. The prisoner's dilemma game is a model of reality and one of many collective action game theory formulations one can envision. The question therefore arises whether it (or any other model) adequately captures human behavior regarding environmental collective action.

10.4 How Do We Test Models of Collective Action?

Models, such as prisoner's dilemma–type games, are useful, but they only predict behavior and do not tell us what people actually do. This is the distinction between *theory*, which is the set of ideas and organizing frameworks to analyze human behavior, and *empirical analysis*, which uses actual data to test behavioral hypotheses.[6]

Let us treat the degree and possibility of collective action as something that might be tested once we understand potential human responses. This approach allows us to understand better what we might expect from humans faced with important and personally costly collective action decisions. Using contemporary research, we might also evaluate "tweaks" that could perhaps improve the functioning of collective action institutions.

How can we gain insight into, for example, whether the prisoner's dilemma result of purely selfish behavior generally holds in real life? The first possible way would be to conduct a focus group. Focus groups of 6-12 people can be very useful for understanding the nuances of how communities operate. The facilitator gives the group a topic to discuss, usually phrased as an open-ended question. The group then discusses the topic, with the responses noted or recorded by someone other than the facilitator. A major advantage of using focus groups to understand cooperative behaviors is that those conducting the analysis do not artificially restrict the topic. It is therefore very possible that issues researchers never considered will come up during the discussion, helping to create nuanced understanding. This unbounded- ness using focus groups also has a downside, because it is difficult to truly test hypotheses.

An alternative and often complementary method is to use surveys that carefully ask people (typically called "respondents") about their cooperative behaviors. For example, suppose we want to understand how to motivate people to avoid littering. We therefore might take a sample of people and ask them (among other things) how many times in the past three days they littered. We might also ask how many times they stopped to pick up trash on the street. These questions could be followed up with others asking why they littered and picked up trash.

Such surveys can be very helpful to understand how people **view** their behaviors and motivations. Often, though, we are interested in *actual* behaviors. With surveys, we need to assume that the responses people give during surveys correspond with their actual behaviors. Collective action involves activities we typically consider prosocial or "good" behaviors and respondents may not want to be perceived as antisocial by researchers or others. They also may think they contribute more than they really do or may not recognize they have done something bad like littering. In either case, respondents may overstate col- lective action behaviors, and reported behaviors and motivations may not correspond very well with reality.

This potential disconnect between reported and real-world behaviors is often called *hypo- thetical bias*. Bias occurs, because what is going on are not merely mistakes that could lead equally to overestimates or underestimates of actual behaviors; respondents are perhaps most likely to make mistakes that cast themselves in favorable lights. We therefore expect that hypothetical bias will cause us to systematically *overestimate* the extent of cooperative behaviors.

We may be able to observe actual behaviors and use this information to infer coopera- tive motivations. For example, we may be interested to test whether new people moving to a city increase environmental collective action. We therefore assemble data on population changes, perhaps by neighborhood. We collect information on collective action behaviors, such as invasive species removal or neighborhood cleanup. Using these data, along with other information (combined with a hefty dose of statistics), we can evaluate whether these two effects are linked, i.e., is this in-migration associated with environmental collective action?

You may notice that I use the words "associated with" rather than "cause." This choice of words is intentional, because particularly using data from one point in time it would be very difficult to convincingly show that the new residents *caused* any uptick or decline in invasive species removals or neighborhood cleanup activities. There are two main reasons it is tough to prove causation. First, and perhaps most obviously, many factors can influence environmental activism. For example, incomes may have increased in the city during the time of in-migration. Higher incomes mean more taxes and donations, some of which will support environmental protection. Environmental organizations may therefore be able to sponsor more trips to local forests and parks to remove invasive species. If we do not take account of the increased donations and taxes when we do the analysis, we risk confusing the effect of the population influx with the (potentially related) effect of generally rising incomes. Indeed, it would not be difficult to mistakenly conclude that the new folks are more environmentally conscious than existing residents, when actually there is just more money around to fund environmental collective action. Clearly, these are very different explanations that are being confused!

The second and more subtle reason why causality may be difficult to infer is that we simply do not know for sure which way causality runs. People, such as newcomers, create desirable places to live, but it is also true that desirable places (e.g., those with healthy environments due to environmental activism) draw people to live there. We therefore might have the idea that our new migrants are more likely than the retrograde existing residents to help the environment, when in actuality it is the environmental work of the existing people that attracted the new people.

It is not so easy to use existing or survey data to answer such seemingly straightforward questions about collective action. To overcome problems with these methods, collective action is now often examined using field experiments or lab experiments. Field experiments involve actual experimental manipulations of the real world for the explicit purpose of testing hypotheses about human behavior. Lab experiments are also interventions, but they are not interventions in the real world. Instead, lab experiments construct an artificial situation where people play games that contain key aspects of the real world. Data from lab experiments are then analyzed to test hypotheses.

One of the many puzzles in environmental economics is why privately profitable and socially beneficial energy efficiency measures are often not adopted. For example, why do people leave lights on when nobody is in the room? Why are televisions left on? Why are cars left idling in parking lots? One hypothesis that has been advanced is that even if costs are low (e.g., the time to walk into the living room to turn off the TV), people have psychological difficulty seeing the benefits they receive later from investments now – there is a present bias.[7]

But recognizing that such a psychological issue exists does not itself get people to save energy. We need policy instruments that encourage the changes in behaviors that reduce energy consumption. Ayers et al. (2012) used a field experiment to analyze one potential policy instrument using data from two electric utilities in California and Washington. The instrument was novel in that they tested whether making people aware of how their energy behaviors compared with their neighbors affected energy consumption. These utilities partnered with a company called Opower to provide a subset of homeowners either monthly or quarterly reports on their energy use. On these reports, energy use was explicitly compared with neighbors to offer relevant comparisons that could potentially spur energy users to reduce their consumption.

Importantly, the subset of homeowners that received the Opower reports was *randomly* selected and then compared to a random sample of those who did not receive the reports. Those receiving the Opower reports are referred to as the *treatment* group and the rest are *controls*. Randomness, combined with a sample size that is sufficient to avoid drawing false conclusions, is extremely important for field experiments, because it is what allows an experiment to draw conclusions about the population. Without a random sample, there are no guarantees that the experimental results are valid in the real world.

Ayers et al. (2012) found that over a one-year period the Opower reports caused (now we can use the word "cause") energy users to reduce their consumption by 1.2%–2.1%, which is a potentially important, but not overly large effect. It also, though, is a very low cost intervention.

Allcott and Rogers (2014) then analyzed the persistence of Opower report effects over a four- to five-year period. They found very strong early efficiency improvements that declined over time. As reports continued to be sent to energy users, over time the decay rate declined. They also found that it takes a long time for energy users to develop habits and full habituation may take up to two years.

Present bias related to energy behaviors may be especially strong when people must invest cash rather than just little bits of time, and cash is especially precious in low-income countries. Mobarak et al. (2012) examined adoption of improved biomass-using cooking stoves in Bangladesh. These technologies use less wood, reducing collection times and costs, and some add chimneys to reduce smoke in homes, potentially improving human health. They may also be quite cheap – generally under $50. Many view adopting such stoves as no-brainers, but Mobarak et al. (2012) found that people basically do not buy at all at full cost and are still reluctant to buy with significant subsidies. They concluded that upfront costs generate oversized responses, with a 1% increase in cost reducing the chance of buying by 8%–10%.[8] Bensch et al. (2015) found similar results for Burkina Faso, where the improved stove in question cost only $7.00! They and many other writers advocate free distribution as the best policy. Whether in high-income countries or low-income settings, present bias therefore appears to be a significant feature of human behavior that must be considered when considering collective action policies.

10.5 Using Laboratory Experiments to Evaluate Collective Action

Those who develop and use laboratory experiments pick out key features of reality and craft behavioral experiments that embody those features. Researchers then run the games to test specific hypotheses. As lab experiments only represent reality and are not themselves real-world contexts, their success depends wholly on how well games capture reality and elicit real human behaviors. The fundamental idea of lab experiments is given in Fig. 10.3.

BASIC IDEA

Understand the actual situation

By seeing how people play a simplified game that is similar to the real situation

Figure 10.3 How are Experiments Used?

The advantages of lab experiments are that they are laser-focused on the hypotheses to be tested and can be much cheaper than field experiments or surveys. A key pitfall of lab experiments is that if poorly constructed, they may not appropriately capture the key elements of reality and therefore may not offer much insight about the nature of real-world human behaviors. Making sure that experimental results are valid in the real world is called *external validity*.

Second, to be relevant, participants in lab experiments must act as they would outside the game. They must therefore take the game seriously and ideally allow themselves to be as fully immersed in the experiment as they would be in reality. If participants do not take their games seriously, we end up with the same hypothetical bias that is the bane of surveys. For this reason, externally valid games offer monetary payments or other economically mean- ingful rewards to participants. Those who take the game seriously earn more than those who do not.

Any lab experiment representing environmental collective action behaviors must include three key elements. First, because collective action involves groups, there should be more than two players. Second, there should be interdependencies between players. As is the case with real-world environmental investments, individual benefits depend not only on one's own actions, but also on the actions of other players. Third, as is true in reality, in the game private benefits should not perfectly line up with social benefits. This mismatch offers the potential for free riding on the efforts of other group members, which is the feature of reality we have identified as particularly inhibiting environmental collective action.

Figure 10.4 presents a simple game (courtesy of Astrid Dannenberg of University of Kassel in Germany) that has nothing explicitly to do with the environment, but nicely captures the tension between private and public interests. Suppose there are 25 people in a group playing the game, where the only choice is the color of a playing card to hand in. Players can hand in a black card, which has a large private payoff and no social payoff, or a red card that offers social benefits.

Which card would you hand in? Red card? Black card? What is the best play for the whole group? Before reading the discussion in the next paragraph, please try to

A Game Where Private and Public Interests Conflict

Suppose we play a game as a group.
Each person has two playing cards, a red card and a black card, then is asked to hand in one card.

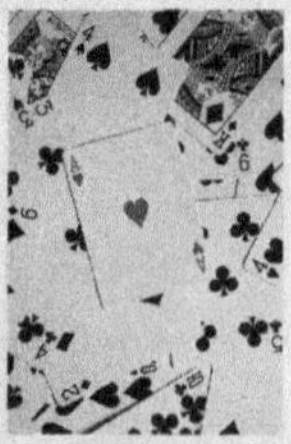

- If you hand in your **red card**, you get $1. In addition, you get $1 for every **red card** handed in by anyone else in the group.

- If you hand in your **black card**, you get $5. In addition, you get $1 for every **red card** handed in by anyone else in the group.

These rules apply to everyone in the group.

Which card do you hand in?

Figure 10.4 The Red Card-Black Card Game

understand the game and make the six calculations, two related to social benefits and four focused on private benefits, that are listed in Box 10.2. Try to understand the best strategy for the group and to what degree that play corresponds with the best private strategy.

Box 10.2 Calculations to Make in the Red Card–Black Card Game

1. Social Benefits: Amount the Group Gets if....
   a. Everyone turns in Red
   b. Everyone turns in Black.
2. Private Benefits: Amount You Get if You Turn in...
   a. Red and everyone else does the same
   b. Red and everyone else turns in Black
   c. Black and everyone else turns in Red
   d. Black and everyone else turns in Black.

The best strategy for the group as a whole is for everyone to play their red card, but playing the black card is the strategy to maximize private returns.[9] No matter what others in the group choose to do, from an individual perspective, playing black always dominates playing red. There is therefore a tension between private and group decisions that is similar to the prisoner's dilemma game. The Red Card–Black Card game merely extends these incentives beyond two players and, critically, offers a framework for testing collective action.

Playing red is socially beneficial behavior because benefits from the action multiply to others. On the other hand, playing black is selfish, because the returns from playing black are not shared and anyone who plays black free rides on the efforts of all those who have played red.

Using a standard deck of playing cards, try this game with your friends and see what happens! Just give each friend one red card and one black card and clearly explain the game using Fig. 10.4.

There are two related classes of games that are often relevant for analyzing environmental collective action problems. These are listed in Fig. 10.5. We are already acquainted with the prisoner's dilemma game, which can be useful for analyzing two-player situations. The common pool resource game is particularly useful when a resource is renewable, but depends on the stock, and the decision is how much of the resource to harvest in each period. The common pool resource game is especially applicable to hunting, fishing and forestry situations.

The most generally applicable game is the Voluntary Contribution Mechanism (VCM) experiment. These types of games attempt to simulate cases where groups must contribute valuable resources (usually effort or money) to a project that improves the welfare of the group. As we have already discussed, collective action can be very complicated if players do not agree on the value of the project. For this reason, most VCM experiments are phrased as generic projects that benefit everyone in known ways.

Two Games are Especially Relevant to Environmental Collective Action

- **Voluntary Contribution Mechanism** - players contribute to a public good, which is set up as a multi-person prisoner's dilemma game
- **Common Pool Resource Game** - players cooperate by not extracting too much from a resource that is accessible to all players, but is also depletable.

Figure 10.5 Environmental Collective Action Games

Free riding can be a key feature of all games that try to represent collective action and this is true whether we are talking about restrained hunting, reduced fuelwood collection, using fewer antibiotics to reduce antibiotic resistance, reducing air pollution emissions, using less packaging or picking up litter on the street. The #1 question to be addressed is how to reduce free-riding behaviors that can undermine collective action. But how bad is free riding? Obviously, if more than 70% of game players are free riders, we have a serious problem, but if it is only 6%, perhaps we can suppose that people basically cooperate.

Some colleagues and I ran a generic VCM game with 327 players in 21 communities in rural Nepal (Bluffstone et al., 2020). All players were part of formal or informal forest management groups. In each of the 21 sites, there were about 10–15 villagers who participated, but such a group size is really too large for people to operate strategically. We therefore randomized people into groups of three anonymous players. That is, each person was in a group with two others, but nobody knew the other group members' identities. Our particular VCM game was called a Public Goods Game or more precisely a Linear Public Goods Game.

The payoff to each player took the form shown in Eq. 10.1, where w = an endowment of money given to each player (100 Nepal Rupees in our case, which in 2013 was the pay for about 4 hours' work by a rural laborer), n is the number of players, g_i is the contribution of player i to the public good and α is the marginal return per person on total investment (i.e., the sum of the contributions of all n players)[10] in the public good (in our case, the marginal return was 0.50, indicating that each Nepali Rupee invested in the public good generated a return of 1/2 Rupee).[11]

As long as $1 > \alpha > 1/n$, if players assume Nash strategies by other players, there is a conflict between private and public benefits. The Nash equilibrium was therefore $g_i^{NE} = 0$, and the Social Optimum was $g_i^{SO} = w$. In other words, the best outcome for the group was that everyone contributed their whole w to the public good, but Nash play would suggest that nothing should be contributed.

This setup is shown in Eq. 10.1 and the full game instructions are included in the appendix to this chapter. Try it for yourself with your friends!

Equation 10.1

$$\pi_i = w - g_i + \alpha \sum_{j=1}^{n} g_j$$

So, what was the result of the game? Are people purely selfish? Purely altruistic? We found that the average contribution to the public good was 62.5 Rupees out of $w = 100$, which indicates that people are definitely not purely selfish (e.g., see Monbiot, 2015). Indeed, though not fully altruistic, the average person tends more in that direction. This is a very common result in the literature (Ostrom, 2000); people seem to have altruistic tendencies, which perhaps helps preserve common pool ecosystem services and may go a long way toward explaining why we have not already fully destroyed the environment!

The results of our game (and similar VCM games) turn out to be quite nuanced and are presented in Fig. 10.6. The figure is called a histogram. On the vertical axis is the fraction of respondents who contributed to the public good at each contribution level. On the horizontal axis are the various contribution amounts in Nepali Rupees. Figure 10.6 therefore tells us how common are various contribution levels.

We see two particularly common contributions: 100% and 50%. The first likely represents preferences of some respondents for altruism (100%) in which they commit to doing the right thing even at potential personal loss. The second spike is right at 50% and is perhaps driven by preferences for moderation and maybe equality and fairness; even though these

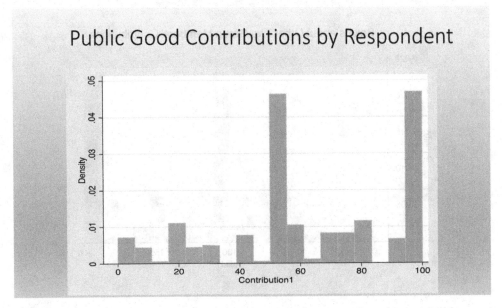

Figure 10.6 Public Good Contributions
Source: Bluffstone et al. (2020).

players may know that 100% contribution is best for their group, they may be a bit doubtful that their fellow players will fully contribute. They therefore contribute half. We also see minor contribution spikes at round numbers 20, 40, 60 and 80, with a wide distribution. That responses vary so much suggests players bring diverse preferences and experiences to the contribution decision.

We also asked respondents to predict the contributions of their two partner players. If they guessed correctly (and surprisingly, a number did guess right!), they received a 50 Rupee reward. Offering a monetary prize is standard practice because it provides incentives for respondents to try to predict others' contributions as accurately as possible. As shown in Fig. 10.7, the pattern is almost identical to Fig. 10.6 and our statistical analysis indicated that players' beliefs corresponded almost identically to their own contributions. Participants on average believed that their fellow group members would contribute about 60.2 Rupees, which was just below the actual average contribution.

This link between beliefs about others' contributions and one's own contributions is a second very robust finding of the VCM literature. It is certainly true that some people stubbornly calculated the Nash equilibrium of the game and contributed nothing. It is also the case that others thought it was right to give 100% and therefore did it. Most people, though, were *conditional cooperators* who contributed based on what they thought others would do.

Early in this chapter, we considered group projects in which we have been involved. Most of us can recall group efforts that for some reason got on track and worked well, and others that were total disasters. It turns out that collective action is highly *path* dependent. That is,

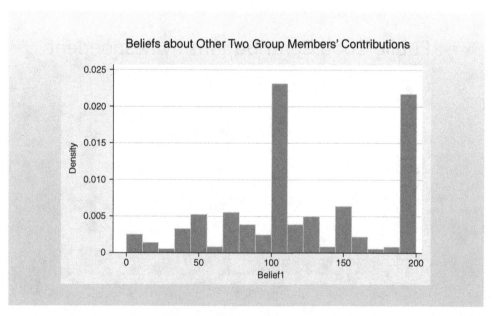

Figure 10.7 Beliefs about Others' Contributions
Source: Bluffstone et al. (2020).

a lot can depend on how projects get started. If things start well, the whole project can go well, but if the start is shaky, collective action can fall apart.

Thinking in terms of conditional cooperation, the reason for path dependence becomes a bit clearer. If conditional cooperators believe others are cooperating, they will cooperate, but if over time they become convinced that others in their group are not pulling their weight, they give up on collective action. That most people are conditional cooperators helps explain why collective action is so fragile.

This insight leads us to the third main result of most VCM experiments. Many researchers have found that when public good games are played over multiple rounds, average contributions decline and end up close to zero – the Nash equilibrium (Ostrom, 2000). Fischbacher and co-authors in a path-breaking paper published in 2001 said that this result occurs because the world is largely made up of conditional cooperators.

Conditional cooperators take cues from others' behaviors before settling on their own choices. Of course, if the world were made up only of conditional cooperators who were willing to try working together if others did the same or, better yet, mainly altruistic people who always stepped up, successful collective action would be the norm. Unfortunately, though, free riders who mooch off the contributions of others and conditional cooperators who don't quite fully step up and give what they should, are part of most games and real-life groups. When conditional cooperators get a whiff that they are being played by others, Fischbacher et al. found that they bolt and start to act like free riders. It is therefore the coexistence of many types of players along with a large percentage of conditional cooperators that make collective action unstable.

Challenge Yourself

Can you name three human behaviors related to the environment that offer evidence of conditional cooperation?

What does experimental economics see as the solution? In short, some sort of sanctions for bad behavior must exist to keep free riders and weak conditional cooperators in line. It turns out that tendencies to punish the antisocial are quite strong. Fehr and Gächter (2000) found, for example, that people are eager to hold laggards to account even if it is personally costly for them. They used an anonymous linear public goods game, such as the one we used in Nepal, except the game was played 20 times. They played the game with and without the possibility of punishment for bad behavior, where sanctioning others was personally costly for those imposing the penalties. They showed that contributions remained high only when people could punish free riders. Typical of their results is the graph in Fig. 10.8, which shows that voluntary contributions increase with the possibility of punishment but decline when no sanctions are possible.

Furthermore, they found that the people who were sanctioned the most were the ones whose contributions were the most outrageously below what average participants contributed. They concluded that it is the violation of *norms* created by participants in the

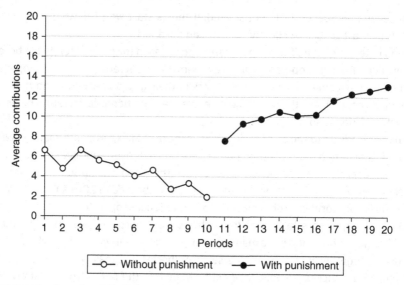

Figure 10.8 Voluntary Contributions with and without Punishment
Source: Fehr and Gächter (2000).

group – even if people do not know each other – which cooperative individuals are sanctioning. People find violating these norms objectionable and are willing to sanction others even if it costs them money to do it.

10.6 Chapter Conclusions

Collective action is at the very heart of environmental protection, but it is also extremely difficult. People may not agree on collective action goals and interest groups may have ideological or self-serving reasons why they support non-cooperative approaches to managing environmental assets. As common pool ecosystem services like climate stability are shared, each person advocating for them gains little compared with those who gain from private ecosystem services that conflict with them; common pool ecosystem services may therefore lack strong, focused advocates.

Even without differing objectives, collective action is difficult, because there are often private incentives to not cooperate. Probably we have all felt dumped on by slackers in groups and such behaviors can make us less willing to cooperate ourselves. Such interactions are often modeled as two-person prisoner's dilemma games and it is this game that forms the intellectual underpinning for predicting the so-called "tragedy of the commons" in collectively held natural resources.

Fortunately, it turns out that the Nash equilibrium of the prisoner's dilemma game and multi-player extensions like the VCM game is not the norm – at least in one-period games. Most people seem to have natural propensities for imperfect cooperation, but things can still go south when interactions repeat, and conditional cooperators feel they are being taken for fools by free riders. Collective action is therefore unstable and mechanisms, such as sanctions, are critical to sustaining cooperation.

We should not forget that national, regional and local governments are our main institutions for collective action. It is in the public sphere, through discussions of public funding and environmental policies, where the rubber hits the road on environmental cooperation. Non-cooperation therefore in practice means avoiding paying taxes, fighting environmental regulations and seeking special favors that have environmental implications. It is to the issue of seeking special favors in the form of subsidies that we turn our attention in the next two chapters.

Issues for Discussion

1. Can you think of someone who always says environmental protection should never be sacrificed? Is there someone you know who is openly hostile to doing anything for the environment? Why do you think they have such strong and perhaps emotional views?
2. Experimental evidence suggests that most people are conditional cooperators. Do you think this really translates to the real world or are most people out for themselves?
3. How important are interest groups in the environmental debate? Given that governments are our main institutions for collective action, do interest groups serve a useful purpose?
4. To what degree do you think that people are really conditional cooperators? Can you give two examples of conditionally cooperative behaviors?

Practice Problems

1. Below please find a two-by-two normal form prisoner's dilemma game of the type discussed in the chapter. This game shows the number of tons of mushrooms harvested in the long run depending on current harvest decisions. Mushroom harvesting using selective techniques is assumed to damage forests less, yielding more harvests in the long run, than harvests using non-selective harvest methods.

		Mushroom Harvester #2 Strategy (2nd number in pair gives their long run harvest)	
		Harvest using selective harvest techniques	Harvest using non-selective harvest methods
Mushroom Harvester #1 Strategy (1st number in pair gives their long run harvest)	Harvest using selective harvest techniques	20, 20	0, 30
	Harvest using non-selective harvest methods	30, 0	10, 10

a. What is the social optimum of the game? What is the Nash equilibrium? Please explain clearly how you know these are the answers.
b. Challenge yourself: Alter the matrix above so that the social optimum is the Nash equilibrium of the game. Please discuss what types of policies or other coordination mechanisms could help the mushroom harvesters achieve the socially optimal equilibrium.

2.   Get at least five of your friends together and try the red card–black card game. What are your results? To what degree are your friends cooperative and how many are defectors? To what degree do your results support the predictions of the prisoner's dilemma model?

Notes

1   See Hickman (2011) or https://www.youtube.com/watch?v=vAA2sLtzXJM from the Cornwall Alliance.
2   In addition, if environmental protection laws and regulations exist, consultancy firms are likely to arise to help implement those policies. These firms have financial stakes and can be powerful private sector environmental protection advocates.
3   See, e.g., Revkin (2009), Bach (2017) and Brulle (2014) for information on financing for individuals and organizations who produce information that seeks to deny the existence of climate change. See Busch and Judick (2021) for a discussion of the messaging of these organizations in the US and Germany.
4   See https://pr.princeton.edu/pr/news/95/q1/0126tucker.html
5   John Nash's main Ph.D. dissertation advisor at Princeton was Albert Tucker, who developed the prisoner's dilemma game. It was for this and related work that Nash won the Nobel Prize.
6   According to Merriam-Webster, a hypothesis is an assumption or concession made for the sake of argument.
7   This is one of the many topics covered in the fascinating book *Thinking Fast and Slow* by 2002 Nobel prizewinner Daniel Kahneman.
8   As we discussed in Chapter 7, such comparisons of dissimilar things (e.g., probability of buying and price) are done using percentage comparisons expressed as a ratio called *elasticities*. In the Mobarak et al. (2012) paper, the lowest elasticity is -8.0 = 8% reduction in probability of adoption for a 1% increase in price.
9   With 25 players, if everyone plays Red the group gets €1 * 25 + €1 * 24 * 25 = €625. If everyone plays Black, the group gets €5 * 25 = €125. If you play Red, you get €25 if all others play Red and €1 if all others play Black. If you play Black, though, you get €29 if all others play Red and €5 if all others play Black. Playing Red is therefore the social optimal strategy, but playing Black is the privately dominant strategy.
10   The Greek capital letter sigma denotes that all the g_j's are added up from the first player ($j = 1$) to the last ($j = n$).
11   I am indebted to my co-author Peter Martinsson for designing the game.

Further Reading and References

Allcott, H. and T. Rogers. 2014. "The Short-Run and Long-Run Effects of Behavioral Interventions: Experimental Evidence from Energy Conservation." *American Economic Review* 104 (10): 3003-3037.

Ayers, I., S. Raseman and A. Shih. 2012. "Evidence from Two Large Field Experiments that Peer Comparison Feedback Can Reduce Residential Energy Usage." *The Journal of Law, Economics, and Organization* 29 (5): 992-1022.

Bach, M.S. 2017. "Is the Oil and Gas Industry Serious about Climate Action?" *Environment: Science and Policy for Sustainable Development* 59 (2): 4-15.

Bensch, G., M. Grimm and J. Peters. 2015. "Why do Households Forego High Returns from Technology Adoption? Evidence from Improved Cooking Stoves in Burkina Faso." *Journal of Economic Behavior and Organization* 116: 187-205.

Bluffstone, R.A., A. Dannenberg, P. Martinsson, P. Jha and R. Bista. 2020. "Cooperative Behavior and Common Pool Resources: Experimental Evidence from Community Forest User Groups in Nepal." *World Development* 129: 104889.

Brulle, R.J. 2014. "Institutionalizing Delay: Foundation Funding and the Creation of U.S. Climate Change Counter-Movement Organizations." *Climatic Change* 122 (4): 681-694.

Busch, T. and L. Judick. 2021. "Climate Change–That is not Real! A Comparative Analysis of Climate-Sceptic Think Tanks in the USA and Germany." *Climatic Change* 164 (1-2): 18.

Fehr, E. and S. Gächter. 2000. "Cooperation and Punishment in Public Goods Experiments." *The American Economic Review* 90 (4): 980-994.

Fischbacher, U., S. Gächter and E. Fehr. 2001. "Are People Conditionally Cooperative? Evidence from a Public Goods Experiment." *Economics Letters* 71 (3): 397-404.

Hardin, G. 1968. "The Tragedy of the Commons." *Science* 162 (3859): 1243-1248.

Harrison, D. and D. Rubinfeld. 1978. "Hedonic Housing Prices and the Demand for Clean Air." *Journal of Environmental Economics and Management* 5 (1): 81-102.

Hickman, L. 2011. "The US Evangelicals Who Believe Environmentalism is a 'Native Evil'." *The Guardian*, May 5. Retrieved from https://www.theguardian.com/environment/blog/2011/may/05/evangelical-christian-environmentalism-green-dragon 10 July 2018.

Lutzenhiser, M. and N. Netusil. 2001. "The Effect of Open Space on a Home's Sale Price." *Contemporary Economic Policy* 19 (3): 291-298.

Mobarak, A.M., P. Dwivedi, R. Bailis, L. Hildemann and G. Miller. 2012. "Low Demand for Nontraditional Cookstove Technologies." *Proceedings of the National Academies of Sciences* 109 (27): 10815-10820.

Monbiot, G. 2015. "We're Not as Selfish as We Think We are: Here's Proof." *The Guardian*, October 14. Retrieved from https://www.theguardian.com/commentisfree/2015/oct/14/selfish-proof-ego-humans-inherently-good 27 March 2018.

Netusil, N., S. Chattopadhyay and K. Kovacs. 2010. "Estimating the Demand for Tree Canopy: A Second-Stage Hedonic Price Analysis in Portland, Oregon." *Land Economics* 86 (2): 281-293.

Olson, M. 1965. *The Logic of Collective Action: Public Goods and the Theory of Groups*. Harvard University Press: Cambridge, MA.

Ostrom, E. 1990. *Governing the Commons: The Evolution of Institutions for Collective Action*. Cambridge University Press: Cambridge, MA.

Ostrom, E. 2000. "Collective Action and the Evolution of Social Norms." *The Journal of Economic Perspectives* 14 (3): 137-158.

Revkin, A. 2009. "Industry Ignored Its Scientists on Climate." *New York Times*, April 23.

Runciman, W.G. and A.K. Sen. 1965. "Games, Justice and the General Will." *Mind*, New Series, 74 (296): 554-562.

APPENDIX
Public Good Game Denominated in Thai Bhat (about 33 THB/1$US)

We will now give you instructions and examples for the game. This is followed by a test in which we will check if you have understood the game or not. Once we are sure that you have understood the game, we will begin playing the game.

You have been divided into groups of three players based on the IDs we have given you. You will not come to know to which group you belong. Likewise, you will not come to know the identity of the other players in your group. Similarly, the other players will not come to know your identity.

At the beginning of the game, each player will receive THB 30 from us. Now you have to decide how many of the THB 30 to put into your pocket and keep for yourself and how many to put into a group project. You may put any amount between 0 and THB 30 into the project. Now we will show you how this is done. Please note that since this is an example, we will tell the player how many Bhat to put into the project. But when we play the actual game, you will have to decide this on your own, without any help from us. (*Randomly select a player* and give him THB 30. **Please make sure that each time YOU tell the person how much he should put into the project.** *Do not allow the player to take a decision because this may influence the decision of other potential players.*)

Suppose you are a player in this game. As mentioned before, you receive THB 30 from us. Now let us assume that out of THB 30, you put zero Bhat into the project. Please put zero Bhat into the project. *Ask the group*: Can you tell me how many Bhat there are in the project? How many Bhat does the player have in his pocket? Have you understood this?

Now, let us assume that out of 30, you put 5 Bhat into the project. Please put 5 Bhat. How many Bhat are in the project? How many Bhat does the player have in his pocket? (A: 25) I will now give you another example. Now, let us assume that out of 30, you put 18 Bhat into the project. Please put 18 Bhat. How many Bhat are in the project? (A: 18) How many Bhat does the player have in his pocket? (A: 12) *(Carry on this procedure for X = 10, 20, and THB 30.)*

Have you understood this part? Do you need additional examples? *(If yes, select another person and repeat the examples in the same order.)*

Any amount in the project will be increased by 0.5 Bhat for each Bhat in the project. For example, if you put 0 Bhat into the project, the project amount will be increased by 0 Bhat. Now, the final amount of money in the project is 0 Bhat. If you put 1 Bhat into the project, the project amount will be increased by 0.5 Bhat. Now, the final amount of money in the project is 1.5 Bhat. *(Carry on until THB 30 using THB 5, 10 and 30.)* I repeat, the project amount will be increased by 0.5 Bhat for each Bhat in the project. Have you understood this? Do you need additional examples? *(If yes, select another person and repeat the examples in the same order.)*

After the project money has increased, it will be divided equally between you and your two partner players, irrespective of how much you have put into the project (**Please repeat this again**). For example, if the project contains 0 Bhat, it will be increased by 0 Bhat and then divided equally between you and your partner players. However, since zero does not increase, both you and your partners will get zero Bhat from the project. For example, if the project contains 1 Bhat, it will be increased by 0.5 Bhat. Now the total value of the project is 1.5 Bhat, and the three of you get 0.5 Bhat each from the project *(Carry on until THB 30 using THB 10 value (A: THB 5 each), 20 value (A: THB 10 each) and 30 value (A: THB 15 each).* Have you understood this part? Do you need additional examples? *(If yes, select another person and repeat the examples in the same order.)*

Please remember that any money that you put into the project is first increased and then divided equally among the players in your group. Any amount that you put in your pocket remains the same. If you put 1 Bhat in your pocket, it remains 1 Bhat. It neither increases nor is it divided.

Your final earning from the game is the sum of the amount you have in your pocket and the amount you receive from the project.

We will now give you three examples. Please note that since now we are learning how to play this game, you can see the identity of each player as well as the decisions made by them. When we play the actual game, you will not come to know of this. **Do you understand this?**

We will now select three people and tell them to take the following decisions in the game. You are player I and you are player II and you are player III (*look for participants with weak comprehension and always give them a chance to act as player I and player II and player III*). We give you THB 30 each at the start of the game. Those who play together in this practice round will not play together in the game.

Example 1: Now we will see what happens if all three players put zero Bhat into the project.

Player I and II and III: Please put zero Bhat into the project. Now, can you tell me how many Bhat did player I put into the project? How many Bhat does he have in his pocket? (A: 30) How many Bhat did player II put into the project? How many Bhat does he have in his pocket? (A: 30) How many Bhat did player III put into the project? How many Bhat does he have in his pocket? (A: 30) How many Bhat are in the project? We have zero Bhat in the project. Since zero Bhat does not increase and cannot be divided, each player gets zero Bhat back from the project.

Player I has put zero Bhat into the project, so he has THB 30 in his pocket. He gets zero Bhat from the project. Can you tell me, what is his income? Since player I has 30 Bhat in his pocket and he gets zero Bhat from the project, his final income is THB 30.

(*Please repeat the procedure to calculate the income of the second player and the third player.*)

Example 2: Now we will show you the second example. You are player I, you are player II and you are player III (*select three other participants*). You get THB 30 from us at the beginning of the game. Now we will see what happens if all three players put THB 30 each into the project. Player I and II and III, please put THB 30 into the project. Now, can you tell me how many Bhat did player I put into the project? How many Bhat does he have in his pocket? (A: 0) How many Bhat did player II put into the project? How many Bhat does he have in his pocket? (A: 0) How many Bhat did player III put into the project? How many Bhat does he have in his pocket? (A: 0) How many Bhat are in the project?

We have THB 90 in the project. The project amount will now be increased by 45 Bhat since 0.5 Bhat is added for each Bhat we have in the project. In the project we have THB 90 and if we give 0.5 Bhat for each Bhat there is additional 45 Bhat. The final amount in the project is THB 90 + 45 Bhat = 135 Bhat. Now 135 Bhat is divided equally among the three players. So, each player gets 45 Bhat. Now, can you tell me, how many Bhat does player I have in his pocket? (A: THB 0) How many Bhat does he get from the project? (A: THB 45) What is his final income? (A: THB 45) We repeat, since player I has zero Bhat in his pocket and he gets 45 Bhat from the project, his final income is 45 Bhat. (*Please repeat the procedure to calculate the income of the second player and the third player.*)

Example 3: Now we will show you the third example. You are player I, you are player II and you are player III (*select three other participants*). We will see what happens if player I puts zero Bhat into the project and player II puts 15 Bhat into the project and player III puts 30 Bhat into the project. Player I, please put zero Bhat into the project. Player II, please put 15 Bhat into the project. Player III, please put 30 Bhat into the project.

Now can you tell me how many Bhat did player I put into the project? (A: 0) How many Bhat does he have in his pocket? (A: 30) How many Bhat did player II put into the project? (A: 15) How many Bhat does he have in his pocket? (A: 15) How many Bhat did player III put into the project? (A: 30) How many Bhat does he have in his pocket? (A: 0) How many Bhat are in the project? (A: 0 + 15 + 30 = 45)

We have 45 Bhat in the project. The project amount will be increased by 22.5 Bhat.

So, the final amount in the project is 45 Bhat + 22.5 Bhat = 67.5 Bhat. Now 67.5 Bhat is divided equally among the three players. So, each player gets 22.5 Bhat. Now, how many Bhat does player I have in his pocket? (A: 30) How many Bhat does he get from the project? (A: 22.5) So, what is his final income? (A: 52.5) We repeat, since player I has THB 30 in his pocket and he gets 22.5 Bhat from the project, his final income is 52.5 Bhat.

How many Bhat did player II put into the project? (A: 15) How many Bhat does he get from the project? (A: 22.5) So, what is his final income? (A: 37.5) I repeat, since player II has 15 Bhat in his pocket and he gets 22.5 Bhat from the project, his final income is 37.5 Bhat. How many Bhat did player III put into the project? (A: 30) How many Bhat does he get from the project? (A: 22.5) So, what is his final income? (A: 22.5) I repeat, since player III has 0 Bhat in his pocket and he gets 22.5 Bhat from the project, his final income is 22.5 Bhat.

We will now summarize the key results from these examples:

a. If all three players put zero Bhat into the project, they earn THB 30 each.
b. If all three players put THB 30 into the project, they earn 45 Bhat each.
c. If the first player puts zero and the second player puts 15 Bhat and the third player puts 30 Bhat into the project, the first player who puts zero Bhat earns 52.5 Bhat, the second player who puts 15 Bhat earns 37.5 Bhat, and the third player who puts 30 Bhat earns 22.5 Bhat.
d. If all players put the same amount into the project, they all earn the same income.
e. If you put less than what your partner players put into the project, you earn a higher income than they do.
f. If you put more into the project than your partner players, you earn a lower income.

If you have any questions, you may ask them now.
(If yes, repeat the examples in the same order.)

First Decision

Hello! Have a seat please. I hope you have understood the game. Your baseline public good game identity card, please? Here are your THB 30. Now you have to decide out of THB 30 how much you would like to put into the project. Please put that amount here. Put the money that you would like to put in your pocket here. *When participant has put their money then ask*: So, you would like to give X Bhat to the project and keep Y Bhat for yourself? *(Important: Do **not** allow the participants to take the money they put in their pocket with them when they leave the room. All the money remains in the room. Participants get their money only at the end of the experiment.)*

How many Bhat do you believe your two partner players **in total** will put into the project? (*Important: the participants should guess the **sum** of the other players' contributions.*) **Please think before you answer the question. If you guess correctly, you will receive additional 15 Bhat**. Please say a number between 0 Bhat and 60 Bhat. Thank you. Please do not discuss this with the other players.

11 Governments Messing with Prices and Agricultural Subsidies

---

What You Will Learn in this Chapter

- The ways governments affect market prices
- The consequences and challenges associated with governments influencing market prices
- How is water allocated in the US?
- About irrigation water subsidies that support agriculture in the arid American West

---

11.1 Recap and Introduction to the Chapter

We know from previous chapters that natural resource management is fraught with a variety of bad incentives that make it tough to sustain flows of common pool ecosystem services. There are several frameworks we can use to analyze these problems, including focusing on property rights or negative externalities, such as excessive use of sink ecosystem services that affect others. The most flexible way to think about the loss of critical ecosystem services is as a collective action problem. People have difficulties working together on joint projects – or may not even agree on what *is* the project – and our most important environmental problems are all big joint projects with a capital "P." Habitat and watershed protection, biodiversity conservation, reducing ocean acidification and stemming climate change are all fundamentally collective action problems; in some cases, it may be difficult to holistically think about these issues in any other way.

In the previous chapter, we considered challenges to environmental conservation viewed through a collective action lens. We found that some sort of accountability, such as group sanctions or comparative information, can be very useful for getting people to do the right thing. Lab and field experiments tell us that over time people are more likely to free ride on the efforts of others. Due to myopia (i.e., excessively short-run viewpoints) and other cognitive limitations, we may even not adopt measures that are in our own interests.

Of course, national, regional and local governments play the most critical role in facilitating collective action. They create and implement rules for businesses, individuals and households that try to curb the worst of our environmental moral hazard tendencies and they collect money for environmental investments.

DOI: 10.4324/9781003308225-13

But governments can also sometimes throw the environment under the bus. Governments have numerous priorities and sometimes natural resource protection does not even make the list. As a result, a variety of choices are made that can negatively affect environmental assets and especially the common pool ecosystem services they produce. One set of policies that can negatively affect natural resources is subsidies, which use public funds to benefit specific parts of economies. Subsidies often reduce prices of goods and services, including for provisioning ecosystem services.

As we have discussed on several occasions in this book, reducing prices increases quantity demanded. But how do subsidy policies operate and how can they affect natural resources? The remainder of this chapter and Chapter 12 try to address this question with regard to agriculture and energy. To develop our theoretical understanding of subsidy policies, we first turn our attention to how governments can influence or set prices outside the world of natural resource management. After considering these more general applications, we will discuss how governments subsidize agriculture, sometimes to the detriment of natural resource quality.

11.2 Messing Directly with Prices

In previous chapters, we emphasized the importance of prices, which steer resources around economies. High prices for consumers, especially combined with low costs for producers, tell businesses that an activity is profitable, a good area for investment and perhaps it is time to start hiring workers. Prices also give information about scarcity. In Chapter 3, we asked whether a candy bar or a Mercedes-Benz was on average scarcer and more valuable and, of course, concluded that the Mercedes was both scarcer and more valuable. We know this because both goods are transacted in markets and the price of a Mercedes is maybe 30,000 times more than a candy bar. This high price reminds us every day that the Mercedes is highly desirable and contains components and engineering knowledge that are very scarce.

In general, we have problems when prices do not reflect scarcities. People get wrong signals and either excessively avoid buying goods and services, because the prices are high, or buy too much because prices are low. When price signals are wrong due to underpricing, businesses react and do not conserve subsidized inputs. They may also not choose the most appropriate mix of inputs. For example, businesses may substitute capital and energy for labor, if for some reason those inputs were cheap.

Challenge Yourself

Can you think of three markets you are part of that might be distorted? What evidence can you point to?

If prices are distorted, businesses can be expected to produce products that may not reflect economic scarcities. Consumption patterns will also differ from what scarcities suggest should be the right consumption mix. For example, if gas prices were below those suggested by scarcity, people would probably buy more SUVs, pickup trucks and Hummers instead of electric cars. They would also drive more, take the train less and let their bikes gather dust. Such twisting of economic behaviors compared with what scarcity would suggest people should be doing is called *economic distortion*. As distortions generally affect and work through prices, they are often referred to as *price distortions*. These distortions create economic inefficiencies, such as producing the wrong products with the wrong mix of inputs. Though we can and will point to cases where distortions could make sense, they should in general be viewed with skepticism.

One rather extreme way governments distort prices is by requiring minimum prices or setting maximum prices. If prices are held below the competitive equilibrium, we will have what is typically called a *price ceiling* - a maximum price above which governments say prices cannot go. With price ceilings, not only will people want a lot, but the low prices will starve that industry of financial resources. As a result, firms would be expected to not want to supply as much as people want to buy, creating the possibility for shortages. We don't generally observe market shortages in places with free markets, good supply chains (which bring together all the inputs needed to produce goods and services) and good institutions to deliver finished goods to markets.[1] When governments rather than markets dictate private sector prices, however, we might see persistent shortages. Figure 11.1 shows a price ceiling situation based on Fig. 3.3 in Chapter 3.

Let us remind ourselves of our simple, but potentially powerful model of human behavior, which assumes buyers try to get good deals and sellers attempt to earn high profits. Getting a good deal means that buyers' values of market goods are at least as high as the price they have to pay. In other words, if price > a buyer's marginal benefit, consumers are not getting good deals! On the supply side, sellers are only able to make a profit selling a market good if prices are at least as large as the marginal costs of supplying goods and services offered for sale.

At the price ceiling (P_2) in Fig. 11.1, all quantities up to quantity B have buyer marginal benefits > P_2. We therefore predict that at price P_2, buyers will want to buy quantity B. At that same price, though, sellers are predicted to want to sell a quantity of only A. As B > A, we have a shortage in the market. As we discussed in Chapter 3, because of this shortage, there will be upward pressure on prices. At quantity A, for example, which is expected to be the amount sellers are willing to offer for sale at P_2, the marginal benefit (as read off buyers' marginal benefit curve) is substantially above P_2. The only problem, though, is that with

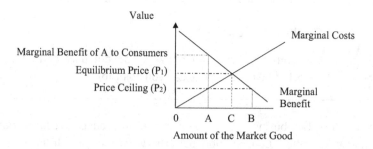

Figure 11.1 Price Ceilings

the legal price ceiling, prices cannot rise to the equilibrium of P_1. Furthermore, the quantity will not increase to the equilibrium at C, where there is no shortage or surplus; with price ceilings, shortages can persist.

An example of such government-mandated prices is rent control, which seeks to make housing affordable by controlling rents. Under strict rent control, municipalities set prices for different categories of apartments that are below equilibrium. This means the controlled prices are *binding* on the market. Usually though, rent-controlled apartments are difficult to get (Gyourko and Linneman, 1989) and relatively poorly maintained (Downs, 1988), which calls into question whether strict price-setting rent control actually helps renters.

An episode of the television show *Seinfeld* illustrates the point. Jerry mentions to Elaine that an older couple is moving from New York to Florida. Elaine immediately dials into the question of what happens to the couple's large, well-located, rent-controlled apartment. Her efforts to get the apartment are frustrated, though, when (after many laughs for the audience) she learns that the couple is passing it on to their son. The lesson is that with a deal that sweet, outsiders don't get access.

San Francisco and New York City in the US have systems of rent control, as did Cambridge, Massachusetts prior to 1995 (Autor et al., 2014). In 2019, Oregon passed first-in-the-nation statewide rent control, but the law does not specify prices. Instead, it limits increases in rents and includes other protections for tenants (State of Oregon, 2019). A key reason the law does not directly set prices is the government did not want to excessively distort rental markets, potentially creating shortages.

Box 11.1 Rent Control in Practice

Until the mid-1990s, San Francisco exempted small multi-family rental units from rent control. This exemption applied to nearly half of San Francisco housing and was exploited by large rental businesses. In 1994, San Francisco removed the exemption, bringing rent control to a new stock of housing. Stanford researchers found that while the policy did lower rents below market prices, it also limited the mobility of renters who feared their rent would increase if they moved.

Source: Diamond et al. (2019).

It is relatively unusual for governments to simply mandate minimum prices, which are called *price floors*, because they cannot force people to buy at above-equilibrium prices. For example, the US government is not going to specify a minimum price of gasoline to help the oil companies, just as it is not going to set a minimum price for California almonds.

The exception is minimum wages, which seek to help people working in low-wage markets or what the Bureau of Labor Statistics (BLS, 2018) refers to as the "working poor." Current policy proposals in many cities in the US, including Portland, Oregon (State of Oregon, undated), Seattle (City of Seattle, undated), New York (New York City Department of Labor, undated) and San Francisco (City of San Francisco, undated), have minimum wages moving to $15.00, if they are not there already.

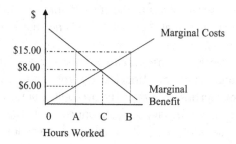

Figure 11.2 Price Floor in a Low-Wage Labor Market

Though higher wages in low-wage labor markets to some could themselves be contro-versial, many people would be in favor of increasing the wage rate per hour from $7.25 (the current US government federal minimum wage) to $15.00 if it would lift workers out of pov-erty. Rather than taking issue with the goal of higher wages, critics often instead focus on *unemployment*, which is the term we typically use for surpluses of labor, and which could potentially be increased by a minimum wage.

The situation is shown in Fig. 11.2, where the minimum wage is set at $15.00 per hour, but the equilibrium wage rate in the low-wage labor market is $8.00. $15.00 turns out to be the willingness to pay by firms at hiring level A, as read off the marginal benefit curve (remember, firms are the "buyers" of labor and workers are "selling" labor). Looking at what workers give up to enter the labor market (i.e., their marginal costs), at hiring level A workers are giving up time and effort valued at $6.00 per hour, which is much less than $15.00. Indeed, at a $15.00 wage rate, they are willing to supply B hours. As B > A, our model predicts that this binding price floor will cause oversupply of labor and unemployment, because firms will only want to hire A workers. The amount B minus A is the number of labor hours people want to work, but cannot work, because of unemployment.

Box 11.2 The Extent of Minimum-Wage Work in the US

The US government federal minimum wage is $7.25 per hour. The US Bureau of Labor Statistics (BLS) estimates that in 2022, about 1 million workers were paid the minimum wage or less. This estimate does not count overtime, tips or commissions.

Source: BLS Reports, August 2023.

Workers are, of course, not goods or services. They may be motivated to do good jobs or stay with a company if wages are high and firms are often well-regarded for paying living wages. Firms may, therefore, get real business benefits, including higher worker productivity and public relations value, from paying above-equilibrium wages.

By contrast, firms do not get such benefits from paying more for other factors of produc-tion. For example, nobody is going to congratulate a firm for paying a lot for rent or overpaying for machines. Unlike other factors of production, workers can also be "motivated," which

helps explain why it is unclear to what degree minimum wages actually create unemployment. Not surprisingly, the topic is hotly debated in the academic literature and in the
popular press (Scheiber, 2018).

11.3 Agricultural Price Supports

Even if governments do not go so far as to directly set prices, they certainly are interested
in what goes on in markets and the prices that result from them! Because of this interest,
governments support economic sectors all the time. They try to boost agricultural production, bolster economic sectors that are in decline, support green technology, promote renewable energy, encourage energy efficiency and much more.

Using government resources (either borrowed or from taxes), other resources or
regulations to boost an industry is called *subsidization* and is an important method for
promoting activity. Subsidization can be done in ways that increase or reduce key prices,
twisting them away from what scarcities might dictate. Modern governments do not want
to create shortages and surpluses when they distort prices, so rather than price-fixing, they
often use policies such as taxes, direct subsidies, trade barriers, low-priced natural resources
and government regulations to try to help parts of their economies.

Subsidies affect prices and therefore we want to keep in mind the important scarcity
information prices offer buyers and sellers. We might also want to recall our discussion of the
externality view of environmental problems from Chapter 9, as well as our externality model
in Fig. 9.2. This is a great place to start, because the bottom line of that discussion was that
when environmental (or other) negative externalities occur, a tax can *internalize* those externalities. If set correctly, which is admittedly a big "if," the tax reduces the quantity bought
and sold to a level that truly balances benefits and costs, including non-marketed environmental costs. As was discussed in Chapter 9, this human welfare- or net benefit-maximizing
quantity is known as the *efficient* level. The tax also yields a price that shows the true scarcity value of the good or service, which "gets the prices right" in the market.

A critical role of governments is to use public resources to support collective action
goals. Public expenditures are therefore not necessarily subsidies. For example, assistance
to the underprivileged uses public funds to give people a hand. We can think of some public
expenditures as supporting *positive externalities*. For example, public libraries, transit, public
schools, infectious disease control and roads all have positive externalities that could justify
public support. Viewed through a collective action lens, these are all also potential examples
of public expenditures that support collective action.

Whether some other expenditures are distorting subsidies may be a bit more debatable.
For example, there may be promising technologies that are underutilized (e.g., green technologies, LED light bulbs, heat pumps, LPG cylinders in rural areas and fertilizers in some
countries in Africa) that governments may want to jumpstart. Some may see such moves as
legitimate policies to scale up technologies and move them to market, but others could view
them as distorting otherwise well-functioning markets.

**Going forward in this chapter and the next, please make your own judgements about
whether using public funds for the purposes discussed makes sense. Consider what
would be your criteria for deciding when governments distort markets.**

At their most basic level, subsidies do the opposite of getting prices right. They use governments' regulatory and budgetary authority to try to achieve goals by moving competitive market prices away from scarcity values. Governments have many ways, both *explicit*, using tax revenues and borrowed money, and *implicit*, using other policies, to subsidize the private sector. The sheer number of possibilities indeed makes measuring subsidies challenging. In higher-income countries, where well-functioning markets are often thick with many buyers and sellers, governments typically (but not always, as we will see!) focus on implicit subsidies. Outside the high-income world, explicit subsidies may also be very important, sometimes with important environmental implications.

As we learned in Chapter 3, our supply-demand model suggests that equilibrium prices can be increased by either increasing demand or reducing supply. Between 1949 and 2014, the US federal government operated several programs with the goal of supporting the price of milk (Blayney & Normile, 2004, p. 43). One program to increase demand was the Dairy Product Price Support Program, which set a fixed price for milk that was typically slightly above the market equilibrium price. The government committed to buy any surplus milk at this price, so producers no longer had to worry about low and/or highly variable prices. This policy likely led to larger and more stable milk markets.

The US government supported the price of milk by essentially entering the market as a big buyer offering a guaranteed price. This meant that the private milk market had to match the government's price. Milk bought by the government was refrigerated, dehydrated and made into other products, such as butter and cheese, which were stored and donated in the US (hopefully to people who otherwise would not buy milk) and abroad. US Department of Agriculture (USDA)-administered dairy price supports cost US taxpayers about $1.3 billion per year during the first half of the 2000s and are estimated to have increased the incomes of US dairy farmers by $1.9 billion per year during the same period. The system helped ensure price stability, because the US government guaranteed the price floor, reducing volatility by an estimated 15% (Blayney & Normile, 2004).

Figure 11.3 incorporates policies to increase milk prices into our standard model of markets using actual numbers from the US in the 2000s. The government entered the milk market and committed to buy any surplus milk at $1.17 per gallon. This policy on average drove up retail milk prices by 1.5% (Blayney & Normile, 2004), which was the intention of the policy.

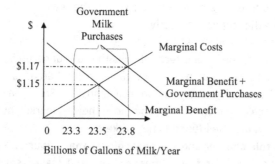

Figure 11.3 Demand-Side Dairy Price Support Policy

Notice that the equilibrium quantity of milk production ("equilibrium" because there was no shortage or surplus) was 23.8 billion gallons per year rather than 23.5 billion if the market were subsidy-free. Private milk purchases declined from 23.5 billion gallons to 23.3 billion gallons (Andreyeva et al., 2010) as the price rose due to the subsidy. Note that we can easily approximate the cost of the subsidy program to taxpayers as the number of gallons the government purchased times the final retail price.

Before reading on, ask yourself whether the dairy price support policy made sense. It helped dairy farmers, but do you see those results as sufficient? What other distortions could this policy have caused? In other words, other than buying and selling more milk, how might this policy affect related markets and natural resources? How might the policy affect the mix of ecosystem services produced by environmental assets? What common pool ecosystem services might especially be affected?

Challenge Yourself

What are some government services that may be given up when the US government uses resources to subsidize dairy and other farmers?

The high price of milk combined with the guaranteed government purchases made the dairy industry more financially attractive than without the subsidy. This should have increased herd sizes and led to technical change in the dairy industry, such as better milking technologies and cows that produced more milk. The USDA estimates that dairy price support programs increased the number of dairy cows by 1.5%–2%. With these additional cows came increased environmental effects. The Food and Agriculture Organization of the United Nations (FAO) notes that cows are an especially large source of greenhouse gas (GHG) emissions, because they emit methane, a particularly powerful GHG. Aside from the cows themselves, manure processing and milk transportation result in additional GHG emissions. The amounts are not necessarily small. Dairy cows account for 2.9% of *all* human-related GHG emissions (FAO, undated).

The effects don't end with the dairy industry itself. All those cows need to be fed, so the program also increased the area of land dedicated to growing corn, hay and other feed stocks. Increased cattle feed production in turn increased water use and irrigation. Alfalfa, a common hay crop, requires large amounts of water. In 2000, California provided $70 million in water subsidies to the alfalfa industry (Kuminoff et al., 2000).

The Dairy Product Price Support Program was an example of a demand-side subsidy program, but many crop-based agricultural subsidies focus on the supply side. Production quotas are an important supply-side policy that limits the production of farmers. Early US programs that paid farmers to leave land fallow, known as acreage reduction programs, intended to raise prices by reducing supply. But such programs suffered from low participation rates, were difficult to enforce and often failed to stabilize prices.

The primary acreage/crop supply reduction program in the US today is the Conservation Reserve Program (CRP). The CRP is an explicit agricultural subsidy that aims to take land out of agricultural production and support prices, while increasing the habitat values associated with American agriculture. Established in 1985, the CRP pays farmers to remove environmentally sensitive land from agricultural production for periods of 10-15 years (USDA Farm Service Agency, undated). Unlike previous acreage reduction programs, the CRP takes into account the environmental benefit of removing agricultural land from production - a positive externality - which allows the government to prioritize environmentally sensitive land.

In exchange for enrolling in the CRP, as of the late 2010s the government paid farmers an average of $85 per acre per year. In 2018, a total of $1.9 billion was paid to farmers and 22.4 million acres were under contract (USDA Farm Service Agency, 2019). The environmental benefits of the CRP include improved water quality due to reduced runoff, lower GHG emissions, improved wildlife habitat, enhanced soil productivity and reduced soil erosion and flood damages (Barbarika, 2012).

11.4 Government Subsidies to Agriculture: Cheap Irrigation Water

Not surprisingly, many people think taxpayer dollars should not be used to pay farmers not to produce (Lauinger, 2018) or to overproduce (Baur, 2019), and some even take issue with the CRP (Ogburn, 2011). Implicit subsidies do not directly pay producers or buy or sell their products, but instead use other ways to help them. In the US, corn ethanol is common in gasoline because of regulations that generally require that 10% of the volume of motor vehicle gasoline consumed in the US be made from ethanol (EIA, 2020). The regulation helps the US meet requirements in the Clean Air Act, but it also helps corn farmers, mainly in the US Midwest, by increasing demand for their product, driving up equilibrium prices and quantities.

Box 11.3 Low-Cost Logging Rights in the Global South

All countries face a range of potentially conflicting objectives in managing forest resources. Governments may seek to maximize timber revenue, maintain ecosystems or foster economic development. In much of the world, economic development is prioritized - sometimes at the expense of environmental sustainability. This focus may result in selling timber rights at below competitive prices. The short-term effects of such policies are increased timber production and lower prices. The long-term effects can include deforestation, timber scarcity and higher prices.

Source: Gray (1983).

Another way governments can implicitly subsidize the private sector is by giving them sweet deals on government-owned natural resources. Governments often control key resources and therefore can offer businesses and indirectly consumers below-market prices, which is an important type of implicit subsidy.

Examples of such subsidies related to natural resources include offering below-market prices for rights to harvest timber, graze cattle on public lands, drill for oil or extract minerals.

All these marketable provisioning ecosystem services come from land and they conflict in important ways with other land-based ecosystem services. For example, when logging companies take advantage of low-cost or no-cost rights by clear-cutting timber, this reduces common pool habitat values. Low-cost rights to graze cattle on US government land in the American West subsidize meat production, skewing people's diets and potentially interfering with land-based ecosystem services, such as soil conservation, camping and endangered species protection.

Water is a very important natural resource that governments often control. In most countries, waterways are publicly owned and often governments can decide who receives water and at what prices. They may also provide critical infrastructure for delivering water to users and get to decide how to price that infrastructure. Both the pricing of the resource and water infrastructure offer the possibility for implicit and even explicit subsidies to farmers and others who need water.

In the US, and most countries around the world, the largest user of delivered water is agriculture, mainly for irrigation. The arid American West may get only 10 inches (25 cm) of rain or fewer per year, and a key complication with promoting agriculture in the western US has always been that so much of the landscape is arid. General John Wesley Powell famously predicted after his exploratory mission to the southwest US between 1867 and 1872, "many droughts will occur; many seasons in a long series will be fruitless; and it may be doubted whether, on the whole, agriculture will prove remunerative" (Powell, 1879, p. 3). Developing agriculture in the arid US West requires potentially complicated and expensive diversions of surface water. If water in the US West is so scarce and costly, why is so much food grown there? Answer: Government subsidies.

Before talking about irrigation subsidies, let us first explore how water rights are allocated in the arid US. In US states, such as California, Colorado and New Mexico, going back to the 19th century, water rights were allotted to European immigrants who claimed them first and systematically excluded Native Americans, Asians and other groups. That is, water was allocated on a *first-come-first-served* basis to those of European ancestry and those rights were then sold, resold and passed down to contemporary users. For example, orange growers in Southern California may have claimed water rights in the 1860s, but those rights could now be owned by the City of Riverside, California, which needs to supply over 320,000 people with water every day.

Water rights are long-lived and, depending on the US state, either can be renewed in perpetuity or need not be renewed at all. In some states, water rights are secure only if water is put to "beneficial use," which is often interpreted as human-focused uses, such as agriculture, landscaping and drinking water. While the intent of this "use-it-or-lose-it" system was perhaps to encourage economic development and discourage hoarding, it also resulted in overuse by those who feared losing their rights. This incentive can, for example, leave some fields flooded with more water than needed, while farms and towns farther downstream face water shortages.

The above discussion relates to US water rights, which probably have important incentive effects, but they are not subsidies. Irrigation subsidies in the arid US West go back to the 1860s and predate Powell's predictions. During the Civil War, the international cotton market was seeking new suppliers in light of the Union blockade of Southern exports. This situation caused local governments in Arizona to offer cash and water subsidies to new cotton farmers (Lustgarten & Sadasivam, 2015).

Related subsidies remain in place today and include below-cost irrigation water, low-interest government loans, loan forgiveness during bad crop years and insurance against increased costs. These policies help existing farmers, and draw new producers into the market. Irrigation subsidies in particular may create important environmental effects due to lower in-stream flows.

California provides an important example of agricultural subsidies via cheap government-supplied irrigation water. Not long after the passage of the US Reclamation Act of 1902, the Central Valley Project (CVP) was created. The CVP is the largest water storage and delivery system in California and the largest federal water project in the US. It covers nearly 500 miles from north to south in California's Central Valley, transforming millions of acres of desert into fertile farmland (EWG, 2004).

While it is often hard to find good data on irrigation water use,[2] and water pricing is heavily dependent on time, place and quality, a 2002 report by the non-profit research and advocacy group Environmental Working Group (EWG)[3] can help us understand something about the irrigation subsidies offered via the CVP. According to EWG, in 2002 California taxpayers and the CVP provided more than 6800 Central Valley farms with water through subsidies worth more than $700 million in 2023 dollars. CVP farmers were consuming about 1/5 of all the raw water used in California at a small fraction of the prices others paid (EWG, 2004).[4]

EWG (2004) notes that the CVP has a limit of 960 acres per farm,[5] but by subdividing large farms into smaller farms the largest 10% of farm owners received over 67% of the water. The water subsidies make planting water-intensive crops, such as rice, cotton and alfalfa, economically feasible in a region that would naturally be very dry. According to EWG (2004), CVP water diversions reduced or eliminated the ecosystem services from more than 700 miles of rivers and lakes and over 55,000 acres of wetlands and estuaries in the Central Valley.

Table 11.1 presents EWG (2004) data on the top ten CVP water and irrigation districts[6] that provided water to farms, ranked by the amount of subsidies given to them in 2002. All costs have been updated to 2023 values using the US consumer price index. As shown in the second column, the unit of irrigation water in the American West is the acre-foot, which is the amount of water that is spread over one acre of land one foot deep.[7] Column 3 shows the amount paid by these water districts to the US government and column 4 shows the EWG estimate of the subsidy provided to the water district, which is always positive, meaning that all irrigation/water districts paid less for water than it cost to provide it to them. Annual subsidies ranged from $16 million to $40 million per district and were sometimes much more than the amount districts paid for their water.

As shown in Table 11.2, which presents a selection of water prices, based on actual prices paid in 2022, residential and commercial users pay the most for water.[8] We see that the two irrigation districts paid about $20 per acre-foot of water, while the two water districts that supplied non-agricultural users paid 20–40 times that amount per acre-foot (Poole, 2023). Notably, members of the Glenn-Colusa Irrigation District, which as shown in Table 11.1 received over $25 million in subsidies in 2002 (in $2023), paid only about $22 per acre-foot.

Table 11.1 Top Ten Central Valley Project Water Districts Ranked by Subsidy Value (2023 $US)

Water District	Amount of CVP Water Purchased (Acre-Feet) in 2002	Number of Farms (2002)	Amount Paid to the US Government for Water (2023 $US)	Cost to Supply the Water Minus the Amount Paid (i.e., the Subsidy Amount) (2023 $US)
Westlands Water District	721,258	422	$22,793,197	$40,630,000
Madera Irrigation District	130,154	541	$3,932,935	$33,830,000
Delano-Earlimart Irrigation District	125,156	93	$4,042,114	$32,300,000
Sutter Mutual Water Company	105,391	73	$850,794	$29,750,000
Lower Tule River Irrigation District	112,727	188	$3,379,668	$29,240,000
South San Joaquin Municipal Utility District	105,549	155	$3,514,267	$27,030,000
Glenn-Colusa Irrigation District	90,207	388	$521,856	$25,670,000
Chowchilla Water District	93,611	209	$2,683,312	$24,480,000
Del Puerto Water District	90,693	115	$3,215,657	$23,120,000
Panoche Water District	63,637	45	$1,692,554	$16,745,000

Costs updated to 2023 $US using the US consumer price index between July 2002 and July 2023 of 1.70 (i.e., consumer prices rose by 70% between 2002 and 2023), available at https://data.bls.gov/cgi-bin/cpicalc.pl

Source: EWG (2004).

Table 11.2 Selection of California Water Prices in 2022

Water District	Price Per Acre-Foot	Notes
Glenn-Colusa Irrigation District	$21.79	Base rate
Central California Irrigation District	$20.00	Rate for first 2.5 acre-feet per acre
Metropolitan Water District of Southern California	$799	Tier 1 full service untreated water
Westland Water District	$432.37	Wholesale rate for untreated water for low-income communities

Source: Poole (2023).

Water subsidies are certainly not unique to the US. In the 1960s, present-day Uzbekistan, which was then part of the Soviet Union, began a centrally planned system of irrigation projects to promote cotton farming. A critical element of the initiative involved diverting water from the Amu Darya River, which naturally flows into the Aral Sea (Gurney, 2012). This set of projects resulted in a major expansion of cotton production, from 2 million tons in 1950 to more than 6 million tons in 1980 (Djanibekov et al., 2010). After the dissolution of the Soviet Union in 1991, the network of cotton producers was privatized, but the government retained significant control over water resources and water continues to be provided to cotton farmers (Yuldashbaev, 2010).

The environmental effects of these publicly provided irrigation projects have been significant. The Aral Sea has lost much of its primary water source and between 1960 and 2000 the water volume fell by 75% (Encyclopedia Britannica, undated). The size of the Aral Sea declined dramatically over time and as of 2014 was just a fraction of its original area. Reduced water volume, combined with agricultural chemical use, generated high levels of water toxicity and potentially important human health effects tied to pesticides in dust left after the Aral Sea receded. As O'Hara et al. (2000) noted in their article published in the medical journal *The Lancet*, "The Aral Sea region is one of the world's foremost ecological disaster zones and there is increasing local concern for the health of millions of people living in this region" (p. 627).

Let us now consider the effects of water subsidies using our standard marginal benefit – marginal cost model. As shown in Fig. 11.4, subsidizing an important production input like irrigation water reduces the cost of agricultural crops, such as almonds, at all quantities produced. Our market model then predicts that such a policy would increase the equilibrium quantity of almonds. It should also reduce the price compared with what otherwise would have occurred. As shown in the figure, we predict that the quantity of almonds produced would be B rather than the unsubsidized market equilibrium = C. Further, equilibrium price per pound is, for example, $4.25 rather than $5.00 per pound. Our market model analysis therefore suggests that water subsidies will distort agricultural markets, over-use water and increase equilibrium quantity.

But how do subsidies affect the mix of ecosystem services produced by natural resources? We recall that environmental assets typically produce a number of ecosystem services, some of which conflict with each other. For example, when river water is diverted for irrigation, it will change the riverside habitat,[9] reducing biodiversity values like birdwatching opportunities. To the extent that almond production is supported by subsidies, it affects competing ecosystem services.

Subsidizing water for agriculture would be expected to tilt the balance of ecosystem services in the direction of provisioning ecosystem services, such as crops. Suppose the first-best market-focused alternative use of water diverted from a river to almond orchards was

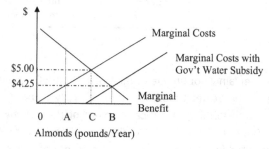

Figure 11.4 Government Irrigation Subsidies

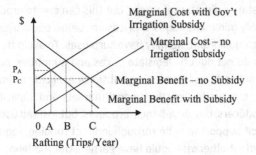

Figure 11.5 Government Irrigation Subsidies and Rafting Services

in-stream flows that support rafting tourism. Let's look at irrigation subsidies from the perspective of river rafting. River rafting requires sufficient water and beautiful riparian scenery. As more water is diverted to irrigation due to subsidies, at some point offtake reduces the ability of the river to supply rafting services. For example, rafting companies may not be able to offer trips on all days or in all seasons. They may also need to offer trips in different, more costly areas. Water subsidies create a less-than-level marketed ecosystem service playing field that could increase costs for rafting companies.

If the quality of the river declines because of increased irrigation, it can also make each rafting trip less enjoyable. For example, with lower water flows trips may be less thrilling and the scenery not as attractive. Due to irrigation subsidies, rafters may get lower marginal benefits and therefore may be less willing to book trips at all possible prices.

Figure 11.5 models the marginal benefit and marginal cost effects of an irrigation subsidy. First, the increase in cost to rafting companies is modeled as a shift in the marginal cost to the left, meaning that costs are higher at all quantities. This effect reduces our equilibrium quantity from C to B and increases the price of rafting trips. Drawing down the river flow for irrigation can also make rafting trips less exciting and impact valuable riparian habitat. This effect would reduce the benefit that rafters get from a trip on the river. We include this effect by shifting the marginal benefit curve to the left. This effect further decreases equilibrium quantity (from B to A) and pulls rafting trip prices down.

Please note that we do not know how the final equilibrium price – with both marginal benefit and marginal cost shifts included – compares with the initial price, because it depends on whether the supply- or demand-side effect is stronger. The bottom line, though, is when environmental assets are degraded – as they often are – due to extractions and/or pollution, ecosystem services that are sensitive to natural resource quality can be hit with the double whammy of both increased costs and reduced benefits.

11.5 Chapter Conclusions

Governments are our #1 institution for addressing the important moral hazard problems discussed in this section of the book. Governments have many priorities and environmental protection may not always be #1. As a result, natural resources may be harmed in the process of pursuing other interests.

Governments sometimes directly set prices, but this can create problems as in the case of rent control and possibly minimum wages. If prices are below equilibrium, we can end up with shortages and if above equilibrium, we may have surpluses. To avoid these types of problems, governments typically do not directly legislate prices and quantities, but instead affect them indirectly through subsidies. Governments can subsidize explicitly or implicitly, where explicit means using budget resources to directly support markets and implicit means governments do not directly pay producers, buy or sell their products, but instead use other policies to help producers. Such implicit support may be through indirect funding, regulations, and by giving up returns the government otherwise could have earned. Our modeling suggests that policies such as subsidies affect costs, change prices, can generate economic distortions and potentially damage natural resources and reduce some types of ecosystem services.

A key way to subsidize agriculture when rainfall is insufficient is by providing water below its scarcity value and even below cost. Such policies distort decisions and cause overuse of irrigation water. They can also tilt the balance of ecosystem services provided by rivers and streams dramatically in favor of agriculture.

Agriculture is a favorite target for government largess because farming is such an important economic sector and securing the food supply is critical. It is hardly the only priority, though. Chapter 12 focuses in on levels and effects of fossil fuel subsidies around the world. It also looks at the case of the Soviet Union, where all prices, including those for energy, were set by the government. This pretty much unlimited government control of prices yielded enormous economic distortions and some disastrous environmental consequences. The Soviet Union can therefore offer potentially important insights into the economic and environmental effects when governments own all capital and can freely set prices.

Issues for Discussion

1. What is subsidization and can you name two cases where you receive government-subsidized pricing?
2. The US Conservation Reserve Program is one type of US farm program. What argument might you use to support the program? Are there any possible economic distortions the CRP might introduce?
3. The arid US West produces a lot of food and dominates production of many vegetables, fruits and nuts. To what degree do you think that food security is a legitimate argument in favor of US irrigation water subsidies?
4. Governments sometimes subsidize agricultural inputs other than irrigation water, including chemical fertilizers and pesticides. In low-income countries, including in sub-Saharan Africa, use of agricultural chemicals can be very limited. Under such circumstances do you think it is OK to promote fertilizers and pesticides using subsidies? Why or why not?
5. The chapter made the case that irrigation water offtake can affect freshwater ecosystem services other than agriculture by both increasing the cost of those services and reducing demand. Can you name two marketed or non-marketed freshwater services other than rafting that might be similarly affected?

Practice Problems

Please use the description below for the following three questions.
Suppose the price of taxi services in Bangkok is capped by the city government at 10 THB per kilometer (about $0.30). The marginal benefit of taxi trips is given by MB = 40 - 2Q and the marginal cost of taxi services is MC = 3Q, where Q in this case stands for kilometers (km) traveled.

1. Without the price ceiling, what is the equilibrium price and how many kilometers do taxi drivers drive?
2. How many km are driven at the price ceiling and at what price?
3. Suppose the price was set at THB 30/km. How would that change your answer and is it now a price ceiling (i.e., a maximum price) or a price floor (a minimum price)?

Please use the description below for Questions 4-8.
The government in an arid region is considering building an irrigation canal to serve a group of large rice farmers operating in a competitive market. The project is large enough that it is expected to affect prices in the region. The marginal benefit of rice is MB = 150 - 4Q and the marginal cost is constant at MC = 70. In this problem, Q = tons of rice. MB and MC are expressed in local currency. The project will make it much easier and cheaper to produce rice and is expected to reduce the MC for farmers by 30 at all levels of production.

4. How will this project affect the equilibrium price and quantity of rice in the region? Please round to the nearest ton and local currency unit.
5. Now consider the problem from the perspective of water rather than rice. Suppose the project costs 15 per acre-foot of water. As suggested by Fox (2015), each ton of rice uses five acre-feet of water. Assume that a unit of local currency is equally valuable in the hands of the government, which will use it for education, health, roads etc., or rice farmers. Considering only the opportunity cost of local currency, does this project make economic sense?
6. In addition to the monetary opportunity cost, the diversion of water increases the temperature of the surface water, making it difficult for fish to spawn in rivers. Valuation experts use methods discussed in Chapter 4 and find that this common pool ecosystem service effect adds another 10 to the social cost per acre-foot. Considering all monetary costs to the government and benefits to buyers and sellers, as well as environmental costs, does this project make economic sense?
7. Using your results from Questions 4-6 and any additional arguments, please write three sentences that make your argument **for not** continuing with this project.
8. Using your results from Questions 4-6 and any additional arguments, please write three sentences that make your argument **for** continuing with this project.
9. Desalination is a technology to remove salt from water and produce fresh water that can be used for drinking, cooking and other domestic uses. Assume a large desalination project is planned to serve the city of Phoenix, Arizona in the US. It will be located in Puerto Peñasco, Mexico on the Gulf of California and the fresh water will be pumped through a piping system to Phoenix.[10] The project is expected to increase water costs for Phoenix

residents, but the city argues that it has no choice as groundwater is insufficient for its growth needs.

The Gulf of California is known for its variety of marine wildlife, including sharks, whales, marine turtles and porpoises. The project is expected to affect marine wildlife, because salty brine removed from the desalinated water will need to be discharged into the Gulf of California. The Gulf of California is bordered on three sides by Mexico, and therefore does not flush out very well. These salty discharges are expected to reduce marine mammal habitat and tourism (e.g., whale watching and sailing). It will also force tourism operators to move to new areas farther away from harbors, therefore increasing their costs.

Please use a clearly labeled graph to analyze the effects on the marine tourism industry. Please include clearly labeled marginal benefit and marginal cost curves and show any shifts and new equilibria. Please write a few sentences that explain your graphical analysis.

Notes

1 This situation is often referred to as one of "thick" markets and is the norm in the higher-income world. Occasionally, though, even in high-income countries there are shortages at market-clearing prices. For example, it is not unheard of for certain vaccines to be in short supply, forcing patients to wait for their immunizations. Typically, though, such shortages are due to short-run mismatches between supply and demand rather than below-equilibrium prices.
2 See New York Times Company v The Superior Court of Santa Barbara County (available at https://scholar.google.com/scholar_case?case=14928417565641588270&q=New+York+Times+Co.+v+Goleta+Water+District&hl=en&as_sdt=4,5), which discusses blocked publication of water usage data.
3 Advocacy organization analysis may not be peer reviewed and could focus on "proving" particular points rather than analyzing objectively.
4 "Raw" water is unprocessed water.
5 Acre is a unit of land area used in the US and is equal to about 0.40 hectares. It is also approximately the size of an American football field.
6 Irrigation districts service only farmers, but water districts may also supply water for non-agricultural uses.
7 Apologies to all readers not familiar with feet and acres as units. A foot is about 0.30 meters and an acre of irrigation water one foot deep would contain about 326,000 US gallons or approximately 1.23 million liters.
8 Not all water is equal. Water value depends on a variety of factors, including quality and location.
9 Typically called *riparian* habitat.
10 Note that this example focuses on a real project that is being considered. See Flavelle (2023) for an overview.

Further Reading and References

Andreyeva, T., M.W. Long and K.D. Brownell. 2010. "The Impact of Food Prices on Consumption: A Systematic Review of Research on the Price Elasticity of Demand for Food." *American Journal of Public Health* 100 (2): 216-222.

Autor, D.H., C.J. Palmer and P.A. Pathak. 2014. "Housing Market Spillovers: Evidence from the End of Rent Control in Cambridge, Massachusetts." *Journal of Political Economy* 122 (3): 661-717.

Barbarika, A. 2012. *Conservation Reserve Program: Annual Summary and Enrollment Statistics*. US Department of Agriculture, Farm Service Agency. Downloaded from https://www.fsa.usda.gov/Assets/USDA-FSA-Public/usdafiles/Conservation/PDF/summary12.pdf

Baur, G. 2019. "The Best Way to Help Dairy Farmers is to Get Them Out of Dairy Farming." *Washington Post*, June 12.

Blayney, D. and M. Normile. 2004. *Economic Effects of U.S. Dairy Policy and Alternative Approaches to Milk Pricing*. US Department of Agriculture, July. Downloaded from https://www.ers.usda.gov/webd ocs/publications/83460/ap-076.pdf

City of San Francisco. Undated. "Office of Labor Standards Enforcement." Retrieved from https://sfgov. org/olse/minimum-wage-ordinance-mwo August 18, 2020.

City of Seattle. Undated. "Minimum Wage." Retrieved from http://www.seattle.gov/laborstandards/ord inances/minimum-wage August 18, 2020.

Diamond, R., T. McQuade and F. Qian. 2019. "The Effects of Rent Control Expansion on Tenants, Landlords, and Inequality: Evidence from San Francisco." Downloaded from https://web.stanford.edu/~diamo ndr/DMQ.pdf August 9, 2020.

Djanibekov, N., I. Rudenko, J.P.A. Lamers and I. Bobojonov. 2010. "The Pros and Cons of Cotton Production in Uzbekistan." In P. Pinstrup-Anderson, ed. *Food Policy for Developing Countries: The Role of Government in the Global Food System*. Cornell University Press: Ithaca, NY.

Downs, A. 1988. *Residential Rent Controls*. Urban Land Institute: Washington, DC.

Encyclopedia Britannica. Undated. "Aral Sea." *Encyclopedia Britannica*. Retrieved from https://www.bri tannica.com/place/Aral-Sea May 10, 2019.

Energy Information Agency (EIA). 2020. "How Much Ethanol is in Gasoline, and How Does it Affect Fuel Economy?" June 24. Retrieved from https://www.eia.gov/tools/faqs/faq.php?id=27 September 1, 2023.

Environmental Working Group (EWG). 2004. "California Water Subsidies." December 15. Retrieved from https://www.ewg.org/research/california-water-subsidies August 15, 2020.

Flavelle, C. 2023. "Arizona, Low on Water, Weighs Taking it from the Sea. In Mexico." *The New York Times*, June 10.

Food and Agriculture Organization of the United Nations (FAO). Undated. "Key Facts and Findings." Retrieved from http://www.fao.org/news/story/en/item/197623/icode/ September 1, 2023.

Fox, J. 2015. "Amid Drought, California Fields to Grow Rice: Bloomberg View." *The Oregonian/Oregon Live*, May 17. Retrieved from https://www.oregonlive.com/opinion/2015/05/amid_drought_califo rnia_fields.html September 5, 2023.

Gray, J.W. 1983. *Forest Revenue Systems in Developing Countries*. Forestry Paper 43. United Nations, Food and Agriculture Organization: Rome.

Gurney, V. 2012. "From the Field: Travels of Uzbek Cotton through the Value Chain." Responsible Sourcing Network. Downloaded from https://www.fergananews.com/archive/2012/fieldcottonreport.pdf July 28, 2020.

Gyourko, J. and P. Linneman. 1989. "Equity and Efficiency Aspects of Rent Control: An Empirical Study of New York City." *Journal of Urban Economics* 26 (1): 54–74.

Kuminoff, N.V., D.A. Sumner, G. Goldman and H.O. Carter. 2000. *The Measure of California Agriculture, 2000*. University of California Agricultural Issues Center: Davis, CA.

Lauinger, J. 2018. "Poll Indicates Some Support for Farm Subsidies." *Politico*, July 30. Retrieved from https://www.politico.com/story/2018/07/30/poll-indicates-some-support-for-farm-subsidies-749 870 August 17, 2020.

Lustgarten, A. and N. Sadasivam. 2015. "Federal Dollars are Financing the Water Crisis in the West." *Scientific American*, May 28.

New York City Department of Labor. Undated. "Minimum Wage." Retrieved from https://www1.nyc.gov/ nycbusiness/description/wage-regulations-in-new-york-state August 18, 2020.

Ogburn, S.P. 2011. "Farmland Conservation Program May be Plowed Under." *High Country News*, September 21. Retrieved from https://www.hcn.org/issues/43.15/farmland-conservation-program-may-be-plowed-under

O'Hara, S.L., G.F. Wiggs, B. Mamedov, G. Davidson and R.B. Hubbard. 2000. "Exposure to Airborne Dust Contaminated with Pesticide in the Aral Sea Region." *The Lancet* 355 (9204): 627–628.

Poole, K. 2023. "How California's Water Rights System Gouges You and Me." Natural Resources Defense Council. February 6. Retrieved from https://www.nrdc.org/bio/kate-poole/how-californias-water-rig hts-system-gouges-you-and-me#:~:text=The%20systematic%20exclusion%20of%20Native,criti cal%20water%20resources%20even%20today June 29, 2024.

Powell, J.W. 1879. *Report on the Lands of the Arid Region of the United States with a More Detailed Account of the Land of Utah with Maps*. US Geographical and Geological Survey (2nd ed.). Government Printing Office: Washington, DC. Downloaded from https://pubs.usgs.gov/unnumbe red/70039240/report.pdf

Scheiber, N. 2018. "They Said Seattle's Higher Base Pay Would Hurt Workers. Why Did They Flip?" *The New York Times*, October 22. Retrieved from https://www.nytimes.com/2018/10/22/business/econ omy/seattle-minimum-wage-study.html

State of Oregon. 2019. S.B. 608, 80th Oregon Legislative Assembly, 2019 Reg. Sess. https://olis.oregon legislature.gov/liz/2019R1/Downloads/MeasureDocument/SB608/Enrolled

State of Oregon. Undated. "Oregon Minimum Wage." Retrieved from https://www.oregon.gov/boli/work ers/pages/minimum-wage.aspx August 18, 2020.

US Department of Agriculture (USDA) Farm Service Agency. 2019. "Conservation Reserve Program." Downloaded from https://www.fsa.usda.gov/Assets/USDA-FSA-Public/usdafiles/Conservation/PDF/ One-Pager%20Jul%202019.pdf

US Department of Agriculture (USDA) Farm Service Agency. Undated. "Conservation Programs." Retrieved from https://www.fsa.usda.gov/programs-and-services/conservation-programs/ August 18, 2020.

US Department of Labor, Bureau of Labor Statistics (BLS). 2018. "A Profile of the Working Poor, 2016." *BLS Reports*, July. Retrieved from https://www.bls.gov/opub/reports/working-poor/2016/home.htm

US Department of Labor, Bureau of Labor Statistics (BLS). 2023. "Characteristics of Minimum Wage Workers, 2022." *BLS Reports*, March. Retrieved from https://www.bls.gov/opub/reports/minimum-wage/2022/home.htm

Yuldashbaev, N. 2010. "Uzbekistan - Republic of: Cotton and Products Annual Report." *GAIN Report*. US Department of Agriculture, Foreign Agricultural Service. Downloaded from https://apps.fas. usda.gov/newgainapi/api/report/downloadreportbyfilename?filename=Cotton%20and%20Produ cts%20Annual_Tashkent_Uzbekistan%20-%20Republic%20of_3-15-2010.pdf

12 Energy, Energy Subsidies and Economic Planning Gone Wild

What You Will Learn in this Chapter

- How much energy is used in the world and across countries
- What are the sources of energy?
- About CO_2 emissions from energy
- What is an energy "subsidy?"
- What are the main types of energy subsidies?
- Where is energy subsidized?
- What was the USSR and what were the environmental implications of comprehensive economic planning in the USSR and its allied states?
- The nature of economic planning in the USSR and its allied states
- Why and how energy subsidies were important in the planned economies and what were the economic and environmental implications of those subsidies

---

12.1 Recap and Introduction to the Chapter

The previous chapter introduced us to the idea that governments mess with markets to achieve public policy goals and to provide benefits to stakeholders. Governments occasionally fix prices, sometimes they buy or sell products to affect prices and they may sell goods and services they control below market value. Governments often own and control natural resources on behalf of their populations and so selling, renting or simply giving away natural resources is an especially important class of economic subsidy.

Subsidizing natural resources often has environmental implications. Irrigation subsidies have a number of potential environmental effects. These include degradation of freshwater habitats due to reduced stream flows, increased stream temperature and loss of biodiversity. Other agricultural subsidies, such as those for fertilizers and pesticides, may also have important environmental downsides.

Agriculture is hardly the only area in which governments offer sweetheart deals on natural resources. Energy is another example. As has been discussed in a number of previous chapters, production and use of fossil fuel energy have very important local air pollution

DOI: 10.4324/9781003308225-14

and climate change implications. We now consider the importance of energy for modern economies, energy subsidies and their effects. We also examine the interesting and unique historical case of energy (and other environmentally harmful) subsidies in the Soviet Union and its allied states.

12.2 Energy Consumption and Economies

Energy is enormously important for economies and without fossil fuels, such as coal, oil and natural gas, it is arguable that the progress humans have made during the past 150 years would not have been possible. The importance of energy to prosperity is demonstrated in Fig. 12.1, which shows energy use (measured as tons of oil equivalent per person) graphed against GDP per person[1] in each country. Colors indicate the region of each country and the sizes of the dots represent the total population of each country, with larger dots meaning bigger populations.

The horizontal axis is labelled "2011 USD PPP." This means that the US dollar figures are from 2011 rather than another year. This is important, because prices for goods and services in 2011 were lower than they are now (prices almost always rise over time), which means that money buys less stuff in the future. For example, the US income per capita in 2011 was about $43,000 and at the end of 2023 was approximately $69,400,[2] which means that income per capita increased by about 61%.

The growth in the overall price level is called *inflation*.[3] Between 2011 and 2023, the overall price level rose by about 36%. Real income per person therefore rose by only about 61% – 36% = 25% rather than 61%. Because adjusting measures like income for inflation tells

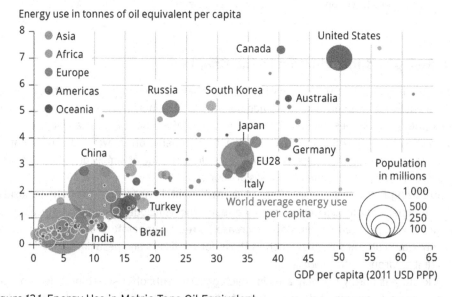

Figure 12.1 Energy Use in Metric Tons Oil Equivalent
Source: European Environment Agency of the European Union, available at https://www.eea.europa.eu/data-and-maps/figures/correlation-of-per-capita-energy

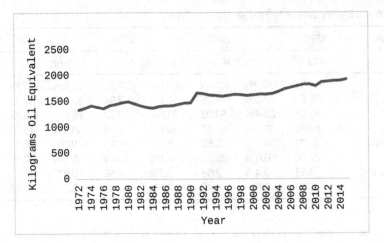

Figure 12.2 World Energy Use per Capita
Source: World Bank, available at https://data.worldbank.org/indicator/EG.USE.PCAP.KG.OE

us how much actual stuff can be purchased, often inflation-adjusted monetary measures are called *real* (e.g., *real* GDP, *real* incomes, *real* stock values etc.). The other adjustment noted on the horizontal axis of Fig. 12.1 is PPP, which stands for Purchasing Power Parity, and accounts for differences in overall price levels across countries. For a reminder about PPP adjustments, please see Chapter 4.

Figure 12.1 shows very clearly that countries with higher GDP per person also use more energy per person. Countries in Africa and South Asia have much lower energy use per person than higher-income countries, such as Japan, South Korea, Australia and the US. Though high incomes and energy use clearly go together, this graph does **not** present a function, because GDP/capita both causes **and is caused by** energy/capita.

As shown in Fig. 12.2, average global energy use per person is about 2 metric tons of oil equivalent per person per year. In the 1990s, this value was roughly 1.6 tons and in 1980 about 1.4 tons. As shown in Table 12.1, as of 2021, about 80% of all energy comes from fossil fuels, roughly evenly divided between coal, oil and natural gas. The other 20% comes from a variety of sources, but fully half of this remaining 20% is biomass, such as fuelwood, crop residues and wastes in low-income countries, where such fuels dominate the energy mix. For example, in sub-Saharan Africa about 2/3 of households cook with biomass.

As we have mentioned in previous chapters, biomass fuels are used for cooking and are often burned indoors. Cooking with biomass fuels also sometimes causes very high levels of indoor air pollution (Jeuland et al., 2016) and perhaps 4 million excess deaths per year (Lim et al., 2012). In rural Ethiopia, I and my co-authors found that traditional biomass burning stoves generated $PM_{2.5}$ air pollution concentrations that averaged over 50 times the 24-hour World Health Organization then guide value of 25 micrograms per cubic meter (LaFave et al., 2021).

Hydropower and other renewables, such as wind and solar power, each make up a small portion of global energy production and consumption, but especially non-hydro renewables have recently grown very rapidly. We should not overweight the scope of the clean energy

Table 12.1 World Energy Supply by Year and Energy Type

Energy Type	1990		2000		2010		2021	
	MTJ	% Total	MTJ	% Total	MTJ	% Total	MTJ	% Total
All Sources	366.51	100.0%	419.69	100.0%	537.95	100.0%	617.79	100.0%
Oil	135.54	37.0%	154.25	36.8%	173.94	32.3%	182.23	29.5%
Coal	93.05	25.4%	97.03	23.1%	153.32	28.5%	168.15	27.2%
Gas	69.60	19.0%	86.57	20.6%	114.45	21.3%	145.99	23.6%
Biofuels/Wastes	37.07	10.1%	41.62	9.9%	49.12	9.1%	58.49	9.5%
Nuclear	22.00	6.0%	28.28	6.7%	30.09	5.6%	30.66	5.0%
Wind/Solar etc.	1.53	0.4%	2.53	0.6%	4.62	0.9%	16.83	2.7%
Hydro	7.70	2.1%	9.40	2.2%	12.41	2.3%	15.46	2.5%

MTJ = million terajoules.

Source: International Energy Agency Energy Statistics Data Browser, available at https://www.iea.org/data-and-statistics/data-tools/energy-statistics-data-browser?country=WORLD&fuel=Energy%20supply&indicator=TESbySource.

transition. Total energy production rose by 68% between 1990 and 2021 and energy produced from coal increased by about 81%.[4]

Between 1961 and 2018, *real* world GDP grew by an astounding six times, from about $11 trillion ($11,000 billion) to $90 trillion measured in 2015 US dollars. Growth in *real* world GDP averaged about 3.5% per year, with low- and middle-income countries growing a bit faster at 3.7% and 4.6%, respectively, and high-income countries growing more slowly (3.1%). As we learned in Chapter 6, these are very fast growth rates, implying that global GDP doubles roughly every 20 years.[5]

Over time, humans have generated more and more production from energy. For example, in 1990 it took about 7.5 megajoules[6] of energy to produce one dollar of 2011, PPP-adjusted output. By 2015, generating a dollar of production on average required only 5 megajoules, which is a 1/3 decline. Output and incomes have been growing faster than energy demand for many years.

Challenge Yourself

Table 12.1 suggests that between 1990 and 2021 energy production from coal increased faster than overall energy production. Which sources grew more slowly than coal? Which grew faster?

As we have discussed in a number of chapters, energy not only powers economies, but is also a major source of air pollution. In addition to pollutants that directly affect human health, such as particulate matter, sulfur dioxide, carbon monoxide and nitrogen oxides, energy production generates about 72% of the world's greenhouse gas emissions.[7] Global energy-related CO_2 emissions hit a record of about 36.8 billion metric tons in 2022, which was a 0.9% increase over the previous year (IEA, 2023).

Table 12.2 Energy-Related Greenhouse Gas Emissions

	Total CO_2e Emissions (Mt), 2021	Growth Rate (%), 2016-2021
China	11,349	17.2%
US	5018	-4.4%
India	2427	9.3%
European Union (27)	2560	-8.7%
Rest of World	16,047	-1.2%
World	37,401	3.3%

Source: International Energy Agency Energy Statistics Data Browser, available at https://www.iea.org/data-and-sta
tistics/data-product/greenhouse-gas-emissions-from-energy-highlights

Table 12.2 presents energy-related GHG emissions in millions of metric tons of CO_2e emissions for the top four emitting countries and the rest of the world. The top four countries produce about 57% of world energy-related CO_2e emissions, with the rest of the world (over 190 countries) generating the remainder. China and the US together produce about 44% of global energy-related CO_2 emissions.

During the period 2016-2021, world energy-related emissions grew by 3.3%, which was about 2/3 of 1% per year. The CO_2e increase appears to have been moderated by declines in emissions in the US, the rest of the world and especially the EU, which reduced emissions by about 9% during the five-year period. Emissions by China and India grew rapidly leading up to 2021.

12.3 Energy Subsidies around the World

Given the importance of energy for economic prosperity, it is not surprising that governments look after their energy sectors. Like irrigation water, some governments subsidize energy to provide lower end-use prices for consumers and, to a much lesser extent, reduce input costs for energy producers.

Equation 12.1 International Energy Agency Energy Subsidy Calculation Method

Total Energy Subsidy = (Reference price - Domestic Price) * Energy Units Consumed

The International Energy Agency (IEA) is an independent agency created by the Organisation for Economic Co-operation and Development (OECD, which we have mentioned periodically throughout this book and introduced in Chapter 2). As we have discussed in previous chapters, the OECD is a consortium of 36 higher-income member countries that supports research and policy analysis in support of its members. The IEA has 30, again mainly higher-income, country members, plus eight associate members, notably China, India, Brazil and Indonesia. It focuses on a variety of energy policies and is an important source of information on energy subsidies around the world.

> **Box 12.1 Does the US Government Subsidize Oil Companies?**
>
> The answer to this question depends on how one interprets the word "subsidy." In analyzing the fossil fuel market, the International Monetary Fund distinguishes *pre-tax subsidies* from *post-tax subsidies*. Pre-tax subsidies are interventions in markets *before* externalities are considered. Post-tax subsidies refer to differences between market prices and *efficient* prices, which are those that would have occurred if all social costs of fuel production and consumption were included. In the US, post-tax subsidies outweigh pre-tax subsidies 20:1. This means that the US is not actively intervening in markets, but is allowing producers and consumers to shift the hidden environmental costs of production and consumption onto the public.
>
> Source: Coady et al. (2019).

The IEA measures energy subsidy value – virtually all of which is used to subsidize fossil fuels – by comparing world prices, which it calls "reference prices," with average domestic prices. If average domestic prices are below reference prices, the difference is called a "price gap" and the IEA registers an energy subsidy. They then just multiply this price gap by the amount of energy consumed, which gives the total energy subsidy.

Often the total subsidy is expressed as a percentage of GDP to give us an idea how big the energy subsidy burden is on economies. Burdens are created by energy subsidies, because country governments must spend or give up state revenues to fund lower prices of energy within their borders. If governments buy the energy on the market to sell at a loss, the cost is obvious. Also, fairly clear is the cost if it uses its taxing and spending powers to reduce energy production costs (e.g., by giving cheap loans for natural gas pipelines). If government-owned companies produce fossil fuels, as in Mexico and Venezuela, and then sell them to their populations at reduced prices, they incur opportunity costs equal to the extra amount they could have earned selling at IEA reference prices. Even countries that produce energy therefore incur costs when they choose to subsidize it.

Using the IEA price gap measure, many countries who subsidize energy are oil producers who use policies to reduce gasoline/petrol prices for domestic consumption. Based on the IEA subsidy definition, none of the highest-income countries have energy subsidies (Black et al., 2023).[8]

The world average retail gasoline price at the pump was $1.22 per liter in 2022 ($4.62/gallon) including all taxes and subsidies.[9] In Iran, though, gas costs only $0.32 per liter and in Indonesia only $0.45 due to heavy subsidies. In Saudi Arabia, the average pump price was about $0.60/liter. The 2022 average price in the US was $1.17 and in Russia $0.75. The highest gasoline prices in the world in 2022 were in Norway at $2.30/liter.[10]

Box 12.2 Fuel Subsidies in Ecuador

Energy subsidies have been used for decades to reduce the cost of energy for Ecuador's consumers. Starting in the 1970s, as Ecuador became an important oil exporter, the subsidies were applied to gasoline, diesel fuel primarily used for trucks, and liquefied petroleum gas (LPG).

LPG is sold in metal cylinders and is widely used for cooking. Consumers pay only about 20% of the actual cost of an LPG cylinder and as of 2018, the price of LPG had not changed since 2001. To the extent that biomass fuels would have been otherwise used for cooking, this LPG subsidy likely reduces indoor air pollution, which is an important externality, and saves lives.

Though premium gasoline prices have edged up since 2015, the price of regular gas has been the same since 2003. These energy subsidies are extremely costly, gobbling up about 10% of the total government budget in 2015. The subsidy on diesel fuel, which like LPG is mainly imported, alone cost the government almost $1.8 billion. The currency of Ecuador, which is called the sucre, is not fully exchangeable with other currencies. To import, Ecuador therefore needed to buy so-called *hard* currencies like US dollars, European Union euros, Japanese yen etc.

All this was roughly manageable – if very costly – until international oil prices, which are generally bought and sold in US dollars, declined after 2014. It then became difficult for Ecuador to meet its international hard currency obligations and in 2019 the government led by President Lenín Moreno negotiated a loan for $4.2 billion from the International Monetary Fund (IMF). The IMF is an international organization of member country governments that focuses on international financial stability. A condition of that loan was to reduce or eliminate energy subsidies. President Moreno announced elimination of the subsidies in an October 1, 2019 decree, causing gasoline and diesel prices to roughly double. These costs rippled through the economy, leading to widespread hardship and large protests. In response to these protests, after two weeks the decree was withdrawn and subsidies were restored.

Sources: Jara et al. (2018) and McCoy (2019).

Though some countries' energy subsidies are quite small as shares of GDP, some are very significant. For example, Ecuador, which in October 2019 succumbed to pressure from protestors to reinstate energy subsidies (see Box 12.2), subsidizes energy costs by 37%, costing the country $3.4 billion or 3.2% of GDP. Subsidies for gasoline in Ecuador alone cost $1.3 billion and kept prices at about $1.00 per gallon (McCoy, 2019). To give an idea of scale, if the US subsidized energy at that level, it would cost about $600 billion.[11]

The Ecuador example is not an especially extreme one. Algeria, Azerbaijan, Iraq, Kuwait and Saudi Arabia, all oil producers, subsidize 50%-70% of energy costs in their countries and spend anywhere from 4% to 10% of GDP on subsidies (IEA, undated a).

As shown in Fig. 12.3, some countries have subsidy levels that are amazing and are no doubt very burdensome for those economies. For example, Iran is estimated to spend about

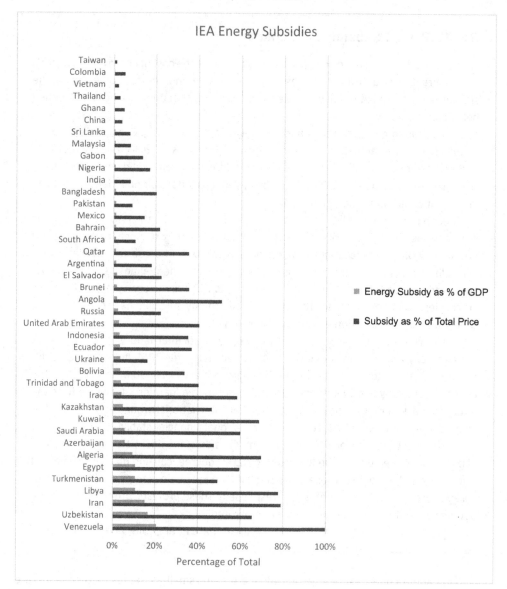

Figure 12.3 2018 World Energy Subsidies

Source: International Energy Agency Energy Statistics Data Browser, available at https://www.iea.org/data-and-statistics/data-product/fossil-fuel-subsidies-database

$70 billion or over 15% of its GDP subsidizing energy. These costs are relatively equally spread over oil, electricity and natural gas. Venezuela spends about $20 billion per year, which is about 20% of its GDP. Of course, rather than subsidizing gasoline – with all the environmental damages burning gasoline implies – those financial resources could have been spent on other things, such as private investments, schools, nutrition, health and environmental protection. In total, the IEA estimates that in 2018 governments spent $427 billion per year to subsidize fossil fuels.

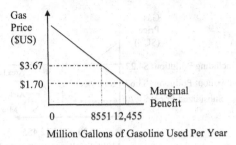

Figure 12.4 Effect of Subsidies on Gasoline Consumption in Indonesia*
*Assumes a 10% increase in gasoline prices reduces consumption by 5%.

We should not forget that, just as with water subsidies, when prices go down, incentives to conserve go down and consumption rises. This situation is shown in Fig. 12.4 using Indonesia, which is one of the countries that has a consumer subsidy for gasoline, as an example. Let us make some rough estimates and see what we can learn, assuming that demand for gasoline is linear and that the price elasticity of demand is -0.5. That means that a 10% increase in the retail price reduces consumption by 5%.[12]

In 2017, according to the IEA, Indonesia used about 1.75 million 42-gallon barrels of oil per day (IEA, undated b). Assuming all oil was used for gasoline (i.e., no diesel) and supposing gasoline refining efficiency was similar to the US at about 19.5 gallons per barrel (US EIA, undated), that means Indonesia used about 34 million gallons of gasoline per day or 12,455 million gallons per year.

How much less would it have used if its average gasoline price was the same as the 2018 world average ($3.67/gallon) rather than $1.70 per gallon? Assuming that each 10% increase in prices reduces consumption by 5%, which is roughly in line with short-run estimates, domestic gasoline consumption in Indonesia would have been roughly 1/3 less or about 8551 million gallons per year. Perhaps not surprisingly, a gasoline subsidy that is so large has a big effect on consumption!

But what if we do not only want to know the increase in gasoline consumption due to the subsidy? What if we also want to estimate the ecosystem services that are lost as a result of the subsidy? In other words, we might want to estimate by how much - in monetary terms - the world is worse off due to the policy. In order to make such an estimate, we refer back to the negative externality model of environmental problems discussed in Chapter 9. Thinking of environmental problems as externalities starts from adding up all social costs and then asking which of those costs are missing from market calculations. The difference between the total cost and the cost markets incorporate into prices is the externality. To estimate the externality cost of the Indonesian gasoline subsidy, we need to first assume how marginal externalities change with gasoline consumption. Let us assume for simplicity that the marginal external cost is constant, meaning that each gallon of gas burned has a fixed effect on the environment.

Davis (2017) estimates that local air pollution and climate change costs of burning gasoline are about $0.50 per gallon. That is, pollution costs at that time were about 15% of the retail price of gasoline. He also estimates that traffic accidents generate costs that are higher at about $0.68 per gallon and congestion costs are a whopping $1.02 per gallon.

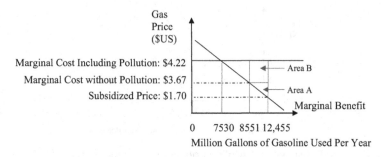

Figure 12.5 Welfare Effects of Gasoline Subsidies in Indonesia*

*Assumes a 10% increase in gasoline prices reduces consumption by 5%.

Figure 12.5 assumes, as is probably reasonable, that the 2018 average worldwide gasoline pump price of $3.67 per gallon includes little or no pricing to cover environmental externalities like air pollution and climate change and is roughly equal to the private marginal cost of production.[13] The social marginal cost, including pollution, but not accidents and congestion, would therefore be about $4.22/gallon. As we saw in Fig. 12.4, the subsidy policy in Indonesia makes gasoline consumption more attractive to consumers, which is of course the point of the policy.

Let us just look at the estimated increase in quantity demanded in Indonesia from approximately 8551 to 12,455 million gallons per year, which is caused by the subsidy. We see there are two parts to the social loss associated with the policy. Assuming $3.67 is the private marginal cost of gasoline, the first loss comes from producing gasoline where the marginal benefits are less than the marginal private cost. This type of loss due to over- or underproduction is called the *deadweight welfare loss*. Assuming as we are that MB is linear, we can calculate this welfare loss that really has nothing to do with the environment but is due to producing gasoline that is not worth producing. Using the formula to calculate the area of a triangle (Area = ½ base*height), we calculate that the deadweight loss in Fig. 12.5 shown as Area A is about {($3.67 - $1.70)*3904 million gallons}/2 = $3845 million per year.

We also need to take account of the air pollution costs from the overuse of gasoline. Using the estimate from Davis (2017) of $0.50 per gallon, we calculate Area B in Fig. 12.5 as $0.50*3904 million gallons = $1952 million per year. We therefore estimate that the Indonesia gasoline subsidy generates approximately $2 billion per year in air pollution-related costs, which is just over half the value of the deadweight welfare loss. With the assumptions noted, the total social welfare loss due to the gasoline subsidy in Indonesia is $3845 million + $1952 million = $5797 million.[14]

An interesting question is who benefits from gasoline subsidies. We said earlier that in Iran gasoline prices are subsidized by the government so that the average price was a little over $0.32 per liter at a time when the worldwide average was about four times higher. Ettehad and Sterner (2012) analyzed which Iranians benefit from the subsidies and found that whether people live in urban or rural areas, subsidies benefit the rich. In fact, they found that in 2008 the richest 20% in Iran burned through about 50% of the gasoline subsidies. The poorest 50% ended up with approximately 20% of the benefits.

And these consumer subsidies can be difficult to stop. As shown by the Ecuador example and the persistence of explicit energy consumption subsidies, especially in oil-producing countries, once subsidies are put in place they are hard to withdraw. People make decisions, including long-run investments, presuming that subsidies will remain in place, and can be angry when they are taken away.

The US and other higher-income countries do not appear on the IEA list of energy-subsidizing countries, because they do not try to reduce prices below reference prices. This observation does not mean they aren't interested in the status of energy supply and industry profitability. Instead of using direct consumption subsidies, as in Iran, Turkey, Russia and Saudi Arabia, among others, most high-income countries use indirect or "off-budget" sub-sidies to support fossil fuel production.

Table 12.3 from Steiner (2008) discusses some of the many possibilities to *implicitly* sub-sidize fossil fuel energy. These opportunities include special tax benefits, trade restrictions that protect domestic industries, loan programs, direct government investments and offering low-cost insurance to energy companies. One example of an implicit subsidy is preferential tax treatment. Table 12.4 shows the estimated 2012–2022 savings to the US federal govern-ment updated to $2023 values associated with eliminating industry-specific tax deductions and credits (Aldy, 2013). We see that subsidies exist but are small compared to the very large federal budget and the enormous US economy.

It turns out that even more costly than the government expenditures associated with fossil fuel subsidies are the environmental damages that the excess consumption causes. Coady

Table 12.3 Potential Indirect Energy Subsidy Mechanisms

Subsidy Type	Example	Subsidy Mechanism		
		Reduce Production Cost	Increase Producer Price	Reduce Price to Consumer
Financial payments	To producers	•		
	To consumers			•
	Preferential loan terms	•		
Tax breaks	Rebates on taxes, royalties or tariffs	•		
	Tax credits	•		•
	Accelerated depreciation	•		
Trade restrictions	Quotas, technical restrictions and trade embargos		•	
Energy services provided by the government below market rates	Investments in infrastructure	•		
	Public research and development	•		
	Liability insurance and facility decommissioning	•		
Energy regulation	Demand guarantees and mandated deployment	•	•	
	Price controls		•	•
	Market access restrictions		•	

Source: Steiner (2008).

Table 12.4 Provisions of the US Tax Code that Subsidize Fossil Fuel Extraction

Tax Code Provision	10-Year Opportunity Cost (2023 $US Billions)
Depletion of oil, gas and coal	$17.6
Tax deductions for oil, gas and coal production	$15.8
Intangible drilling costs	$18.5
Independent operator accelerated depreciation	$1.9
Other	$1.3
Total	**$55.1**

Inflated from 2012 $US to 2023 values using the Federal Reserve Bank of Minneapolis Inflation Calculator.[a]

a   Indicative only, as price inflation may differ by economic sector.

Source: adapted from Aldy (2013).

et al. (2017) estimated that the biggest costs associated with fossil fuel subsidies (almost all of which promote consumption) are the air pollution costs. By eliminating both explicit and implicit fossil fuel subsidies around the world, they estimated that world air pollution deaths, which at that time were estimated to be about 4 million per year (more now), would fall by more than 50%. They also found that climate change-creating CO_2 emissions would fall by over 20%. They then valued the environmental damages using methods that we discussed in Chapter 4 and found that the total cost of global fossil fuel subsidies was over $5 trillion. By 2023, that cost had increased to $7 trillion, which was roughly 7% of world GDP. This enormous cost, due to un-internalized air pollution externalities, is roughly equal to the GDP of all Latin American and Caribbean countries combined!

12.4 Energy (and Other Environmentally Harmful) Subsidies Gone Wild: The Soviet Union

In 1917, the Union of Soviet Socialist Republics, known as the USSR or Soviet Union, was established in what was previously the Russian Empire. Over time, it expanded to include present-day Ukraine, the Baltic countries, most countries in Central Asia and others.

Countries like Poland, the eastern part of present-day Germany, Czech Republic, Slovakia, Romania and Bulgaria were also heavily under the influence of the USSR. A key feature of the USSR, which ended as a country in late 1991, triggering enormous political and economic changes, was comprehensive economic planning. Imagine whole economies without any private businesses. No private factories, no private farms, no private stores, houses, apartments, restaurants or even coffee shops. Everything most people think of as "businesses" were divisions of governments. It was like the Ministry of Economics owned and ran your apartment building, grocery store, Walmart, mobile phone company and McDonalds! As shown in Table 12.5, in most countries in the so-called Soviet Bloc the government produced huge majorities of the economic output.

Centralizing production with the government could possibly have allowed governments to sustainably manage economies, because they might have been better able to balance management of the ecosystem services produced by environmental assets. This does not seem to be what happened in the former Soviet Union and allied countries. Little was known in the

Table 12.5 Size of the Government Sector by Output and Employment for Planned and Market Economies (% of Total)

	Output	Employment
Planned Economies		
Czechoslovakia (1986)	97.0%	Not available
East Germany (1982)	96.5%	94.2%
Poland (1985)	81.7%	71.5%
Hungary (1984)	65.5%	69.9%
Soviet Union (1985)	96.0%	Not available
Market Economies		
France (1982)	16.5%	14.6%
West Germany (1982)	10.7%	7.8%
Italy (1982)	14.0%	15.0%
Portugal (1976)	9.7%	Not available
Spain (1979)	4.1%	Not available
US (1983)	1.3%	1.8%

Source: Sachs (1994).

democratic world about environmental issues in the Soviet Bloc prior to the mid-1980s. After the Berlin Wall, which separated East and West Berlin, came down and East Germany began to be reintegrated with western Germany, and after the Soviet Union officially collapsed, researchers and others began to understand the depth of the environmental problems.

What they found shocked the world. The Soviet Bloc had water and air pollution emissions that were often much, much higher than what would have been expected for countries with similar output and population levels. In some cases, much larger countries or similar countries just a few hundred kilometers west had significantly lower pollution levels than the countries in Central and Eastern Europe and the Soviet Union. These high levels of pollution created large environmental damages that amazed and worried outsiders, and helped generate support for rebellions in several countries, including Czechoslovakia (now two countries, Czech Republic and Republic of Slovakia), Poland, Estonia, Latvia and Lithuania.

Table 12.6 offers a comparison of sulfur dioxide emissions of market and planned economies. It is astounding to consider that East Germany had total SO_2 emissions that were almost 5 times those of France and West Germany, and a full 1/4 of those of the US, with an economy that was much less than 10% the size of those in all three countries![15] The same amazing excess pollution could be found in all the countries of the Soviet Bloc – relatively small economies had enormous levels of pollution.

We see from the table that this environmental inefficiency translated into high SO_2 emissions per unit of output (measured as GNP, which includes production by countries abroad rather than only within borders as for GDP). Even if we measure emissions per capita, Central and Eastern Europe and the Soviet Union did not fare any better, with 1.5–15 times more SO_2 emissions per person. Measuring emissions per mile of land area also does not change the conclusion that the Soviet Bloc had much too much pollution.

Why were the planned economies so polluting and what happened after they became market economies? The answer to the first question is complicated, but overuse of fossil fuel

Table 12.6 Sulfur Dioxide Emissions per Person and per $US of GNP in 1988

	Emissions per Person (Kilograms)	Emissions per $US 1988 of GNP (Grams)
Planned Economies		
East Germany	317	31
Czechoslovakia	179	24
Poland	110	20
Hungary	115	17
Bulgaria	114	21
Market Economies		
US[3]	84	4
UK	64	5
West Germany	21	1
France	22	1
Sweden	25	1

Source: French (1990).

energy and other materials are likely at the center of the story. The Soviet Bloc countries were astoundingly energy intensive compared with Western countries. Poland, Czechoslovakia, East Germany and virtually all other planned economies in 1988 used anywhere from 30% more (Hungary) to almost two to three times more energy (Romania and USSR) than an average OECD country. Furthermore, the energy sector typically generated about 3/4 of SO_2 and over half of particulate emissions (Bates et al., 1994), making these differences in energy use very important from an air pollution standpoint.

Box 12.3 Life-Threatening Air Pollution in Czechoslovakia in the Early 1990s

The north of what used to be the Czech part of Czechoslovakia is a mountainous area with large lignite coal reserves. Lignite is especially dusty and of significantly lower quality than harder coals. To take advantage of these resources, prior to 1991, the government built three large power plants and two district heating plants in a valley close to the coal mines. Many households throughout what in 1995 became the Czech Republic also burned coal in their homes for heating.

   The result of these decisions was a major air pollution problem, even in the capital, Prague, and especially in winter, when the country was subject to temperature inversions.

These countries wasted a lot of energy, which generated huge amounts of pollution, because the economies were planned in ways that emphasized industry, and energy was heavily subsidized. In Chapter 5, we discussed how most countries have economies dominated by services. In the Soviet Union, though, services were not even recognized as legitimate economic activities and were instead described as the "unproductive sector." Steel, energy, cement, metal fabrication and glass were politically preferred economic sectors.

These preferences tilted the economic structures in those directions. For example, by the end of the 1980s, steel production capacity in the USSR was almost twice that of the US. This observation implied that the Soviet Union could produce 15 times more steel per dollar of GDP than the US, with an economy only 1/8th as big (Sachs, 1995)! In most Soviet Bloc countries, 40%–60% of people worked in either energy production or one of five highly energy-intensive sectors (iron and steel, metal fabrication, glass/ceramics, building materials and wood pulp/paper production) (Wilczynski, 1990).

What the economist Jeffrey Sachs of Columbia University called the "worst kind of obsession with heavy industry" (Sachs, 1995) resulted in economies that were heavily distorted in favor of industry and against services, which even in 1979 made up over half of market economies. Whereas a typical higher-income market economy had only a bit over 1/3 of its economy in industry, in the planned economies the percentage was 50%–70%. Agriculture made up much higher portions of economic output than in the West. Anywhere from 8% to 19% of output in the planned economies was agriculture, versus 2%–9% in market economies. For example, 71% of Czechoslovakia's GDP was in industry, 8% in agriculture and only 18% in services. By comparison, Italy, which is less than 800 kilometers away from Prague, the former capital of Czechoslovakia, had only 43% of its economy in industry, 7% in agriculture and 50% in services (Wilczynski, 1990).

Planners in the planned-economy countries set all prices for goods and services because there were no markets. Indeed, an important way economic planners favored or penalized economic sectors was to change prices. Price setting, of course, included energy, where energy prices were typically only 20%–25% of energy *production costs* (Sachs, 1995). We should let this soak in for a moment. **These governments were selling energy to their populations at 1/5–1/4, not of what IEA would call *reference prices*, but of their *production costs*!** This is an astounding level of subsidy, which meant that prices were barely 1/3 of those in Western Europe (Bates et al., 1994).

Once economic planning collapsed, governments no longer controlled prices and these quickly started to climb to market levels. For example, at the end of 1989, coal prices in Poland were about $6.00 per ton. By May 1991, though, prices had increased sevenfold to $37 per ton. As most coal was burned to produce electricity, this meant that electricity prices also rose very quickly. For example, household electricity prices in Poland increased by an enormous 12-fold between 1989 and 1992 and thereafter continued to rise until they reached reference levels (Bates et al., 1994).

These huge increases created hardships for people, but also enormous incentives for conservation. I personally witnessed thousands of old, leaky windows being replaced during the three years I lived in Vilnius, Lithuania in the 1990s. Such investments were occurring throughout the region as price increases affected behavior. The result was a major decrease in energy consumption in the countries transitioning from economic planning to markets. Between 1989 and 1996, consumption of solid fuels (largely coal) fell by 30%. As petroleum- and gas-rich Russia could no longer afford to subsidize those fuels, consumption declined by an enormous 69% and natural gas use fell by 26%. In just six years during the 1990s, energy use fell by 43% (Zamparutti, 1999).

Because air pollution is so closely tied to energy, throughout the 1990s air pollution declined. Between 1989 and 1996, depending on the air pollutant (SO_2, particulates and NO_x) and country, air pollution emissions fell by anywhere between 50% and 70%. Concentrations of air pollutants in urban areas declined by 5%-7% per year throughout this same period, which is an extremely fast rate. These rapid declines in pollution meant that by mid-decade people in urban areas were breathing substantially better air than at the beginning of the decade (Zamparutti, 1999).

Water pollution fell even faster. The volume of wastewater, which is typically called *effluent*, declined by an average of 12% per year between 1990 and 1996, which means during those six years wastewater volume was approximately cut in half. Industrial water use fell at similar rates (Hughes and Lovei, 1999).

Pesticides and fertilizers are important sources of water pollution, because if over-applied they can run off into waterways, disrupting aquatic ecosystems. These agricultural inputs, which can be very energy-intensive to produce and therefore contribute to overuse of energy, were heavily subsidized before 1990. Once those subsidies were reduced and ultimately withdrawn, between 1990 and 1996 applications of both agricultural chemicals fell by anywhere from 8% to an almost unbelievable 24% per year. Water quality also dramatically improved throughout the region. The Baltic Sea, which borders Denmark, Estonia, Finland, Germany, Latvia, Lithuania, Poland, Russia and Sweden, saw especially large improvements during this period (Hughes and Lovei, 1999).

In Slovakia, subsidies for nitrogen, phosphorus and potassium fertilizers were phased out between 1989 and 1992, with retail prices basically doubling by 1995. In the process, because fertilizers were more expensive, total tons of fertilizer used in the country fell by 82%. Applications of phosphorus and potassium per hectare of agricultural land dropped by 90% or more during this period[16] (Statistical Office of the Slovak Republic, 1995).

12.5 Chapter Conclusions

Just like water, which was discussed in the previous chapter, energy is enormously important for modern economies. Energy consumption has grown substantially more slowly than economic output during the past decades, and high-income countries use a lot more energy than lower-income countries.

Because of its economic and social importance, the production and end-use cost of energy are of keen interest to governments around the world. Governments will therefore sometimes explicitly or implicitly subsidize energy. Measuring subsidies is tricky, but where they exist, they reduce incentives for conservation and promote energy use, with sometimes enormous budget expenditures and pollution. In general, energy subsidies reduce social welfare, though subsidies that address externalities, such as indoor air pollution from biomass fuels, may increase human welfare. All energy subsidies must be analyzed carefully because of potential lock-in effects.

Comprehensive economic planning in the so-called Soviet Bloc was an extreme example of what can occur when subsidies are substantial. All economic activities in the Soviet Union and most allied countries were under government control. In principle, this structure could have made it possible to better protect the environment. In practice, though, government control resulted in substantially more environmental problems. Though we may not know for sure how much of this extra environmental damage was directly due to energy subsidies, we know that energy in the Soviet Bloc was much cheaper than in the West. When the system fully ended in 1991, energy and other prices shot up, consumption of energy fell and pollution problems came to look much more like those in Western Europe. Now 11 of the 27 members of the European Union are former Soviet Bloc members. Three current EU members (Estonia, Latvia and Lithuania) were actually part of the Soviet Union.

Issues for Discussion

1. What is subsidization and can you name two cases where you receive government subsidies? How do you know you get subsidies?
2. Suppose you are an advisor to the Ministry of Environment in Ecuador, which regulates economic activity to improve the environment (like the US EPA). The minister is convinced that the gasoline subsidy in Ecuador is bad for the environment *and* for the economy. He asks you for arguments he can use to try to convince his counterpart at the Ministry of Energy and Nonrenewable Resources. Unfortunately, she has been very reluctant to increase gasoline prices because, "Ecuador is rich in oil and it should be used for the benefit of the people." What is your counter-argument?

Practice Problems

1. Suppose the average rush hour traffic on a stretch of Highway 271 in Cleveland, Ohio was 10,480 cars per hour in 2023 and 13,490 in 2024. What is the percentage change in average rush hour traffic?
2. Based on monitoring data from the same stretch of Highway 271 mentioned in Question 1, in 2023 average daily particulate matter (PM) concentration in the atmosphere was 75 micrograms per cubic meter ($\mu g/m^3$). In 2024, the average concentration was 98 $\mu g/m^3$.

   a. Which grew faster, traffic or PM? What is the percentage point difference (i.e., by how many percentage points did traffic grow more or less quickly than PM)? What is the percentage difference in these growth rates?
   b. Assume that the increase in traffic *caused* the increase in PM. Further, assume that the relationship between rush hour traffic and PM concentrations is linear. Suppose in 2025 the average rush hour traffic increased to 16,200 cars per day. Using your previous calculations, what is your prediction of the average daily PM concentration in $\mu g/m^3$?

3. Suppose the retail market for electricity in a region is in equilibrium at a price of $0.12 per kilowatt hour (kwh). A kilowatt is 1000 watts, which is a measure of energy usage. Suppose that the marginal private cost of producing electricity is also $0.12/kwh. The amount of energy produced and consumed is 1000 kwh. The government uses some of its tax revenue to pay electric companies to reduce the price of energy to $0.10/kwh.

Assume that each 1% decrease in the price of electricity causes the quantity demanded to rise by 0.7% and that demand is linear.

a.   What is your prediction of the quantity demanded after the electricity subsidy?
b.   Assuming that the marginal private cost of producing electricity continues to be $0.12/kwh, how much is the welfare loss due to overproduction of electricity that is caused by the subsidy?
c.   Electricity in the region is produced by burning coal in large power plants, which emit air pollutants like SO_2, CO_2 and PM. A clever specialist estimates that the monetary value of the damages due to producing 1 kwh of electricity is $0.02. What is the estimated monetary value of environmental damages due to the electricity subsidy?

4.   **Challenge Problem**: Figure 12.5 presents the estimated demand for gasoline in Indonesia at various prices. The associated discussion goes through how to calculate social welfare losses due to over-consumption of gasoline from the subsidy. The text also addresses the air pollution damages due to more use of gasoline. I have included in the figure an estimate of the *efficient* level of gasoline consumption, which would occur if the price were $4.22 per gallon, the estimated full social marginal cost of consuming a gallon of gasoline.

a.   Using the graph, what would be the additional social welfare gain by pricing gasoline at $4.22 per gallon?
b.   Using the information provided in the graph and the text, derive the linear demand function used to make the estimates. Just a reminder that linear demand will be of the form $P = a - bQ$.

Notes

1   Often, as in the table, this is phrased as per capita, which is Latin for "by heads."
2   Source: https://fred.stlouisfed.org/series/A939RX0Q048SBEA
3   The US Federal Reserve Bank, which is the central bank of the US, targets annual inflation at 2%.
4   % change in energy from coal = {(168.15 - 93.05)/93.05}*100 = 80.7%.
5   These data are from the World Bank and can be found at and downloaded from https://data.worldbank.org/indicator/ny.gdp.mktp.kd.zg
6   For better or worse, there are many energy measures, including calories, British thermal units (BTU), tons oil equivalent and electron volts, but fortunately they are easily converted to each other. A megajoule is 1,000,000 joules, which is about 0.000024 tons oil equivalent. To convert tons of oil equivalent to megajoules, just multiply by 41,868.
7   https://www.c2es.org/content/international-emissions/
8   This is not the only way to define energy subsidies. As we discuss later in this chapter and as mentioned in Box 12.1, the International Monetary Fund (IMF) includes unpriced pollution externalities in one class of subsidy measures. Using this metric, high-income countries do subsidize fossil fuels.
9   There are 3.785 liters/US gallon. To convert price/liter to price/gallon, just multiply (price/liter)*(liters/gallon). Notice that liters cancel, leaving price/gallon.
10   For information on gasoline prices around the world, see the IEA website: https://www.iea.org/data-and-statistics/data-tools/end-use-prices-data-explorer?tab=Overview

11 Analyzing the LPG share of the fossil fuel subsidy is a bit tricky. LPG subsidies may be welfare-increasing, particularly if they sufficiently reduce indoor and outdoor air pollution externalities and human mortality from cooking with biomass fuels.

12 If you need a refresher on elasticities, please see Chapters 7, 9 and 10.

13 Due to competition as discussed in Chapter 3.

14 I have also calculated the efficient level of gasoline consumption in which all costs – including environmental externalities – are included in the price of $4.22/gallon. This value turns out to be 7530 million gallons. Reducing consumption from 8551 to 7530 million gallons would further increase social welfare. In the practice problems, you will have the chance to calculate this additional welfare increase.

15 Though because of poor information and very different national accounting methods estimating GDP in the Soviet Bloc was challenging, East Germany's GDP was approximately $79 billion in 1988. West Germany's GDP was about $1400 billion, France's GDP was approximately $1100 billion and the US GDP was $5200 billion (World Bank, undated; Protzman, 1989).

16 Between 1918 and 1992, Slovakia was joined with the Czech Republic in what was called Czechoslovakia. One hectare is 10,000 square meters or about 2.5 acres.

Further Reading and References

Aldy, J. 2013. "Proposal 5: Eliminating Fossil Fuel Subsidies." In *15 Ways to Rethink the Federal Budget*, ed. M. Greenstone, M. Harris, K. Li, A. Looney and J. Patashnik. February. The Brookings Institution: Washington, DC.

Bates, R., S. Gupta and B. Fiedor. 1994. "Economywide Policies and the Environment: A Case Study of Poland." World Bank Environment Working Paper No. 63. January. World Bank: Washington, DC.

Black, S., A. Liu, I. Parry and N. Vernon. 2023. "IMF Fossil Fuel Subsidies Data: 2023 Update." International Monetary Fund Working Paper WP/23/169.

Coady, D., I. Parry, L. Nghia-Piotr Le and B. Shang. 2019. "Global Fossil Fuel Subsidies Remain Large: An Update Based on Country-Level Estimates." IMF Working Paper WP/19/89. May. International Monetary Fund.

Coady, D., I. Parry, L. Sears and B. Shang. 2017. "How Large are Global Fossil Fuel Subsidies?" *World Development* 91: 11-27.

Davis, L.W. 2017. "The Environmental Cost of Global Fuel Subsidies." *The Energy Journal* 38 (S1): 7-27.

Ettehad, S. and T. Sterner. 2012. "Distributional Effect of Reducing Transport Fuel Subsidies in Iran." In *Fuel Taxes and the Poor: The Distributional Effects of Gasoline Taxation and Their Implications for Climate Policy*. RFF Press: New York.

French, H. 1990. "Green Revolutions: Environmental Reconstruction in Eastern Europe and the Soviet Union." Worldwatch Paper No. 99. Worldwatch Institute: Washington, DC.

Hughes, G. and M. Lovei. 1999. "Economic Reform and Environmental Performance in Transition Economies." World Bank Technical Paper No. 446. World Bank: Washington, DC.

International Energy Agency (IEA). 2023. *CO_2 Emissions in 2022*. Retrieved from https://www.iea.org/reports/co2-emissions-in-2022 March 2024.

International Energy Agency (IEA). Undated a. International Energy Agency Energy Statistics Data Browser. Retrieved from https://www.iea.org/data-and-statistics/data-product/fossil-fuel-subsidies-database March 2024.

International Energy Agency (IEA). Undated b. *Indonesia*. Retrieved from https://www.iea.org/countries/indonesia/oil March 2024.

Jara, H.X., P.C. Lee, L. Montesdeoca and M. Varela. 2018. "Fuel Subsidies and Income Redistribution in Ecuador." WIDER Working Paper 2018/144. November. United Nations University World Institute for Development Economics Research.

Jeuland, M., S. Pattanayak and R.A. Bluffstone. 2016. "The Economics of Household Indoor Air Pollution." *Annual Review of Environmental and Resource Economics* 7: 81-108.

LaFave, D., A.D. Beyene, R.A. Bluffstone, S.T.M. Dissanayake, Z. Gebreegziabher, A. Mekonnen and M. Toman. 2021. "Impacts of Improved Biomass Cookstoves on Child and Adult Health: Experimental Evidence from Rural Ethiopia." *World Development* 140: 105332.

Lim, S.S., T. Vos, A.D. Flaxman, G. Danaei, K. Shibuya, H. Adair-Rohani, M.A. AlMazroa, M. Amann, H.R. Anderson, K.G. Andrews et al. 2012. "A Comparative Risk Assessment of Burden of Disease and Injury Attributable to 67 Risk Factors and Risk Factor Clusters in 21 Regions, 1990-2010: A Systematic Analysis for the Global Burden of Disease Study 2010." *The Lancet* 380 (9859): 2224-2260.

McCoy, T. 2019. "Celebration, Cleanup in Ecuador as Deal Ends Nationwide Protests - For Now." *The Washington Post*, October 14.

Protzman, F. 1989. "East Germany Losing its Edge." *The New York Times*, May 15.

Sachs, J. 1994. Table 1.4 in *Poland's Jump to the Market Economy*. MIT Press: Cambridge, MA.

Sachs, J. 1995. "Keynote Address. Economies in Transition: Some Aspects of Environmental Policy." Environment Discussion Paper No. 1. February. Harvard Institute for International Development, Harvard University: Cambridge, MA.

Statistical Office of the Slovak Republic. 1995. *Statistical Yearbook (Štatistická ročenka Slovenskej republiky)* . Bratislava.

Steiner, A. 2008. *Reforming Energy Subsidies: Opportunities to Contribute to the Climate Change Agenda*. United Nations Environment Programme: Nairobi.

US Energy Information Agency (US EIA). Undated. *Refined Petroleum Product Consumption (Mb/day)*. Retrieved from https://www.eia.gov/international/data/world/petroleum-and-other-liquids/annual-refined-petroleum-products-consumption October 2024.

Wilczynski, P. 1990. "Environmental Management in Centrally Planned Non-Market Economies of Eastern Europe." World Bank Environment Working Paper No. 35.

World Bank. Undated. "GDP (Current US$)." *World Development Indicators*. Retrieved from https://data.worldbank.org/indicator/NY.GDP.MKTP.CD November 2023.

Zamparutti, A. 1999. "Environment in the Transition to a Market Economy: Progress in Central and Eastern Europe and the Newly Independent States." Organisation for Economic Co-operation and Development Nonmembers Branch: Paris.

SECTION III

Solving Environmental Problems around the World

13 Social Institutions for Better Environmental Collective Action

What You Will Learn in this Chapter

- About governments as collective action institutions
- How governments get resources to do collective action activities
- Privatization as a mechanism for addressing environmental moral hazard
- Community-level collective action success stories

13.1 Recap and Introduction to the Chapter

The previous section presented a standard economic view of natural resource problems. The part of this book titled *The Challenge of Protecting Natural Resources* started with the notion that environmental assets are inherently difficult to manage. They produce many ecosystem services, which is great, but because some (especially pollution and provisioning/ direct use services) conflict with other services, natural resources can be hard - and political - to manage. As was discussed long ago by Krutilla (1967), not all ecosystem services are under similar levels of threat. Typically, it is common pool ecosystem services, which are depletable/rival and difficult to exclude, that keep us up at night. Examples of these shared services include a stable climate, respiratory health due to clean air and biodiversity in forests.

This combination - important competing and common pool ecosystem services - creates nasty incentives for humans to not take due account of the value of common pool ecosystem services when they make decisions. Examples include driving too much, biking too little, using outdated heating systems, converting forests to agriculture or housing and malls, buying too much stuff, watering too much, running air conditioning when not necessary, buying inefficient equipment, yada yada yada.

To derive appropriate policy instruments to stem the slide in key common pool ecosystem services, we need to have some idea of the weird incentives we are dealing with. Subsidies, which reduce costs of market goods and services courtesy of governments, should always be viewed with suspicion, because of the well-known economic distortions they may create.

DOI: 10.4324/9781003308225-16

Subsidies can twist economies in favor of explicitly or implicitly subsidized markets, often with unintended environmental consequences.

But the problem of antisocial incentives is not even close to being all about government subsidies. Property rights may be poorly defined, creating self-interested incentives for natural resource overuse and under-investment. Of course, if all of us were to act in the collective action interest (whatever that is!), there would be no need to worry so much about property rights. Improving property rights to help partially align private and social interests can therefore be thought of as one, albeit imperfect, way to deal with collective action problems.

We can also look at some problems, especially pollution, productively as negative and unaccounted-for environmental spillovers from markets onto society as a whole. Of course, if firms *chose* to conduct business as collective action enterprises, externalities would be automatically internalized. Internalizing externalities can therefore also be viewed as a collective action tool.

Both these ways of thinking about environmental problems rely on a human view of natural resources that is primarily self-interested, which no doubt has a fair amount of on-the-ground credibility. Property rights and internalized externalities approaches are especially useful, because they have classes of solutions that naturally fall out from them. For externality-type problems, the challenge becomes to "get the prices right," often using tax mechanisms. When property rights are vague, a part of the solution may be to clarify rights. Of course, many, many natural resources generate ecosystem services over which it is hopeless to try to firm up property rights or get prices right – basically all the supporting services that make life possible, and most regulating and cultural services.

Viewing the challenge of conservation through a collective action lens is a versatile and often the most relevant framework for conceptualizing environmentally bad behavior in the 21st century. We can all identify environmentally harmful moral hazard we have personally engaged in despite our environmental *bona fides*. We are all also well-familiar with the difficulties of working with others, including frustrations associated with free-riding, which is again partially due to self-interested behavior. Most of our worst environmental problems are therefore probably best thought of as collective action problems, but unfortunately this conceptual framework leads less cleanly to policy solutions.

Fortunately, environmental economists and others have begun to think about policy instruments to improve environmental collective action and outcomes. This chapter begins the final section of this book, which presents some of the most important, relevant, though still in-progress, policy instruments to deal with the enormous challenges of protecting natural resources. In this chapter, we focus on government and grassroots social institutions, property rights and policies to create group norms, which have been productively used around the world to address collective action failures.

13.2 Governments as Collective Action Institutions

Governments are constituted at national, state/province, regional (e.g., county) and local levels, such as cities, towns and villages. These governments, whether they be in India, Chile, Denmark, Canada or Vietnam, collect resources via taxes and other sources, borrow money and may be allocated funds from national governments. These governments then use money to provide services as agents of collective action.

We know these government services very well. Governments are generally responsible for roads, water, sanitation, macroeconomic policy, national defense and a variety of regulations on the private sector, from banking management to occupational health and safety and environmental protection. Services are provided by governments in some countries, but not others. For example, many countries at least purport to provide health care for all citizens, but some do not (e.g., the US and Chile). In most countries, governments deliver drinking water, but in some places in some countries water systems can be private (e.g., France and the UK). Electricity is often provided by governments, but there are a lot of places where power is produced and distributed by private companies under a variety of government regulations related to services and rates. Though the vast majority of roads around the world are public, it is not hard to find privately financed roads with rates regulated by governments.

There is enormous variation in the levels and qualities of services governments provide to their countries' residents. In some countries, services are extensive, and governments seem pretty responsive. In other countries, much less so. There is also huge variation in quality within countries, with some state/province and local jurisdictions offering very high-quality services – perhaps accompanied by high taxes – and some places doing the minimum or less.

It is interesting to also note what governments typically **do not** do. Though again there are exceptions, governments generally do not sell coffee. They don't develop online games or produce movies. They don't make cars. Basically, governments typically do not provide anything that in Chapter 8 we called "private" goods and services – anything that is fully excludable and depletable/rivalrous. They also may be less involved in excludable, non-depletable/non-rival "club" services than the private sector. Examples of such club services include music, apps, movies and uncrowded fitness clubs.[1]

Basically, governments do things that can only be done if we do them collectively, and they are without question our #1 institution for environmental collective action. Again, with varying degrees of quality around the world and within countries, governments build and run sewage treatment plants and systems. They manage protected areas, preserve wilderness, make sure that people can use beaches and protect migratory birds. Governments are also **the** institutions for pollution control, with key roles in ensuring that humans and other species can breathe the air, that rivers are clean enough to be swimmable, and that climate change is eventually stabilized (fingers crossed), among many other tasks.

How do governments get the money they need to provide services? Answer: They mainly levy taxes and fees on the private sector, pulling out value that is then used for collective services. They also borrow money from the private sector. In countries such as China, publicly owned enterprises, which produce 40% of GDP, can also generate significant revenues (BNP Paribas Bank, 2021). To keep things simple, let us ignore borrowing and public enterprises in our consideration of government revenues.

Taxes include levies on individual and business incomes, property, wealth, sales, value added and foreign trade.[2] The size of tax revenues is typically measured relative to the size of the economy, measured as GDP. Total tax revenues in 2019[3] as a percentage of GDP range from 36% in tiny Nauru in the South Pacific and 33% in larger Denmark, to under 10% in several countries (e.g., China, Ethiopia, Somalia, Bangladesh, Indonesia). The average across all countries is 17%, which means that on average 17% of the value of output is swept up as tax revenues and run through governments. Notably, about 40% of countries do not provide these data - or indeed any data about their tax revenues - perhaps because they do not want to report them, do not know them or do not have functioning formal tax systems. Basically, all these non-reporting countries are in the lower-income world (e.g., Afghanistan, Egypt, South Sudan, Tanzania, Tajikistan, Zimbabwe).

What do countries mainly tax to get resources? Taxes on incomes, capital gains (i.e., increases in value of certain assets) and business profits lead other sources, with an average country getting 38% of its 2019 revenues from those sources. The leader here is the US with over 90% of revenues, but several countries get under 10% from these categories (e.g., Latvia, Russia, Saudi Arabia, Belarus).

Taxes on goods and services run a close second, with the average country receiving 31% of its 2019 revenues from such taxes. The leaders are the Bahamas and Mauritius (tourism-dependent island nations) at 60%, lots of countries score over 40% and the US is at the bottom with only 2.6% of revenues derived from taxing goods and services. The European Union is in the middle of the pack at 33%. A total of only 14 countries reporting received less than 15% of tax revenues from goods and services and, as is the case with all these data, about 40% of countries did not report.

Taxes on international trade, such as on imports or exports, on average make up a relatively small share of total taxes. The average country in 2019 derived about 5.5% of revenues from trade taxes, with 40 countries getting less than 1% of tax revenues in this way. The European Union countries received no revenue from trade and high-income countries on average did not tax this sector. A minority of countries, mainly in the low- and lower middle-income world, get larger percentages from trade. Twenty-seven countries, including Ghana, Jamaica, Gabon, Philippines and Fiji, get more than 10% of tax revenues from trade, led by Somalia (32%), Solomon Islands (27%) and Namibia (27%). In these countries, trade taxes are roughly as important as taxes on goods and services are in most other countries. Though trade taxes can be detrimental to the functioning of economies because they make imports and exports more expensive, compared with collecting income and value-added taxes, it is relatively easy to levy taxes at the border. For lower-income countries with weak public administration systems, trade taxes can therefore be important sources of revenue.

In several chapters, we have discussed issues related to the world's poorest countries. These countries often cover significant portions of central government expenditures with foreign aid provided by countries in the Global North. In fact, government budgetary assistance from other countries is sometimes many times more than the total amount spent by governments themselves![4] For example, the Central African Republic gets total foreign assistance that is three times total government spending. Somalia also gets much more than the government spends and 16 countries get foreign assistance (what the World Bank calls "official development assistance" or ODA) that is more than half the total government budget. Thirty-six countries get foreign help to the tune of more than 10% of what the central government spends.

The World Bank also reports ODA revenues as percentages of gross national income (GNI), and several countries get significant support for their economies from ODA. For example, Tuvalu in the South Pacific gets half of GNI from ODA, Syria gets almost half (46%), and nine countries receive over 20% of GNI from other countries. Afghanistan, Burundi, the Central African Republic and Somalia are all part of these nine. Many countries get over 10% of GNI from foreign assistance. Almost 100 countries get 1% of GNI, which is an important portion, from foreign aid.[5]

Though governments are our #1 environmental collective action institution, sometimes they may not be quite right for the job. The next section discusses privatization of natural resources – devolution to individuals, households, firms and perhaps other institutions – followed by local community collective action institutions, as potential ways to neutralize some of the moral hazard problems associated with collective action.

13.3 Privatization to Sort Out Environmental Collective Action Problems

In this subsection, we will talk about policies that have been used to support and improve collective action, but as has been emphasized throughout this book, flexible and practical approaches seem to be necessary to save the planet. As we discussed in Chapter 9, open access to natural resources means free, open and perhaps no-limit availability of provisioning (typically, private) or perhaps cultural (generally, club) ecosystem services that are either bought and sold or used directly.

Open access usually ends up with gross misallocations, overuse and natural resource degradation and has often been the starting point for making available natural resources like forests, fish, water, land and even minerals. This approach can perhaps work when natural resources produce a few extractable products (i.e., there are few conflicts between different provisioning ecosystem services). It also needs to be the case that for whatever reason it is reasonable for everybody to take what they want.

When managing renewable natural resources, it can sometimes make sense to just focus on the private and club ecosystem services (see Chapter 8) that environmental assets produce and concentrate our policy efforts on clarifying property rights. Getting rid of open access to provisioning ecosystem services is an important result, and it is possible and even likely that other services, such as regulating and supporting ecosystem services, will increase as provisioning and cultural services from renewable resources become more available.

13.3.1 Private Forests

In the US, private forest ownership is very important. More than half of all US forest lands are private, almost 40% are owned by families and only about 2% are community controlled.

Private forests are estimated to be the origin for about 30% of drinking water for the country and over 90% of domestically produced wood (USDA, undated). In Europe, as of 2010 more than half of all forest lands were private (Schmithüsen and Hirsch, 2010) and forests cover about 1/3 of Europe (Nichiforel et al., 2018).

> ## Box 13.1 Devolution of Mangrove Forests to Households in Vietnam
>
> A three-author team led by Truong Dang Thuy of the University of Economics in Ho Chi Minh City in Vietnam examined whether devolution of mangrove forests from the central government to households increased mangrove area in the Mekong Delta Region. Mangroves grow in saltwater, supporting fisheries, providing coastal protection and storing some of the highest densities of carbon in the world. Despite their ecological and economic importance, mangrove areas in Viet Nam decreased rapidly during the last century. In the Mekong River Delta, mangrove area fell by about 65% between 1975 and 1995, mainly due to cutting for wood and conversion to shrimp farms.
>
> Like so many countries facing rapid deforestation, Vietnam decided to devolve management to local people. As in China, management was decentralized to households, who were allowed limited access to the provisioning ecosystem services of mangroves. In this case, that meant using some area for shrimp farming.
>
> The authors found that devolving mangroves to households – without allowing households to do whatever they want, but giving them the responsibility for conservation – increased mangrove area. Interestingly, they found that the best devolution policy was to have management contracts with households, which explicitly give them rights – and responsibilities – for mangrove management.
>
> Source: Thuy et al. (2021).

Up until the early 1980s, all forests in China were either government-owned or controlled by villages. Such government ownership and control approximated free access conditions and villagers and state forestry enterprise managers, who used and controlled forests, had limited incentives for sustainable management. In response, starting in 1981 as part of a wide-ranging reform of land use policies, the central government began allowing villages to *devolve* (i.e., move "down") forests to the household level. By taking such steps, the government sought to transfer responsibility for, and the benefits of, planting and managing forests to farmers. These and additional rules clarified property rights around forests, sidestepping many of the problems associated with forest collective action, and created essentially private forest plots. As we have seen, private plots are common in the US, Europe and elsewhere.

By 1986, in several provinces almost 3/4 of forest area was under household management. This process of privatization continued through the 2000s and over time dramatically increased (Xu and Jiang, 2010). These policy changes made villagers responsible for forest health and required permits for logging. Land could not be changed from forests to other uses. This set of policy reforms is now known as the Three Fixes Program and was substantially strengthened by the government in 2003 (Xie et al., 2016).

This and other land tenure reforms in China seem to have had enormous effects. Estimates of changes in forest area depend on the definition of forests, methods used (there are lots of methodological issues!), etc., etc., but the official China national forest inventory reports that total forest area increased by over 2% during 2004-2008 and another 3% from 2009 to 2013 (Li et al., 2017). Nationwide, the portion of the country covered by forests increased from a little under 19% of the total land area in 2000 to over 21% in 2010. As of 2020, forests covered 23% of China's land (World Bank, undated). These are very large increases,[6] especially considering that between 2000 and 2010 world forest area **declined** by 1.5% (D'Annunzio et al., 2017). Using careful methods, Xie et al. (2016) estimated that in their study areas the 2003 tenure reforms, which promoted individual and partnership tenure, almost immediately increased forest cover by about 8%. Most of the increase was due to farmers planting tree seedlings, which hopefully grew into mature forests.

China's forest reforms have been lauded for using the power of private incentives to enhance forests (e.g., see Zhang, 2001; 2019; Xu et al., 2006). Analysts have often viewed the topic through forest area or timber products lenses (e.g., see Ren et al., 2015; Zhao et al., 2022; Hyde, 2019), which may be incomplete and insufficient measures of forest improvement. Several researchers on Chinese forestry have indeed argued that a lot of the new forests are single species, often called *monoculture* plantations (e.g., Zhai et al., 2014; Hua et al., 2016; 2018).[7] Some argued that trees were planted in environments that historically did not grow trees (Xu, 2011), which may not increase ecosystem services such as wildlife or biodiversity. That said, as discussed by Zhang et al. (2000), the original, pre-reform forestry system was certainly no prize and itself relied heavily on monoculture and resulted in widespread loss of natural forests.

13.3.2 Private Fisheries: Aquaculture

FAO et al. (2022) evaluated the state of global fisheries (i.e., "capture" fishing) and found that about 35% of fish and other aquatic animal stocks that humans use for food are at biologically unsustainable levels (up 1.2% from 2017 and up from 10% in 1974). Almost all the 65% of fishery stocks that are within biological limits are fished at their maximum possible levels, which suggests problems in fisheries management. At the same time, aquatic food consumption per capita increased by almost twice the rate of population growth, indicating that they are making up increasing portions of diets around the world (FAO et al., 2022).

Ashe et al. (2018) describe fishing and collecting of other marine species as "the last major hunting and gathering industry" (Fluet-Chouinard et al., 2018). Citing many sources (e.g., Gordon, 1954; Smith, 1969; Clark, 1973), they note what should by now be firmly imprinted in all our minds, that open access to renewable resources, such as fish and other aquatic animals, is the enemy of natural resources, leading to overexploitation and under-investment. They emphasize the role of marine fisheries management, which we will sidestep in favor of focusing on private fisheries as a potential policy to deal with open access.

Box 13.2 Are Oysters Expensive Where You are? Blame Lack of International Trade

Oysters are one of the oldest types of aquaculture and are farmed around the world. Global production has increased over time and has long been dominated by Asian growers. Since 1990, China has by far produced the most oysters, but as is the case in other countries, only a small percentage of Chinese oysters are exported. The global market for oysters is therefore *segmented*, meaning that markets are independent, not linked and consumers in other countries cannot tap into productivity in other countries. Prices therefore differ dramatically across countries. In 2016, for example, prices in the US were about 1/4 of the prices in France and half of the prices in Canada and Japan. South Korean prices were roughly 1/8th of French prices! International trade would tend to equalize prices across countries.

Source: Botta et al. (2020).

According to the US National Oceanic and Atmospheric Administration (NOAA), which is a scientific agency that is part of the US Department of Commerce and focuses on the air and water, aquaculture is the "breeding, raising, and harvesting of fish, shellfish, and aquatic plants. Basically, it's farming in water" (NOAA, undated).[8] In the US, *marine* aquaculture produces species such as oysters, clams, mussels, shrimp, seaweeds, and fish such as salmon, black sea bass, sablefish, yellowtail and pompano. Marine shellfish production can be done by "seeding" small shellfish on the seafloor or by growing them in cages. Marine finfish farming is typically done in net pens in the water or in tanks on land. NOAA (undated) notes that "U.S. aquaculture is an environmentally responsible source of food and commercial products, helps to create healthier habitats, and is used to rebuild stocks of threatened or endangered species."

Environmentally responsible or not, aquaculture is clearly a growth industry, which supplies over half of humans' aquatic foods (Botta et al., 2020). It has also perhaps displaced open access fishing as consumption has increased. In a wide-ranging review, Naylor et al. (2021) showed that global aquaculture production – basically all private – tripled between 1997 and 2017 to 112 million metric tons and FAO et al. (2022) reported that production reached a record 123 million tons three years later. Seventy-five percent of that tonnage was produced in freshwater environments, and almost all of it came from Asia (Naylor et al., 2021); China alone supplied almost 60% of global aquaculture volume.

Farmed shrimp offers examples of some of the potential problems with aquaculture, especially in lower-income countries. Shrimp is an important marine aquaculture product, particularly in Southeast Asia, where it is produced for export. Production ponds are filled with seawater and are therefore typically located in coastal zones or just inland. They are stocked very intensively, requiring significant inputs of antibiotics to control diseases, aeration to keep dissolved oxygen levels up and external feed inputs. Aquaculture often relies on feed from wild-caught fish, which may be essentially open access (Belton et al., 2020). In China, up to 1/3 of smaller schooling fish (sometimes called "bait" fish) are fed directly to aquaculture species (Liu and Su, 2017).

Intensive production of aquatic foods like fish and shrimp creates a lot of wastes that pollute marine and freshwater environments (Naylor et al., 2021), can alter aquatic biochemistry (Liu and Su, 2017) and contribute to low dissolved oxygen conditions (Wartenberg et al., 2017). When shrimp are harvested, production ponds are often simply drained into marine ecosystems, creating significant pollution plumes. As other shrimp producers take in seawater to fill ponds, these large-scale discharges can spread diseases, sometimes destroying shrimp production across large areas.

Mangroves are trees or shrubs that grow in salty soils along tropical and subtropical coastlines. They are important, because they protect the coastlines from storms, sequester carbon and offer habitats for a variety of marine species, including commercially important fish. Unfortunately, mangrove forests are often cut down to create shrimp and other aquaculture farms. Goldberg et al. (2020) used satellite data and found that globally about 3400 square kilometers of mangrove forests were lost during 2000-2016. Most of that deforestation occurred during 2000-2005 and the authors found that humans caused almost 2/3 of the reduction in mangrove area. They also found that about 80% of human-driven mangrove loss occurred in six Southeast Asian countries where aquaculture, including shrimp production, was especially important.

13.3.3 *Private Protected/Conservation Areas*

The International Union for Conservation of Nature (IUCN) defines a protected area as "a clearly defined geographical space, recognized, dedicated and managed, through legal or other effective means, to achieve the long term conservation of nature with associated ecosystem services and cultural values" (IUCN, undated).

A total of 15.3% of global land and freshwater environments were protected in 2019 (up from 14.1% in 2010) and somewhere between 3% and 7.5% of the marine world (Maxwell et al., 2020). This amount is close to, but below, the 17% by 2020 specified in the 2010 Strategic Plan for Biodiversity under the Convention on Biological Diversity (CBD), which was signed by 168 countries in 1992. It is also only half of the 30% of land and ocean area to be conserved by 2030, which was specified in the 2021 meeting of CBD signatory countries (UN, undated). All this is much, much below the 50% of global land and sea area that the noted Harvard University biologist Edward (E.O.) Wilson said should be preserved (https://www.half-earthproject.org/).

Many of us think of habitat conservation as something that only governments do, and to some degree that is true. As of 2014 (the latest data I could find), globally about 77% of protected areas were government owned or managed and another 13.5% were co-managed with other actors in the country (Stolton et al., 2014). But other institutions, including the private sector, also contribute to increasing habitat protection and in some cases, private management can be a successful approach.

Private protected areas (PPAs) are officially called "other effective area-based conservation measures (OECMs)" by the CBD (Maxwell et al., 2020). Though poorly documented (e.g., see Drescher and Brenner, 2018; Mitchell et al., 2018), OECMs are probably preserving significant habitat. Private land is very important around the world, often making up well over 50% or more of countries' land areas (Powell, 2012). It is therefore perhaps not surprising that some of that land area would be used for conservation, effectively sidestepping the problem of collective action.

Box 13.3 Bosque Pehuén Private Protected Area in Chile

Bosque Pehuén is a private forest, which covers 882 hectares that are typical of temperate rainforests in South America. It is located adjacent to the 63,000-hectare Villarrica National Park, offering important connectivity and collaboration benefits. Both protected areas are in the Andean foothills of the Araucanía region, an area of Chile that has high levels of biodiversity and *endemism* (i.e., species not found elsewhere). Located in the popular lake district, it is also one of the most human-affected areas of the country and the landscape is one of the least-protected.

Bosque Pehuén conserves grasslands, old-growth forests, secondary forests and riparian zones. It is also an important area for animals, such as pumas, foxes, woodpeckers and many other birds. It protects endangered species, such as the araucaria (*Araucaria araucana*), Darwin's frog (*Rhinoderma darwinii*) and the little mountain monkey (*Dromiciops gliroides*). In addition to conservation, the PPA supports research, acts as a laboratory for innovative conservation activities and promotes environmental awareness.

Sources: Mitchell (2018) and https://fundacionmaradentro.cl/en/
proyectos/bosque-pehuen/.

The World Database of Protected Areas (https://www.protectedplanet.net/en) reports 22,637 PPAs, most of which are in the US and New Zealand. As of 2017, these lands covered over 161,000 km^2, which is only 0.42% of the total global area. This finding may be partly because reporting on PPAs is weak, and in most countries they are not recognized under law (Bingham et al., 2017). PPAs collectively represent 8.5% of total protected areas, and most are owned by individual landowners (69%), followed by non-profit organizations (27%) (Bingham et al., 2017).

Of course, unless private landowners have primarily altruistic motivations, which no doubt is the case for some, PPAs may only stay protected because they make money. This means that PPAs are producing important indirect-use ecosystem services, such as large mammal viewing, rafting or camping, which tourists value. Shumba et al. (2020) refer to this motivation as "If it Pays, it Stays."

A key question is how the quality of PPAs - or, more generally, protected areas that allow human uses - compares with protected areas that do not allow people to do activities that directly and economically benefit themselves. Elleason et al. (2021) analyzed exactly this question using the World Database of Protected Areas for 19,000 *terrestrial* (i.e., land-based) protected areas. They concluded that "...multiple-use areas are not necessarily less effective than strictly protected areas..." (p. 1070). They found that this result occurs because - not surprisingly - strictly protected areas tend to be in remote regions that people have no plans to use anyway. As Joppa and Pfaff (2009) noted, "Protected areas are biased towards where

Table 13.1 Number and Area of Private Protected Areas in Countries and Territories for which Data are Available

Country or Territory	Number of PPAs*	Area of PPAs (km²)	PPA Area as a Percentage of Country/Territory's Total Protected Area (Marine & Terrestrial)
US	11,877	21,821	0.24%
New Zealand	4694		
Australia	1620	47,756	1.10%
Canada	1192	232	0.02%
South Africa	922	26,045	9.30%
Colombia	912	803	0.45%
UK	690	1396	0.65%
Mexico	336	4036	1.14%
Guatemala	151	7028	19.60%
Peru	87	28,795	10.28%
Chile	19	3725	0.62%
Puerto Rico	17	402	0.24%
Bermuda	16		
Cayman Islands	16	13	10.81%
Kenya	16	1915	2.61%
Honduras	13		
Costa Rica	12		
Kingdom of Eswatini	9		
Virgin Islands	6	1	0.38%
Bonaire, Ste. Eustatius and Saba	5	77	48.95%
Fiji	3	18	0.13%
Nepal	3	11,657	33.40%
El Salvador	3		
Belize	2	42	0.35%
Falkland Islands (Malvinas)	2	6	5.58%
Madagascar	2	2113	5.95%
Marshall Islands	2	98	1.81%
Mauritius	2	3	1.67%
Namibia	2	2899	0.90%
Saudi Arabia	2		
Philippines	2**	0.40	0.00%
Botswana	1	752	0.44%

*In addition to the countries listed, nine countries have one PPA.

**Source: Bingham et al. (2017).

Sources: Bingham et al. (2021) Number of PPAs; Bingham et al. (2017) Other Data.

they can least prevent land conversion (even if they offer perfect protection)." They also noted that protected areas may be poorly resourced and may even merely be "paper parks" in which protection exists in law or regulation, but not on the ground, leading to inconsistent or no protected area results (e.g., also see Blackman et al., 2015).

But what about PPAs specifically? Though much more needs to be known about the effectiveness of PPAs, Shumba et al. (2020) reported that PPAs in South Africa (a total of

992 in 2021, see Bingham et al., 2021) appear to be conserving habitat. They matched PPAs with unprotected areas that had similar characteristics and found that between 1993 and 2013 PPAs lost significantly less natural land cover and biodiversity intactness than unprotected areas. They concluded that in South Africa, PPAs support conservation and within PPAs those with no legal requirements had natural land cover and biodiversity intactness that were similar to strictly protected areas.

Do PPAs serve any special ecological functions? It seems the answer is yes. Palfrey et al. (2022) analyzed data from over 17,000 protected areas around the world. They found that PPAs occupy special niches within the panoply of protected areas. They tend to be in areas with low protected area coverage, likely increasing their conservation impact. They are also generally in areas with lots of existing human disturbance, which is probably not a surprise given that many PPAs are commercial enterprises that can more easily do business if they are close to people. Finally, while they found that PPAs represent only a bit over 3% of protected areas in their database, they increase the connectivity of existing protected areas (i.e., the degree to which protected areas are linked to each other) by more than 7%, again suggesting an outsized footprint.

13.4 Policies that Create or Strengthen Local-Level Natural Resource Collective Action

As I have tried to emphasize throughout this book, saving the world is going to require a lot of flexible thinking and many different tools. Government-led environmental collective action is our go-to collective action institution, representing the vast majority of efforts around the world. Privatization can also sometimes be helpful, but in some cases groups – if they are small enough, united enough and well-defined enough – can be very effective.

We know from a variety of work by 2009 Nobel Prize in Economic Sciences Laureate Elinor Ostrom (see Chapter 10), her collaborators and many who have come after her that collective action related to common pool ecosystem services can work when the conditions are right. In her Nobel Prize address, Ostrom said that previous economic models, for example those based on the prisoner's dilemma, were too simplistic. They had only privately produced and government-produced services and purely self-interested people. She argued for a more nuanced, evidence-based approach. Rather than relying wholly on theoretical models of human behavior to guide, for example, ecosystem service policies, Ostrom said that approaches should be more practical and flexible, with the core goal to "facilitate the development of institutions that bring out the best in humans" (Ostrom, 2010, p. 435).

Ostrom recognized that collective action often falls apart, but also saw important examples of community-level natural resource management success. What were the key on-the-ground differences that distinguished success from failure? What was missing from failed collective action? These were the types of questions Ostrom and her colleagues were trying to answer. They used experiments very much in the mold of those presented in Chapter 10 to analyze collective action and ended up proposing eight general design principles for strengthening and building successful collective action institutions. These principles are presented in Table 13.2 and come from Ostrom (1993; 2010).

Table 13.2 Elinor Ostrom's Collective Action Design Principles

	Design Principle	More Details
1	Clearly Defined Boundaries	**User and Resource Boundaries:** Clear and locally understood boundaries between legitimate users and nonusers. Clear boundaries that separate a specific common pool resource from a larger social-ecological system.
2	Proportional Equivalence between Benefits and Costs	**Congruence with Local Conditions:** Appropriation and provision rules are congruent with local social and environmental conditions and with each other. The distribution of costs is proportional to the distribution of benefits.
3	Resource Rules Can be Changed by Users or Those Affected by Rules	**Public Participation:** Most individuals affected by a resource regime are authorized to participate in making and modifying its rules.
4	Effective Monitoring	**Monitoring Users:** Individuals, who are accountable to or are the users, monitor resources and the appropriation and provision by users.
5	Graduated Sanctions	**Graduated Sanctions:** Sanctions for rule violations start very low but become stronger if a user repeatedly violates a rule.
6	Conflict Resolution Mechanisms	**Conflict Resolution Mechanisms:** Rapid, low-cost, local arenas exist for resolving conflicts among users or with officials.
7	Recognition of Rights to Organize	**Recognition of Rights:** The rights of local users to make their own rules are recognized by the government.
8	Nested Enterprises	**Nested Enterprises:** When a common pool resource is closely connected to a larger social-ecological system, governance activities are organized in multiple nested layers.

These ideas and design principles have been used around the world to inform better natural resource collective action. In several cases, there have been important successes. Some of these group-focused, community-based innovations are discussed in the subsections below.

13.4.1 Community Forest Collective Action[9]

In the US and Europe, community control of forests is unusual and in China community-based forestry is the "old system," which is in the process of being reformed away. But much of the Global South is embracing decentralized, community management of forests. In fact, by one estimate almost 30% of all lower-income countries' forests are now managed by local communities, well over twice the share of land in protected areas (RRI, 2020).

Management decentralization is the transfer of some decision-making authority from higher to lower jurisdictions, including forest user groups, villages and other communities. During the past several decades, devolution and decentralization have arguably been the most important trend in lower-income country forest policy (Blackman and Bluffstone, 2021). Often forests are devolved with significant restrictions on users or they may be co-managed in important ways, with governments, including by jointly developing management plans.

Public finance constraints and public administration issues, demands by communities to manage forests, and external pressures can be significant drivers of decentralized forest management, which is going forward in dozens of countries (FAO, 2015; Agrawal et al., 2008). Management decentralization is also typically prompted by open access-type overuse problems. For example, Ethiopia is estimated to have originally been heavily forested, but by the early 2000s forest cover was estimated to be only 4.6%. During the 1990s, forest cover fell by 20% (EFAP, 1994; Tumcha, 2004; FAO, 2015 cited in Beyene et al., 2016).

In response to the decline in Ethiopian forests, the federal government started to transfer control and management to regional states and then ultimately to communities through a process called "participatory forest management" (Gebreegziabher et al., 2021). The jury is still out on the effects of these reforms - especially as at the time of this writing Ethiopia is engaged in a civil war - but the devolution process is continuing.

Community management can lead to better forests and maybe even less poverty, by giving rights - and incentives - to forest user groups who probably best understand local conditions and constraints. They therefore in principle have the capacity to sustainably manage forests (Somanathan et al., 2009; Ribot, 2008; Barbier and Burgess, 2001). Though I am not aware of evidence that suggests forest area and quality decline after decentralization, there may be equity downsides. For example, decentralization initiatives sometimes fail to promote meaningful shifts in power, leading to recentralization, or they may allow elites to control forests, perhaps at the expense of others (e.g., see Adhikari and Di Falco, 2009; Ribot et al., 2006).

Box 13.4 Participatory Forest Management in Ethiopia

Though collective action in irrigation water and grazing land management is centuries old and an indigenous tradition in Ethiopia, community-based forest management was only introduced in 2000. Participatory Forest Management is an agreement between regional state governments and communities to allow community management and utilization of forests.

The Sokora Forest Management Association in Oromia Regional State was the first forest user group to sign an agreement with the government. As of 2024, there are over 700 user groups in Oromia and a total of 1.6 million hectares of natural forest are under community management. In total, the Ethiopian federal government seeks to sustainably manage and protect over 2 million hectares through participatory forest management.

Source: Gebreegziabher et al. (2024).

Nepal is a country where significant forest devolution experience and research exist. Decentralization of forest management to communities started in the 1980s and was largely in response to forest degradation and deforestation in the Himalayan foothills. There were especially serious concerns about downstream effects, such as flooding, in the Nepali lowlands, India and Bangladesh (Carter and Gronow, 2005; Guthman, 1997; Springate-Baginski and Blaikie, 2007). In the 1980s and early 1990s, pilot projects, legislation and rules were put in place to give forest user groups legal control of forests. In 1993, this trend culminated with the formal establishment of the Nepal Community Forestry Programme, as well as other programs, and currently about 40% of the population is part of one or more community forest user groups (Fox et al., 2018).

Though there is mixed evidence on how these reforms have affected equity in benefit sharing (e.g., see Luintel et al., 2017; Adhikari and Di Falco, 2009), recent rigorous evidence appears to indicate that the Programme has promoted biodiversity and may increase carbon sequestration (Luintel et al., 2018). Remote sensing research suggests that forest cover increased from 26% of total land area in 1992 (at the start of the Nepal Community Forestry Programme) to 45% in 2016 (Fox et al., 2018). It is very likely that a significant part of the forest improvements observed in Nepal during the recent past are attributable to forest decentralization policies launched over 30 years ago.

So, stay tuned on devolution of forests to communities as a collective action-promoting policy. This is a worldwide trend that is if anything accelerating and has become closely linked with other key objectives, such as those related to Sustainable Development Goal 15 (Life on Land – see Chapter 1) and initiatives focused on Indigenous Peoples' rights.[10] There is also emerging evidence that forest management devolution and decentralization, by tightening up forest property rights, may generate ecosystem services like carbon sequestration and increased biodiversity, and help transform low-productivity rural agricultural systems. As I say, stay tuned.

13.4.2 Community Collective Action in Fisheries

Fishing is a critical source of livelihoods around the world, especially for the poor. According to the Marine Stewardship Council, which certifies seafood sustainability, up to 10% of all people on the planet rely on fisheries for their incomes (MSC, undated). If marine capture fisheries are "the last major hunting and gathering industry" (Ashe et al., 2018, p. 11221), it is also an industry that in many cases has a weak history of sustainability.

As shown in Fig. 13.1, large-scale, commercial fish stocks declined dramatically since indus-trial fishing ramped up in the 1960s, with some species (e.g., tuna) rebounding in recent years. Using long-term historical data, Jackson et al. (2001) estimated that regardless of where one looks, large-scale fishing and associated human actions have ecological effects that have made marine ecosystems essentially unrecognizable over time. Myers and Worm (2003) studied predatory fish like tuna, sharks and cod. They found that fish populations fell by 80% within 15 years of the start of industrialized fishing and estimated that the global ocean has lost more than 90% of large predatory fish, often linked to overfishing of smaller forage fish, such as herring, mackerel, anchovies and sardines. Essington et al. (2015) analyzed data from 55 forage fisheries over at least 25 years and found that about half had experienced collapse and most of the observed population crashes were preceded by overfishing. Parallel to the world of large-scale commercial fishing, which supplies much of the world's grocery stores, is enormous dependence on small-scale, local, often *artisanal* (i.e., traditional, often with limited mechanization) fisheries. According to FAO et al. (2022), almost half a billion people depend at least partially on small-scale fishing, which makes up 90% of fisheries employment.

The sector produces about 40% of global fish catches, with the largest small-scale fishery landings in Asia and Africa. In the Americas, almost 1/3 of total fish catches come from

Total fish stocks

Fish stocks are measured by their biomass: the number of individuals multiplied by their mass.

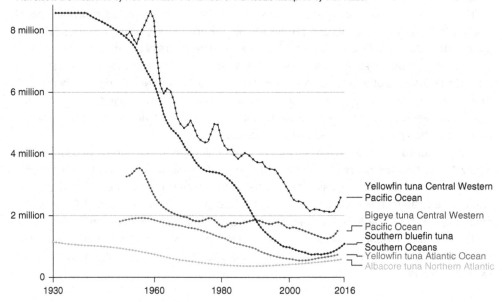

Figure 13.1 Estimated Commercial Fish Populations by Species and Select Region

Source: Ritchie and Roser (2021).

small-scale fisheries. About 2/3 of all landings are from marine (i.e., salt water) ecosystems and about 40% of small-scale fishers are women. Recognizing the importance of such fisheries, the United Nations General Assembly designated 2022 as the International Year of Artisanal Fisheries and Aquaculture.

A significant portion of these important, small-scale, local fisheries have community-level collective management.[11] FAO et al. (2022) estimated that at the local level, about 90% of small-scale fishery catches are by groups with formal collaborative management, and in 40% of the cases fishers are highly involved with management. Fisheries collective action is highest in the Americas, Europe and Oceania[12] and, though still significant, lowest in Asia and Africa.

Coral reefs are highly vulnerable and valuable ecosystems that have seen significant collective action management. Wamukota et al. (2012) conducted a wide-ranging review of community or co-management of coral reefs. They found that of the studies that analyzed various outcomes of fisheries collective action, about 90% highlighted economic and social improvements and at least two-thirds noted positive resource or ecosystem effects.

As we know from prisoner's dilemma–type models, resource degradation and reduced incomes can result from uncoordinated harvests of renewable resource provisioning services. In Bangladesh, 80% of households rely on fisheries – mainly small-scale – for food and/or income, but fishing incomes are low and basically the employment of last resort. Furthermore, fisheries productivity, measured as catch per unit of effort (e.g., per hour fished or money invested), declined substantially during the mid-2010s (Islam et al., 2014). Increasing the productivity and sustainability of small-scale fisheries is therefore potentially important for human livelihoods.

Both Islam et al. (2014) and Khan et al. (2012) analyzed the effects on incomes, expenditures, assets and food intake of a community-based small-scale fisheries collective action project implemented from 2002 to 2006.[13] Both groups of researchers found increases in these measures of human welfare. I would like to focus on the work of Khan et al. (2012), because they used especially useful methods to identify whether collective action actually *causes* increased incomes and expenditures.

In Chapter 8, we talked about the importance of not confusing correlation with causality. Lots of things can cause either higher levels of human welfare or collective action. Some factors may simultaneously cause both outcomes. Figuring out whether collective action really results in better outcomes is therefore a particularly gnarly type of impact evaluation. We suggested in Chapter 8 that experimental methods can help, but a second approach to the problem is to use so-called *quasi-experimental* methods. These approaches try to simulate what would have happened - all else held constant - if an intervention, such as a community collective action project, had not occurred.

Khan et al. (2012) used a quasi-experimental method called *propensity score matching*, which matched comparable fishers in the project with those outside the project. They then measured differences in outcomes depending on whether fishers were included in the project.[14] The authors found that incomes rose due to formal collective action and the increase was roughly equal to the average annual household income in Bangladesh at the time of the study. Expenditures were much higher for households in the project compared with non-participants, including spending on food, health and housing. The authors also found that small-scale fisheries collective action made incomes and expenditures more equal, which are critical effects.

The results from Bangladesh are very encouraging, but, as always, collective action - in fishing or otherwise - has challenges. Collective action projects, such as the one analyzed in Bangladesh, offered participants significant facilitation and finance support. Without outside support, coordination can be much tougher.[15]

13.4.3 Community-Based Wildlife Management

Sub-Saharan Africa has some of the most amazing and tourist-enticing mammals in the world, so it is not surprising that several wildlife management innovations have come from that region. With, of course, large variation across countries, sub-Saharan Africa is home to lions, gorillas, giraffes, elephants, zebras, chimpanzees, cape buffalo and many other species that are international tourist magnets. Translated into the language of Chapters 1 and 4, cultural ecosystem services and indirect use values from African mammals are extremely high. It is also the poorest region in the world, with all that entails, including the need to monetize and funnel ecosystem service values to locals, limited public sector monitoring and enforcement capabilities, and the critical importance of addressing human-wildlife conflicts.

Photo 13.1 Zebra in Eastern Uganda
Source: Author.

It is therefore perhaps not surprising that governments often need local-level buy-in to conserve wildlife. Sub-Saharan Africa is home to some of the most innovative, community-based wildlife collective management programs in the world. So-called community-based conservation has been heavily utilized in Kenya to provide benefits to communities and reduce human–wildlife conflicts from large mammals, such as elephants (Western et al., 2015). The same is true in Tanzania, where the Wildlife Management Areas Program appears to have increased wildlife densities and reduced cattle grazing in wildlife areas (Lee and Bond, 2018). In Namibia, community conservation programs have been found to increase forest cover and elephant populations (Meyer et al., 2021).

Uganda is one of just a couple of countries in the world that hosts mountain gorillas and one park, intriguingly named Bwindi Impenetrable National Park, has about half the world's approximately 1000 mountain gorillas (Bitariho et al., 2020). Established as a forest reserve in 1932 under British colonialism, the forest faced heavy illegal logging, mining and poaching, which in 1991 caused the Ugandan government to create the national park.

The park, which is managed by the Ugandan Wildlife Authority (UWA), is surrounded by farms and densely populated villages. The area near Bwindi has few other natural areas, suggesting what may have happened had the park not been established (Hamilton et al., 2000). Virtually all the people around Bwindi are subsistence farmers and they are often poor by international standards. Many people hunted and collected commercial and subsistence products from the forest reserve, but when the Bwindi Impenetrable National Park was established, harvests were prohibited, which generated a variety of conflicts, violence, arrests and penalties. Most conflicts were with illegal loggers and miners, but farmers also sustained damage from gorillas. Not surprisingly, some were angry.

These conservation-related conflicts appear to have been substantially reduced by the establishment of an integrated conservation and development program in 1994. Villages surrounding Bwindi and other national parks can bring forward complaints, participate in park management and receive significant monetary benefits, including a portion of the $700 per day entrance fee that foreign tourists pay the UWA to go on group gorilla treks with rangers. Villagers also can receive tips from international groups after public cultural programs and can collect agreed quantities of plants for subsistence purposes. Species that are endangered are being propagated and distributed at low or no cost to villagers by organizations, such as the Institute of Tropical Forest Conservation at Mbarara University (Hamilton et al., 2000).

As of 2018, the UWA and related organizations had funded 338 community-level projects in the jurisdictions (called parishes) surrounding Bwindi. Many of these projects focused on water supply and sanitation, household income generation and education, which are all critical needs in eastern Uganda. Communities where such projects are running receive recognizable benefits and perhaps engage in systematically fewer illegal activities in the park (total of 171 events in 2018, down from over 1500 in 2006) (Bitariho et al., 2022).

One of the first and most important community-based wildlife collective action management programs was the Zimbabwe Communal Areas Management Programme for Indigenous Resources (CAMPFIRE). Established in 1989, the origins of the CAMPFIRE program go back to the 1975 Parks and Wildland Act, which decentralized state authority and gave owners or occupiers of land specific rights over wildlife, fish and plants. The law gave responsibility for management to locals, but also gave them incentives to manage sustainably (Taylor, 2009). As of 2009, most of the country was part of CAMPFIRE.

Box 13.5 Does Community-Based Wildlife Management Increase Wildlife in Tanzania?

Tanzania has created 19 Wildlife Management Areas (WMAs) that are managed by communities. Villagers involved in the WMAs share in any revenues generated by wildlife tourism. Lee and Bond (2018) compared populations of giraffe, impala, zebra and dik-dik (a small antelope) in the Randilen WMA with the adjacent Lolkisale Game Controlled Area (LGCA) managed by the government. In 2012, when the WMA was established, the WMA and LGCA were two parts of the same park. The authors found that, as of 2015, in the WMA the numbers of giraffe and dik-dik per square kilometer were statistically greater than in the LGCA. Zebra and impala densities were statistically similar. Interestingly, the WMA had statistically fewer cattle per hectare than the LGCA and about 1/4 of the sheep/goats.

Source: Lee and Bond (2018).

Communities participating in CAMPFIRE are allowed to lease rights to wildlife and wild land to private sector tour operators. These tourist companies then bring in visitors for eco-tourism, but also for trophy hunting, which can be very lucrative (Taylor, 2009). Indeed, the whole idea of CAMPFIRE is to allow harvests of sometimes endangered wild-life to increase its provisioning ecosystem service value, creating incentives for preservation and reducing poaching. CAMPFIRE revenues are then used for local projects and wildlife management.

The program was supported throughout the 1980s and 1990s by various international donors and as of 2006 had generated $30 million in local benefits (Taylor, 2009). The program was also widely touted as a model for market-focused wildlife conservation (Frost and Bond, 2008). As of the early 2000s, the program does not appear to have been funded from the outside, largely because of the economic and political crises in Zimbabwe.

After donors pulled out, the quality of governance seems to have declined and local CAMPFIRE programs were largely captured by elites or outsiders. As a result, in the recent past CAMPFIRE has produced limited benefits for local residents (Tchakatumba et al., 2019; Mutandwa and Gadzirayi, 2007; Balint and Mashinya, 2008; Dube, 2019; Shereni and Saarinen, 2021) and, not surprisingly, there have been demands for improvements (Ntuli et al., 2020). Oduor (2020) found a similar need for improved community wildlife governance in Kenya, suggesting that sometimes outside support and engagement can be critical for such initiatives.

13.5 Chapter Conclusions

Collective action is hard, but a variety of social institutions have been created to move environmental collective action forward. The most important institutions by far are national, state/provincial, regional and local governments, which are responsible – in some cases for better, in other cases for worse – for the vast majority of environmental collective action. Funded typically by taxes of various types, but sometimes by other sources, governments manage environmental assets ostensibly to balance ecosystem service values and avoid deterioration of natural resources.

When the groups most involved with natural resources are reasonably small, it may be possible to exploit the strengths of local-level collective action, such as the potential to marshal diverse expertise, include different perspectives and improve equity and fairness. Examples discussed in the chapter include successful examples of community forestry, communities involved with wildlife management and small-scale fisheries.

Privatization, which may in some cases be effective, strips down collective action and narrows the range of ecosystem services for which assets are managed to strengthen incentives for good management. The whole world of land and water privatization, as well as aquaculture and private protected areas, which have been touted recently by environmental groups such as the IUCN, fit into this category.

Issues for Discussion

1. In some countries, urban water is provided by private companies. Please name two positive features of this institutional arrangement, two negative ones and defend your answers.
2. Governments are our #1 institution for environmental collective action. Governments tend to be funded by taxes, but perhaps there are other financing mechanisms that would be appropriate for funding environmental improvements. Research one non-tax revenue source in your hometown, region or country that is used to fund environmental projects. Does the financing source seem to be appropriate? What do you find are the positive and negative features?
3. Aquaculture appears to have contributed to the increase in marine and freshwater food consumption that has occurred in recent years. Is such a switch from wild capture to farmed a good thing? Why or why not?
4. If you live in the higher-income world, there are probably relatively few community-controlled and -managed forests. Yet, they are important, growing and actively promoted in the lower-income world. Why do we see this difference in forestry management? Do you see scope for increased use of community forestry in higher-income countries? From your experience or a quick internet research, why are there relatively fewer private forests in lower-income countries?
5. Elinor Ostrom's collective action design principles are typically applied to small group settings. Do you see any scope for "scaling up" those principles for larger problems that may be addressed by government or other institution-led collective action? How might those principles be applied?

Notes

1 Governments often can be very involved with parks, uncrowded roads and education. Note that in contrast to music, apps and movies, often such activities do not pay off very well for the private sector.
2 Value-added is the total value minus the cost of intermediate inputs. All tax information in this section comes from the World Bank via https://data.worldbank.org/indicator/GC.TAX.TOTL.GD.ZS. The tax revenue categories mentioned are the World Bank aggregations of all taxes by country and are expressed as percentages of total tax revenues {i.e., (revenue from a particular category/total tax revenues in a country)*100}.
3 2020 is the latest year available as of this writing, but we quote 2019 values due to the 2020/2021 global COVID-19 pandemic.
4 The World Bank quotes foreign aid as a percentage of *central* government expenditures rather than revenues. Sub-national government revenues may therefore not be considered.
5 1% of the US GNI is over $250 billion.
6 The increase in forest area over 20 years was four *percentage* **points**, but the *percentage* **change** in forest area was actually 21% = 0.21 = 4%/19%.
7 These articles analyze the Sloping Lands Conversion Program (SLCP), which, while securing property rights, also provides incentives for farmers to shift land from agriculture or scrub to forests.
8 We encountered NOAA in Chapter 2 when we discussed their work on climate change.
9 Much of this subsection is based on Bluffstone (2022).
10 For example, see https://ukcop26.org/cop26-iplc-forest-tenure-joint-donor-statement/
11 In the small-scale fisheries world, collective action is called co-management, collaborative, joint or participatory management (Pomeroy and Rivera-Guieb, 2006).
12 i.e., Australia, New Zealand, Papua New Guinea and 11 smaller island nations.

13 For more information on the CBFM-2 project and the final project report see https://digitalarchive.
    worldfishcenter.org/handle/20.500.12348/1805
14 For a relatively non-technical treatment of propensity score matching, please see Austin (2011).
15 e.g., see Carillo et al. (2019) for three case studies from Costa Rica.

Further Reading and References

Adhikari, B. and S. Di Falco. 2009. "Social Inequality, Local Leadership and Collective Action: An Empirical Study of Forest Commons." *European Journal of Development Research* 21 (2): 179-194.

Agrawal, A., A. Chhatre and R. Hardin. 2008. "Changing Governance of the World's Forests." *Science* 320 (5882): 1460-1462.

Ashe, F., T. Garlock, J. Anderson, S. Bush, M. Smith, C. Anderson, J. Chu, K. Garrett, A. Lem, K. Lorenzon et al. 2018. "Three Pillars of Sustainability in Fisheries." *Proceedings of the National Academies of Science* 115 (44): 11221-11225.

Austin, P.C. 2011. "An Introduction to Propensity Score Methods for Reducing the Effects of Confounding in Observational Studies." *Multivariate Behavioral Research* 46 (3): 399-424.

Balint, P.J. and J. Mashinya. 2008. "CAMPFIRE during Zimbabwe's National Crisis: Local Impacts and Broader Implications for Community-Based Wildlife Management." *Society and Natural Resources* 21 (9): 783-796.

Barbier, E.B. and J.C. Burgess. 2001. "Tropical Deforestation, Tenure Insecurity, and Unsustainability." *Forest Science* 47 (4): 497-509.

Belton, B., D. Little, W. Zheng, P. Edwards, M. Skladany and S.H. Thilsted. 2020. "Farming Fish in the Sea Will Not Nourish the World." *Nature Communications* 11: 5804.

Beyene, A.D., R.A. Bluffstone and A. Mekonnen. 2016. "Community Forests, Carbon Sequestration and REDD+: Evidence from Ethiopia." *Environment and Development Economics* 21 (2): 249-272.

Bingham, H.C., J.A. Fitzsimons, B.A. Mitchell, K.H. Redford and S. Stolton. 2021. "Privately Protected Areas: Missing Pieces of the Global Conservation Puzzle." *Frontiers in Conservation Science* 2: 748127.

Bingham, H., J.A. Fitzsimons, K.H. Redford, B.A. Mitchell, J. Bezaury-Creel and T.L. Cumming. 2017. "Privately Protected Areas: Advances and Challenges in Guidance, Policy and Documentation." *Parks* 23 (1): 13-28.

Bitariho, R., E. Akampurira and B. Mugerwa. 2020. "Regulated Access to Wild Climbers Has Enhanced Food Security and Minimized Use of Plastics by Frontline Households at a Premier African Protected Area." *Conservation Science and Practice* 2 (10): e275.

Bitariho, R., E. Akampurira and B. Mugerwa. 2022. "Long-Term Funding of Community Projects Has Contributed to Mitigation of Illegal Activities within a Premier African Protected Area, Bwindi Impenetrable National Park Uganda." *Conservation Science and Practice* 4 (9): e12761.

Blackman, A. and R. Bluffstone. 2021. "Decentralized Forest Management: Experimental and Quasi-Experimental Evidence." *World Development* 145: 105509 (Introduction to Special Issue on Forest Governance Decentralization and Devolution).

Blackman, A., A. Pfaff and J. Robalino. 2015. "Paper Park Performance: Mexico's Natural Protected Areas in the 1990s." *Global Environmental Change* 31: 50-61.

Bluffstone, R.A. 2022. *Open Access to Forests and Forest Tenure Reform*. Pathways to Research in Sustainability. EBSCO Publishing: Ipswich, MA.

BNP Paribas Bank. 2021. *China's Public Finances, A Tangled Web*. EcoConjoncture No. 7, September. Downloaded from https://economic-research.bnpparibas.com/pdf/en-US/China-public-finances-tangled-9/28/2021,44528 September 22, 2024.

Botta, R., F. Asche, J.S. Borsum and E.V. Camp. 2020. "A Review of Global Oyster Aquaculture Production and Consumption." *Marine Policy* 117: 103952.

Carillo, I.I.C., S. Partelow, R. Madrigal-Ballestero, A. Schlüter and I. Gutierrez-Montes. 2019. "Do Responsible Fishing Areas Work? Comparing Collective Action Challenges in Three Small-Scale Fisheries in Costa Rica." *International Journal of the Commons* 13: 705-746.

Carter, J. and J. Gronow. 2005. "Recent Experience in Collaborative Forest Management: A Review Paper." Occasional Paper No. 43. Centre for International Forestry Research: Bogor, Indonesia.

Clark, C.W. 1973. "The Economics of Overexploitation." *Science* 181 (4100): 630-634.

D'Annunzio, R., E. Lindquist and K.G. MacDicken. 2017. *Global Forest Land-Use Change from 1990 to 2010: An Update to a Global Remote Sensing Survey of Forests*. Forest Resources Assessment Working Paper 187. Food and Agriculture Organization of the United Nations: Rome.

Drescher, M. and J.C. Brenner. 2018. "The Practice and Promise of Private Land Conservation." *Ecology and Society* 23 (2): 3.

Dube, N. 2019. "Voices from the Village on Trophy Hunting in Hwange District, Zimbabwe." *Ecological Economics* 159: 335-343.

EFAP (Ethiopian Forestry Action Program). 1994. *The Challenge for Development*, Vol. II. Ministry of Natural Resources Development and Environmental Protection: Addis Ababa.

Elleason, M., Z. Guan, Y. Deng, A. Jiang, E. Goodale and C. Mammides. 2021. "Strictly Protected Areas are Not Necessarily More Effective than Areas in which Multiple Human Uses are Permitted." *Ambio* 50 (5): 1058-1073.

Essington, T.E., P.E. Moriarty, H.E. Froehlich, E.E. Hodgson, L.E. Koehn, K.L. Oken, M.C. Siple and C.C. Stawitz. 2015. "Fishing Amplifies Forage Fish Population Collapses." *Proceedings of the National Academy of Sciences* 112 (21): 6648-6652.

Fluet-Chouinard, E., S. Funge-Smith and P.B. McIntyre. 2018. "Global Hidden Harvest of Freshwater Fish Revealed by Household Surveys." *Proceedings of the National Academy of Sciences* 115 (29): 7623-7628.

Food and Agriculture Organization (FAO). 2015. *Global Forest Resources Assessment 2015: How are the World's Forests Changing?* Rome.

Food and Agriculture Organization (FAO), Duke University and WorldFish. 2022. "Small-Scale Fisheries and Sustainable Development: Key Findings from the Illuminating Hidden Harvests Report." FAO: Rome; Duke University: Durham, NC.

Fox, J., S. Saksena, K. Hurni, J. Van den Hock and A. Smith. 2018. "Twenty-Five Years of Community Forestry: Mapping Dynamics in the Middle Hills of Nepal." Report from the East-West Center, University of Hawaii and South Asia Research Institute Conference, Kathmandu, Nepal, November 28-30, 2018. Downloaded from https://www.eastwestcenter.org/sites/default/files/filemanager/Research_pdfs/2018%20Nepal%20tree%20cover%20conference.pdf March 11, 2022.

Frost, P.G.H. and I. Bond. 2008. "The CAMPFIRE Programme in Zimbabwe: Payments for Wildlife Services." *Ecological Economics* 65 (4): 776-787.

Gebreegziabher, Z., A. Beyene, B. Gebremedhin, R. Bluffstone and A. Mekonnen. 2024. "Devolution and Sustainable Management of Forests in Developing Countries: Quasi-Experimental Evidence from Household Level Data in Ethiopia." Unpublished Working Paper.

Gebreegziabher, Z., A. Mekonnen, B. Gebremedhin and A.D. Beyene. 2021. "Determinants of Success of Community Forestry: Empirical Evidence from Ethiopia." *World Development* 138: 105206.

Goldberg, L., D. Lagomasino, N. Thomas and T. Fatoyinbo. 2020. "Global Declines in Human-Driven Mangrove Loss." *Global Change Biology* 26 (10): 5844-5855.

Gordon, H.S. 1954. "The Economic Theory of a Common-Property Resource: The Fishery." *Journal of Political Economy* 62 (2): 124-142.

Guthman, J. 1997. "Representing Crisis: The Theory of Himalayan Environmental Degradation and the Project of Development in Post-Rana Nepal." *Development and Change* 28 (1): 45-69.

Hamilton, A., A. Cunningham, D. Byarugaba and F. Kayanja. 2000. "Conservation in a Region of Political Instability: Bwindi Impenetrable Forest, Uganda." *Conservation Biology* 14 (6): 1722-1725.

Hua, F., L. Wang, B. Fisher, X. Zheng, X. Wang, D.W. Yu, Y. Tang, J. Zhu and D.S. Wilcove. 2018. "Tree Plantations Displacing Native Forests: The Nature and Drivers of Apparent Forest Recovery on Former Croplands in Southwestern China from 2000 to 2015." *Biological Conservation* 222: 113-124.

Hua, F., X. Wang, X. Zheng, B. Fisher, L. Wang, J. Zhu, Y. Tang, D.W. Yu and D.S. Wilcove. 2016. "Opportunities for Biodiversity Gains under the World's Largest Reforestation Programme." *Nature Communications* 7: 12717.

Hyde, W. 2019. "The Experience of China's Forest Reforms: What They Mean for China and What They Suggest for the World." *Forest Policy and Economics* 98: 1-7.

Islam, G.M.N., T.S. Yew and K.K. Viswanathan. 2014. "Poverty and Livelihood Impacts of Community Based Fisheries Management in Bangladesh." *Ocean & Coastal Management* 96: 123-129.

IUCN. Undated. "Effective Protected Areas." Retrieved from https://iucn.org/our-work/topic/effective-protected-areas#:~:text=What%20are%20effective%20protected%20areas,ecosystem%20servi ces%20and%20cultural%20values. October 2024.

Jackson, J.B.C., M.X. Kirby, W.H. Berger, K.A. Bjorndal, L.W. Botsford, B.J. Bourque, R.H. Bradbury, R. Cooke, J. Erlandson, J.A. Estes et al. 2001. "Historical Overfishing and the Recent Collapse of Coastal Ecosystems." *Science* 293 (5530): 629-638.

Joppa, L.N. and A. Pfaff. 2009. "High and Far: Biases in the Location of Protected Areas." *PLoS One* 4 (12): e8273.

Khan, M.A., M.F. Alam and K.J. Islam. 2012. "The Impact of Co-Management on Household Income and Expenditure: An Empirical Analysis of Common Property Fishery Resource Management in Bangladesh." *Ocean & Coastal Management* 65: 67-78.

Krutilla, J. 1967. "Conservation Reconsidered." *American Economic Review* 57 (4): 777-786.

Lee, D. and M. Bond. 2018. "Quantifying the Ecological Success of a Community-Based Wildlife Conservation Area in Tanzania." *Journal of Mammalogy* 99 (2): 459-464.

Li, Y., D. Sulla-Menashe, S. Motesharrei, S. Xiao-Peng, E. Kalnay, Q. Ying, S. Li and Z. Ma. 2017. "Inconsistent Estimates of Forest Cover Change in China between 2000 and 2013 from Multiple Datasets: Differences in Parameters, Spatial Resolution and Definitions." *Scientific Reports* 7: 8748.

Liu, H. and J. Su. 2017. "Vulnerability of China's Nearshore Ecosystems under Intensive Mariculture Development." *Environmental Science and Pollution Research* 24 (10): 8957-8966.

Luintel, H., R.A. Bluffstone and R.M. Scheller. 2018. "The Effects of the Nepal Community Forestry Program on Biodiversity Conservation and Carbon Storage." *PLoS One* 13 (6): e0199526.

Luintel, H., R.A. Bluffstone, R.M. Scheller and B. Adhikari. 2017. "The Effect of the Nepal Community Forestry Program on Equity in Benefit Sharing." *The Journal of Environment & Development* 26 (3): 297-321.

Marine Stewardship Council (MSC). Undated. "The Impact on Communities." Retrieved from https://www.msc.org/en-us/what-we-are-doing/oceans-at-risk/the-impact-on-communities October 15, 2022.

Maxwell, S.L., V. Cazalis, N. Dudley, M. Hoffmann, A.S.L. Rodrigues, S. Stolton, P. Visconti, S. Woodley, N. Kingston, E. Lewis et al. 2020. "Area-Based Conservation in the Twenty-First Century." *Nature* 586 (7828): 217-227.

Meyer, M., E. Klingelhoeffer, R. Naidoo, V. Wingate and J. Börner. 2021. "Tourism Opportunities Drive Woodland and Wildlife Conservation Outcomes of Community-Based Conservation in Namibia's Zambezi Region." *Ecological Economics* 180: 106863.

Mitchell, B. 2018. "Bosque Pehuén: Private, Voluntary Protection in a Chilean Forest." IUCN, March 28. Retrieved from https://www.iucn.org/news/protected-areas/201803/bosque-pehu%C3%A9n-priv ate-voluntary-protection-chilean-forest October 10, 2022.

Mitchell, B.A., S. Stolton, J. Bezaury-Creel, H.C. Bingham, T.L. Cumming, N. Dudley, J.A. Fitzsimons, D. Malleret-King, K.H. Redford and P. Solano. 2018. *Guidelines for Privately Protected Areas.* International Union for Conservation of Nature: Gland, Switzerland.

Mutandwa, E. and C.T. Gadzirayi. 2007. "Impact of Community-Based Approaches to Wildlife Management: Case Study of the CAMPFIRE Programme in Zimbabwe." *International Journal of Sustainable Development & World Ecology* 14 (4): 336-344.

Myers, R.A. and B. Worm. 2003. "Rapid Worldwide Depletion of Predatory Fish Communities." *Nature* 423 (6937): 280-283.

Naylor, R., R.W. Hardy, A.H. Buschmann, S.R. Bush, L. Cao, D.H. Klinger, D.C. Little, J. Lubchenco, S.E. Shumway and M. Troell. 2021. "A 20-Year Retrospective Review of Global Aquaculture." *Nature* 591 (7851): 551-563.

Nichiforel, L., K. Keary, P. Deuffic, G. Weiss, B.J. Thorsen, G. Winkel, M. Avdibegović, Z. Dobšinská, D. Feliciano, P. Gatto et al. 2018. "How Private are Europe's Private Forests? A Comparative Property Rights Analysis." *Land Use Policy* 76 (July): 535-552.

NOAA. Undated. "What is Aquaculture?" Retrieved from https://oceanservice.noaa.gov/facts/aquaculture.html September 1, 2022.

Ntuli, H., E. Muchapondwa and B. Okumu. 2020. "Can Local Communities Afford Full Control over Wildlife Conservation? The Case of Zimbabwe." *Journal of Choice Modeling* 37: 100231.

Oduor, A.M.O. 2020. "Livelihood Impacts and Governance Processes of Community-Based Wildlife Conservation in Maasai Mara Ecosystem, Kenya." *Journal of Environmental Management* 260: 110133.

Ostrom, E. 1993. "Design Principles in Long-Enduring Irrigation Institutions." *Water Resources Research* 29 (7): 1907-1912.

Ostrom, E. 2010. "Beyond Markets and States: Polycentric Governance of Complex Economic Systems." *American Economic Review* 100 (3): 408-444.

Oswalt, S.N., W.B. Smith, P.D. Miles and S.A. Pugh. 2019. *Forest Resources of the United States, 2017: A Technical Document Supporting the Forest Service 2020 RPA Assessment.* Gen. Tech. Rep. WO-97. US Department of Agriculture, Forest Service.

Palfrey, R., J.A. Oldekop and G. Holmes. 2022. "Privately Protected Areas Increase Global Protected Area Coverage and Connectivity." *Nature Ecology and Evolution* 6 (6): 730-737.

Pomeroy, R. and R. Rivera-Guieb. 2006. *Fishery Co-Management: A Practical Handbook.* CABI Publishing: Cambridge.

Powell, L.A. 2012. "Common-Interest Community Agreements on Private Lands Provide Opportunity and Scale for Wildlife Management." *Animal Biodiversity and Conservation* 35 (2): 295-306.

Ren, G., S.S. Young, L. Wang, W. Wang, Y. Long, R. Wu, J. Li, J. Zhu and D.W. Yu. 2015. "Effectiveness of China's National Forest Protection Program and Nature Reserves." *Conservation Biology* 29 (5): 1368-1377.

Ribot, J. 2008. "Building Local Democracy through Natural Resource Interventions: An Environmentalist's Responsibility." World Resources Institute: Washington, DC.

Ribot, J.C., A. Agrawal and A.M. Larson. 2006. "Recentralizing while Decentralizing: How National Governments Reappropriate Forest Resources." *World Development* 34 (11): 1864-1886.

Richmond, L. and L. Casali. 2022. "The Role of Social Capital in Fishing Community Sustainability: Spiraling Down and Up in a Rural California Port." *Marine Policy* 137: 104934.

Rights and Resources Initiative (RRI). 2020. Tenure Tracking Tool. Retrieved from https://rightsandresources.org/tenure-tracking/forest-and-land-tenure/ March 6, 2022.

Ritchie, H. and M. Roser. 2021. "Fish and Overfishing." *Our World in Data.* Retrieved from https://ourworldindata.org/fish-and-overfishing#what-is-the-status-of-global-fish-stocks October 18, 2022.

Schmithüsen, F. and F. Hirsch. 2010. *Private Forest Ownership in Europe.* Geneva Timber and Forest Study Paper 26, ECE/TIM/SP/26, United Nations Economic Commission for Europe/ Food and Agriculture Organization of the United Nations.

Shereni, N.C. and J. Saarinen. 2021. "Community Perceptions on the Benefits and Challenges of Community-Based Natural Resources Management in Zimbabwe." *Development Southern Africa* 38 (6): 879-895.

Shumba, T., A. De Vos, R. Biggs, K.J. Esler, J.M. Ament and H.S. Clements. 2020. "Effectiveness of Private Land Conservation Areas in Maintaining Natural Land Cover and Biodiversity Intactness." *Global Ecology and Conservation* 22: e00935.

Smith, V.L. 1969. "On Models of Commercial Fishing." *Journal of Political Economy* 77 (2): 181-198.

Somanathan, E., R. Prabhakar and B. Mehta. 2009. "Decentralization for Cost-Effective Conservation." *Proceedings of the National Academy of Sciences* 106 (11): 4143-4147.

Springate-Baginski, O. and P. Blaikie, eds. 2007. *Forests, People and Power: The Political Ecology of Reform in South Asia.* EarthScan Forestry Library. Earthscan Press: London and Sterling, VA.

Stolton, S., K.H. Redford and N. Dudley. 2014. *The Futures of Privately Protected Areas.* International Union for Conservation of Nature: Gland, Switzerland.

Taylor, R. 2009. "Community Based Natural Resource Management in Zimbabwe: The Experience of CAMPFIRE." *Biodiversity and Conservation* 18 (10): 2563-2583.

Tchakatumba, P.K., E. Gandiwa, E. Mwakiwa, B. Clegg and S. Nyasha. 2019. "Does the CAMPFIRE Programme Ensure Economic Benefits from Wildlife to Households in Zimbabwe?" *Ecosystems and People* 15 (1): 119-135.

Thuy, T.D., V.Q. Tuan and P.K. Nam. 2021. "Does the Devolution of Forest Management Help Conserve Mangrove in the Mekong Delta of Viet Nam?" *Land Use Policy* 106 (July): 105440.

Tumcha, B. 2004. "The Vision of Ministry of Agriculture on Natural Resources of Ethiopia by 2025." In S. Mengistou and N. Aklilu, eds. *Proceedings of the Public Meetings on Integrated Forest Policy Development in Ethiopia*. Environment and Interchurch Organization for Development Cooperation: Addis Ababa, 1-11.

United Nations (UN). Undated. " A New Global Framework for Managing Nature through 2030: 1st Detailed Draft Agreement Debuts." Retrieved from https://www.un.org/sustainabledevelopment/blog/2021/07/a-new-global-framework-for-managing-nature-through-2030-1st-detailed-draft-agreement-debuts/ September 28, 2022.

US Department of Agriculture (USDA). Undated. "Private Land." US Forest Service. Retrieved from https://www.fs.usda.gov/managing-land/private-land August 1, 2022.

Wamukota, A.S., J.E. Cinner and T.R. McClanahan. 2012. "Co-Management of Coral Reef Fisheries: A Critical Evaluation of the Literature." *Marine Policy* 36 (2): 481-488.

Wartenberg, R., L. Feng, J.J. Wu, Y.L. Mak, L.L. Chan, T. Telfer and R.K.S. Lam. 2017. "The Impacts of Suspended Mariculture on Coastal Zones in China and the Scope for Integrated Multi-Trophic Aquaculture." *Ecosystem Health and Sustainability* 3 (6): 1340268.

Western, D., J. Waithaka and J. Kamanga. 2015. "Finding Space for Wildlife beyond National Parks and Reducing Conflict through Community-Based Conservation: The Kenya Experience." *Parks* 21 (1): 51-62.

World Bank. Undated. "Forest Area (% of Land Area) - China." Retrieved from https://data.worldbank.org/indicator/AG.LND.FRST.ZS?locations=CN March 10, 2022.

Xie, L., P. Berck and J. Xu. 2016. "The Effect on Forestation of the Collective Forest Tenure Reform in China." *China Economic Review* 38: 116-129.

Xu, J. 2011. "China's New Forests Aren't as Green as They Seem." *Nature* 477 (7365): 371.

Xu, J. and X. Jiang. 2010. "Collective Forest Tenure Reform in China: Outcomes and Implications." Proceedings of the World Bank Conference on Land Governance, Washington, DC, USA.

Xu, J., R. Yin, Z. Li and C. Liu. 2006. "China's Ecological Rehabilitation: Unprecedented Efforts, Dramatic Impacts, and Requisite Policies." *Ecological Economics* 57 (4): 595-607.

Zhai, D.-L., J.-C. Xu, Z.-C. Dai, C.H. Cannon and R.E. Grumbine. 2014. "Increasing Tree Cover while Losing Diverse Natural Forests in Tropical Hainan, China." *Regional Environmental Change* 14 (2): 611-621.

Zhang, D. 2001. "Why Would so Much Forestland in China Not Grow Trees?" *Management World* (in Chinese) 3: 120-125.

Zhang, D. 2019. "China's Forest Expansion in the Last Three Plus Decades: Why and How?" *Forest Policy and Economics* 98: 75-81.

Zhang, P., G. Shao, G. Zhao, D.C. Le Master, G.R. Parker, J.B. Dunning Jr. and Q. Li. 2000. "China's Forest Policy for the 21st Century." *Science* 288 (5474): 2135-2136.

Zhao, H., C. Wu and X. Wang. 2022. "Large-Scale Forest Conservation and Restoration Programs Significantly Contributed to Land Surface Greening in China." *Environmental Research Letters* 17 (2): 024023.

14 The Economics of Pollution Control

14.1 Recap and Introduction to the Chapter

The previous chapter reviewed collective action institutions and institutional reforms that can improve environmental collective action. Governments are the #1 environmental collective action institution, but they need to have some tools to organize their efforts; perhaps nowhere do we feel the environmental protection hand of government and perceive the need for instruments more strongly than in the world of pollution control.

For many, it is almost a fact – or at least a maxim – that governments are or should be in charge of pollution control. But what exactly is the role of governments in controlling pollution, so it doesn't interfere too much with other parts of human activity, including the market economy, and more broadly other ecosystem services? Furthermore, how do governments at the national, state and local levels approach the craft of pollution control and how might they decide how much pollution should be allowed? This chapter delves into the analytics that can help us answer these questions, with a focus on cost-effective pollution control policies.

DOI: 10.4324/9781003308225-17

14.2 Optimal Pollution and Abatement

The National Geographic Society, which is a global non-profit founded in the late 1800s in the US, defines pollution as:

> ...the introduction of harmful materials into the environment. Pollutants can be natural, such as volcanic ash. They can also be created by human activity, such as trash or runoff produced by factories. Pollutants damage the quality of air, water, and land.[1]

Basically, pollution includes any materials that are added to the environment that reduce the capability of environmental assets to produce ecosystem services. Examples of pollution include particulate air pollution, which reduces clean air's ability to help breathers stay healthy, and carbon pollution that destabilizes the climate. Garbage reduces the quality of open space or shorelines and may toxify land, reducing its productivity. Suspended solids dumped into rivers and lakes by, for example, food processing companies, stimulate algae. The extra organic material causes algae to multiply, using up oxygen, making waterways less able to support oxygen-using species like fish.

Like many sub-fields in the sciences, the economics of pollution control has some of its own terminology. The key terms are briefly defined in Table 14.1. In the remainder of this chapter, we will talk a lot about pollution *abatement*, which is the process of reducing pollution and is also known as managing the *sink* functions of the environment. Abatement occurs at the pollution *source*, which is where pollution meets the environment. Abatement costs (total, average or marginal) are the costs of pollution reduction. Uncontrolled pollution is just the level of pollution without any abatement, and the pollution we observe at any point in time is the uncontrolled pollution level minus any abatement we've done. Pollution control instruments are the specific policies of pollution control and can include standards, incentive-based or even voluntary instruments. These instruments are introduced at the end of this chapter and a number of them are discussed in detail in subsequent chapters.

We are already well-familiar with efficient or optimal policies, which seek to maximize human welfare by balancing off the benefits and costs, and it turns out these models can be applied to pollution control.[2] Of course, another name for pollution is the sink function

Table 14.1 The Economic Terminology of Pollution Control

Term	Definition
Pollution Source	The point where pollution meets the environment
Abatement/Pollution Control	The process of managing waste disposal (i.e., "sink") ecosystem services
Abatement Costs	Pollution control costs
Uncontrolled Pollution	Pollution when there is no abatement
Pollution Emissions	Uncontrolled pollution minus abatement
Pollution Control Instruments	The specific policies of pollution control
Efficient/Optimal Policy	Policy mix that yields the set of outcomes that maximizes human welfare
Cost-Effective Policy	Policy mix that minimizes the total cost of achieving a predetermined level of pollution control

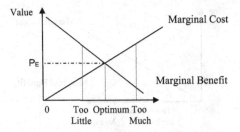

Gasoline (Petrol) Produced and Consumed/Year

Figure 14.1 Example of Optimal, Too Much and Too Little Gasoline

of environmental assets. There is, of course, something useful about the sink function of the environment, because otherwise we'd have no place to put our wastes (i.e., see *Where's the Poop?* by Julie Markes and Susan Hartung). Pollution is therefore "good," but too much pollution is, of course, "bad." I guess in principle too little pollution could also be "bad," but that is not something we worry about too much.

We are no strangers to these ideas of "too much" and "too little." In Chapters 3 and 9, we discussed what it means for an allocation of resources, such as ecosystem services, to be optimal. In Chapter 12 we talked about energy subsidies, including the important effects fossil fuel subsidies can have on pollution, reducing aggregate human well-being. This lost human welfare is typically modeled as too much something – gasoline consumption, electricity consumption, leaky pipelines, high-pollution-content fuels, etc. etc. – in the sense that marginal (i.e., the "extra") costs exceed marginal benefits. The point here, as shown in Fig. 14.1, which gives the example of gasoline (petrol), is that too much (or possibly too little) is relative to an optimal quantity that balances off costs and benefits and maximizes human well-being.

But if we only care about the pollution from these decisions, and we at least partly see pollution as an ecosystem service, why not leave aside products and materials, fuels etc. and do the analysis in terms of pollution itself? Figure 14.2 presents marginal costs and benefits of carbon dioxide (CO_2) air pollution, which is a key pollutant from burning the gasoline noted in Fig. 14.1.

There may be a relationship between the optima in Figs. 14.1 and 14.2, but there is certainly more to vehicles' CO_2 pollution than just petrol consumption. Fuel quality, vehicle maintenance, size of the engine and pollution control equipment all affect the relationship between gasoline use and CO_2 emissions. There are therefore many more handles for policymakers to pull to reduce CO_2 pollution than just reducing petrol consumption! Unless there is some reason not to, why not, therefore, focus directly on carbon pollution?

The first step to analyzing optimal pollution is to make sure we understand what makes up the marginal costs and benefits. Let's start with marginal costs. In this book and in economics more generally, all costs should be thought of as opportunity costs – the next most valued thing that is given up or lost because of a decision. In this case, what is given up when an additional ton of CO_2 is emitted into the atmosphere (e.g., when you drive your car, motorcycle or scooter to work or the university)?

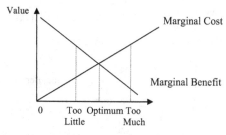

Figure 14.2 Optimal, Too Much and Too Little Carbon Dioxide Emissions

Before reading on, please think carefully about this question.

Of course, when CO_2 is emitted into the atmosphere it causes global warming, which destabilizes the climate and creates all the climate damages that we discussed in Chapter 2. These costs (importantly, depending on the existing total concentration of CO_2 in the atmosphere), such as flooding, excessive heat waves, species extinctions, reduced agricultural production etc., are the costs of carbon pollution. The marginal costs of pollution emissions are therefore the additional *damages* that are created by that pollution.

The marginal costs in Fig. 14.2 are increasing in CO_2 emissions, meaning that as carbon pollution accumulates in the atmosphere the damages from an additional ton of CO_2 are more than when there was less pollution in the air. While the slope of the marginal cost function is an empirical matter, not a theoretical one, meaning basically it is what it is, there is pretty good logic for why it might be upward-sloping. Low pollution concentrations are likely to be easily dealt with by the biosphere, but as emissions increase (or in the case of climate pollution, build up[3]), the environment gets overloaded, making the impact of an additional ton of pollution larger than for previous tons.

For example, as of early 2024 the average concentration of CO_2 in the atmosphere was about 422 parts per million (ppm) (i.e., for every million cubic centimeters or inches, 422 of them were CO_2). The average CO_2 concentration in 1990 was about 354 ppm, which is about 15% lower than 2024. Since 1990, the climate has totally changed, bringing many more heat waves, bigger storms, higher sea levels and droughts. In other words, the effect of *increased* CO_2 emissions has been to cause dramatically *increasing* damage to the environment.

What about the marginal benefits of pollution? Is there such a thing? If we once again think of pollution as the waste-absorbing capacity of the environment – one of the many ecosystem services that make up the magic of nature – we might be able to envision benefits from pollution. If pollution must be reduced, it will come at a cost – an opportunity cost – because pollution abatement is not free. It is costly.

To reduce CO_2 emissions, for example, we have to switch to lower-carbon and eventually carbon-free energy sources, such as solar, geothermal and wind energy. Even if the costs of those energy sources are falling rapidly (which is true) and some are already cheaper than fossil fuels (also true), there are still major switchover costs. For example, how many people in temperate climates would like to switch their home heating systems from a fossil fuel, such as methane, to an efficient electric system like a heat pump,[4] but are not ready to change out their heating systems, because it will cost a lot? If existing heating systems are working fine, this reaction is perfectly reasonable and, due to the environmental costs of manufacturing heat pumps, may even be better for the planet. Allowing pollution to go forward therefore allows polluters, who are part of society too, to avoid abatement costs. These *avoided* abatement costs are the benefits of pollution.

Why might marginal benefits decline as pollution increases? Once again, before reading on please think about this question for a moment.

Recall that the origin (i.e., the 0, 0 point) of Fig. 14.2 represents a situation in which CO_2 emissions are driven to zero. Policymakers, engineers and the rest of us are trying to envision such a situation, but so-called net-zero emissions are not expected to occur in any country until perhaps 2050. It may take so long to reach net-zero because as emissions get closer and closer to zero, more efficient, sophisticated and costly technologies must be used to further reduce emissions.

We can get important CO_2 emissions reductions through energy efficiency measures. Some of these may even be so cost saving they not only help the planet, but also on balance save people money. But eventually we will exhaust these profitable, no-cost and very-low-cost measures and will need to spend real money to further reduce CO_2 emissions. Carbon capture and storage, whereby carbon is extracted from smoke and stored before it can escape into the air, is an example of a costly technology that will likely eventually need to be deployed if we are to stop CO_2 concentrations from rising. Other types of pollution abatement also become more costly as abatement increases, a topic that we will explore in more detail later in this chapter.

The bottom line is that we generally need to use increasingly expensive measures to drive emissions closer to zero. If, on the other hand, we **don't** try to be so ambitious about abatement (i.e., we emit more pollution), polluters don't need to deploy such high-priced measures. The marginal benefit of allowing an additional ton of pollution is therefore expected to be highest when pollution is closest to zero and lowest near the uncontrolled pollution level.

So far, we have phrased our decisions in terms of how much pollution to emit. This approach is very useful, for example, if we are analyzing a pollution quantity-oriented instrument, such as *cap-and-trade*, which we will consider in Chapters 17 and 18. When we want to analyze pollution pricing policies, such as carbon pricing, it may be more useful to think in terms of pollution abatement. We will consider green pricing policies starting in the next chapter.

But switching from pollution to pollution abatement is simple, because if we know the uncontrolled level of pollution and the current pollution level, we can easily calculate abatement. This identity (remember identities from Chapter 5?) is given in Eq. 14.1.

Equation 14.1

Pollution Abatement = Uncontrolled Pollution Level − Actual Pollution Level

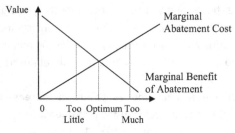

Figure 14.3 Optimal, Too Much and Too Little Carbon Dioxide Abatement

OK, now for the brainteaser part of the chapter. Before reading any further, can you explain the meaning and rationale for the slopes of the marginal abatement cost and marginal benefit functions in Fig. 14.3?

Now the marginal cost is paid by polluters who are reducing pollution. The function is likely to be upward-sloping, because as pollution is reduced, polluters need to marshal increasingly sophisticated and expensive technologies to reduce even more. That is what we mean by increasing marginal abatement cost.

Marginal benefits of abatement are just the pollution damages that are avoided due to abatement. Marginal benefits decline, because as abatement increases, pollution gets driven closer and closer to zero. Marginal damages from pollution are likely greatest when pollution closes in on the uncontrolled level of pollution (i.e., when abatement is at its lowest level) and are lowest when abatement approaches the maximum level, otherwise known as when pollution is as low as possible.

14.3 Do Governments Really Need to Be in Charge of Pollution Control?

The question "Do governments really need to be in charge of pollution control?" may seem a bit ridiculous, but answering this question perhaps delayed passage of the critical battery of US pollution control legislation, which was passed starting in 1970, by at least some years. In Chapter 9 we talked about negative externalities and how a tax equal to a marginal pollution externality can help *internalize* the externality. But what if undefined property rights rather than unconsidered costs are the main reason for excessive pollution? What does that imply about the efficacy of environmental policy instruments, such as environmentally motivated taxes, or for that matter pollution **limits**?

The property rights view implies they are seriously misguided, which is exactly what Nobel prizewinning economist Ronald Coase proposed in his classic article titled "The Problem of Social Cost" (Coase, 1960). In this paper, Coase suggested that it really does not matter whether, for example, a polluter is given the right to pollute, or someone affected by pollution is given the right to be **free** of pollution. They should be able to reach optimal pollution levels if they are able to freely bargain over pollution levels. What is critical from a policy perspective, Coase proposed, is that the rights be well-defined.

Coase argued that clarity about rights over pollution/pollution-free living will allow market-type institutions in environmental degradation and environmental protection to emerge; essentially, people will make deals for the right to pollute and the right to limit pollution. The role of local, state and national governments in environmental protection therefore becomes to define and protect property rights (i.e., eliminate open access), which is very different from what we normally think of as regulation-focused pollution control.

Box 14.1 Grab a Partner to See if the Coase Theorem Works

Background

Suppose we are concerned over the use of a lake near your hometown. The lake is used by fishers, who catch fish in the lake, and a small paper factory, which uses the lake as a sink for its waste. The main pollutant from the factory is biological oxygen demand (BOD), which reduces the productivity of the fishery and the fish catch. The lake lasts one period (call it 100 years) and there are only two uses for the lake: fishing and as a place for BOD pollution.

Instructions

1. This is a two-person game. One person plays the role of the fisher and one is the factory owner.
2. Choose one person who has complete property rights over the lake. This means the owner can do with the lake as they want.
3. Suppose each partner would like to earn as much money as possible from the lake.
4. The non-owner can pay the lake owner to reduce or increase pollution and does so to earn as much money as possible.

Production

Below are the profits each person earns from different levels of BOD emissions into the lake.

BOD (Tons)	0	1	2	3
Paper Factory Profit ($000s)	0	30	50	60
Fisher Profit ($000s)	35	30	20	0

Please answer the following questions:

1. Suppose we would like to maximize the total profits generated by the lake irrespective of ownership or distribution. What level of pollution is best? Zero? Three? Other?
2. If you and your partner bargain freely and compensation can be paid, how much will be paid? By whom? What is the level of pollution the bargainers voluntarily choose?
3. How does this bargained level compare with the optimum?

Coase used an example of a steel producer and a laundry to illustrate the idea that not only can such an institutional arrangement work, but it will lead to the socially optimal level of pollution in Fig. 14.2. Because of Ronald Coase's critical intellectual leadership, this idea that solidifying property rights, allowing the private sector to trade and decide on pollution levels, leads to optimal pollution levels, came to be known as the "Coase Theorem."

OK, so maybe you are thinking that something does not smell quite right, but possibly that is just because typically we do not use market forces to determine pollution levels – we are not used to that institutional arrangement. So, please grab a partner – a classmate, sibling, roommate or friend.

Read the exercise in Box 14.1, work with a friend, do the exercise and see how things work out. Before reading on, what were the results of the pair exercise? What is the best level of pollution and without government oversight were you able to reach it? Who paid whom?

You should have determined that 2 tons of biological oxygen demand (BOD) is the optimal pollution level in the Box 14.1 exercise, because $70,000 maximizes the total value produced by the lake. If you truly bargained with your partner to maximize your own gains, then you found that private bargaining – without any help from any government – led to the socially optimal BOD level. And this result should have held whether the lake was owned by the fisher or the paper factory; ownership does not matter to the final pollution outcome, as long as both sides bargain to earn as much money as possible. In other words, Adam Smith's "invisible hand of the market" from *The Wealth of Nations* goes green!

Now, to reiterate, the Coase Theorem is not the basis for contemporary pollution control policy. Why not?

Before reading on, please write down two reasons why the pair exercise was not fully convincing and in the "real world" the Coase Theorem may not work.

Critically, the pair exercise had only two participants: you and another person. In practice, though, any environmental asset that is of interest would have many more than two stakeholders. Imagine how difficult it would be to reach a bargain if the lake had even five stakeholders (e.g., polluters, aquaculture businesses, recreational fishers, farmers and municipalities getting drinking water from the lake), and each user class had some effect on the lake; it would be very difficult to privately bargain about how much pollution/other ecosystem services to allow.

Challenge Yourself

List two key reasons why the Coase Theorem in practice may not work and therefore perhaps should not be the basis for pollution control policy.

Second, in order for those interested in different ecosystem services to negotiate, they need to know how the actions of others will affect "their" services. For example, if recreational fishers have rights to a lake, only if they know the effect of water pollution on their fish catches and other aspects of their recreation can they decide whether to allow a paper plant to locate on the lake and how much to let it pollute. This decision requires knowledge of the effect of pollution discharges on water pollution concentrations in the lake, the effect of those concentrations on the biology of fish, how those changes link to fish stocks, and the relationship between fish stocks and fish harvests. This is, of course, a very difficult information bar to get over! Likely in many cases no bargain would even be reached, simply because it was so difficult to figure out the effects of others' actions.

Third, those who are considering reducing their pollution need to know how much reducing an additional ton of pollution would cost them. This may seem like something people are more likely to know than the value of pollution damages, and you are probably right, but it turns out that even knowing the marginal costs of pollution control is not that easy.

So, even if in principle the private sector might be able to bargain over pollution and pollution control and achieve the efficient levels, in practice there is a lot that can get in the way of good results. Deals may not be possible and whoever owns natural resources would simply exploit them for "their" ecosystem services and not fully consider others' services. Maybe, therefore, the private sector is not the best institution to deal with pollution and it wouldn't be such a bad idea to have governmental institutions responsible for coordinating things. For example, in the case of our lake, environmental protection agencies, state ecological boards and ministries of the environment could try to balance off the pluses and minuses of pollution through regulation.

14.4 The Economics of Government-Led Pollution Control

But what can we really expect from governments in terms of correctly getting us to optimal pollution levels? Can they really do it and are they any better than the private sector? Governments can certainly help with Coase Theorem problem #1, because now many stakeholders don't need to independently negotiate with each other (they only need to lobby the government). Governments can convene stakeholders by holding meetings, evaluate all the different interests and make a determination in the form of laws, regulations and rules. They are therefore perhaps less likely to be totally stalled than the market,[5] but they may not really be much better on problem #2, which focuses on information related to pollution damages, and may have worse information on marginal abatement costs (point #3) than polluters.

Recall from Fig. 14.2 that to have environmental collective action that achieves the optimal pollution level, governments would need to know the marginal costs (i.e., the marginal damages) and marginal benefits (i.e., marginal abatement costs avoided) of pollution. Because of their access to scientific expertise and potential for a broad stakeholder view, governments may have a better handle on the marginal costs, but as we saw in Chapter 4, estimating the benefits of pollution control (i.e., marginal damages) is difficult and involves tricky methods like hedonic pricing and contingent valuation. Though governments can and do support research into the value of pollution reduction, they can't be expected to do big studies every time they want to set pollution standards; such a lift may be too big and therefore unrealistic even for governments.

What about the marginal benefits of pollution, otherwise known as the avoided abatement costs? Well, in most countries even if polluters have such information, it is private, and polluters do not have to report it to governments. To be concrete, suppose the Pollution Control Department in the Ministry of Natural Resources and the Environment in Thailand[6] was considering proposing a new water pollution regulation on the food processing industry to reduce pollution in rivers, such as the Chao Phraya River, which runs through the capital Bangkok. The Ministry wants to optimally balance off the benefits of reducing BOD pollution against the costs. They conduct a series of studies on the damages of water pollution, such as BOD, and have a pretty good handle on those marginal benefits.

The Ministry, of course, doesn't really know how much it will cost on average to reduce pollutants like BOD, because they are not involved in businesses that produce BOD. They suppose that the food processers would have a better idea, and therefore set up a series of meetings to survey firms on the marginal abatement costs so they can properly set regulations.

What do you think would, on average, be the result of requests by governments for the abatement costs of polluters? Before reading on, please carefully consider this question.

Of course, we do not know for sure what would be the outcome but let us consider the incentives of polluters. If they have as a key goal to earn high profits, understating the marginal costs of pollution reduction would certainly not be in their interests. For example, if they told the Ministry that BOD reduction was no problem and could be done very easily and cheaply (which very well may be true!), the government may use that information to make regulations more stringent and require larger reductions.

Such requirements could turn out to be difficult for firms if their calculations are wrong or something changes, and abatement costs are higher than expected. In their estimates, they would probably, at minimum, err on the conservative side, even at the risk of overstating abatement costs. There is no downside to that strategy and it could offer the benefit of more lenient regulations. Our standard behavioral model assumes producers want to earn as much money as possible. We might therefore expect that when asked about abatement costs, unless there is some other incentive to be aggressive about pollution reduction, polluters will overstate their true costs.

The bottom line is that governments may have a tough time managing pollution based on an efficiency criterion, because both marginal costs and benefits are highly uncertain. The benefits of pollution control typically involve a lot of complicated and important biology and ecology. Even if the science is well-known, there remains the matter of converting environmental information to monetary terms, which is also difficult. On the cost side, firms that cannot be forced to give information on abatement costs may not provide any, and, if asked, would likely have incentives to overstate costs.

How are pollution reduction goals set if not based on a careful balancing of the costs and benefits? There are lots of possibilities, most of which you know very well. So much of pollution control is about protecting human health, so it is not surprising that safe minimum standards are a very important way to set pollution standards. For example, the World Health Organization (WHO) guidelines for annual and 24-hour limits on outdoor air pollution (sometimes called *ambient* concentrations) are based on the concentrations scientists estimate pose limited risks to human health.[7] Many countries adopt or refer to WHO guidelines in their own legislation.

Another possibility is to minimize risk based on the so-called *precautionary principle*. This principle suggests that if an activity raises threats to human health or the environment, precautionary measures should be taken even if cause-and-effect relationships are not fully established. The precautionary principle provides important guidance for environmental decision-making in Canada and several European countries, including Denmark, Sweden and Germany. It is also important for regulating pharmaceutical drugs in the US (Hayes, 2005).

But political feasibility is by far the most important guiding principle for environmental legislation and regulations. Making "reasonable," "ambitious" and "balanced" goals are absolutely critical to contemporary environmental policies.[8] For example, the latest congressional amendments to the 1970 US Clean Air Act, which occurred way back in 1990, require that the Best Available Control Technology (BACT) be used by new sources in regions subject to potentially serious environmental deterioration, but in other situations require Reasonably Available Control Technology (RACT) standards. It also added acid rain deposition standards to the Clean Air Act. For example, it set goals for the year 2000 of reducing annual SO_2 emissions by 10 million tons from 1980 levels and reducing annual NO_x emissions by 2 million tons. The required reductions for large electric-generating facilities were about 50% by 1995 compared with 1980.

Box 14.2 The US Inflation Reduction Act Uses Subsidies to Reduce Greenhouse Gas Emissions

We discussed subsidies in Chapters 11 and 12 in the context of environmentally harmful policies. Subsidies can also be explicitly designed to improve environmental quality. For example, the 2022 US Inflation Reduction Act (IRA) provides significant financial support for renewable energy, decommissioning fossil fuel infrastructure, nature-based climate solutions and climate justice. The IRA allocates a total of $370 billion across a variety of sectors. It includes expanded clean electricity tax credits and funding to promote US manufacturing of solar panels, wind turbines and batteries. It also supports residential installation of low carbon and carbon-free technologies, such as heat pumps and rooftop solar arrays; promotion of electric vehicles and carbon capture and storage; and targeted funding for disadvantaged communities.

The IRA is a major public subsidy program to reduce the carbon footprint of the US. The Rhodium Group consulting company estimates that the law will help the US come much closer to meeting its Paris Agreement Nationally Determined Contribution (NDC) of 50% reduction in greenhouse gases by 2030 compared with 2005. They estimate that the US could achieve a 44% reduction compared with a maximum of 35% without the IRA.

Sources: Barbanell (2022); Larsen et al. (2022).

The Clean Air Act reauthorization of 1990 imposed a variety of requirements on air quality management regions that in 1988 were out of compliance with ground-level ozone (O_3) standards. It established five classes of so-called *non-attainment areas* (marginal, moderate, serious, severe, extreme), which depended on how far regions were out of compliance. The legislation also imposed deadlines for compliance. In 1988, 98 regions were out of compliance at some level. Over time, the ambient standards were tightened (generally by about 50%), but as of July 2022 only 49 regions were out of compliance (CRS, 2022).

The European Union environmental legislation has features that are quite similar to the US. EU laws are called *directives* and once passed by the European Parliament and the Council of the European Union must be written directly into the 27 EU member states' own environmental legislation. The whole body of EU environmental legislation is called the *Environmental Acquis* and is supervised by the European Commission (the executive branch of the EU) Directorate-General for Environment (DG ENV). The Environmental Acquis encompasses 500+ laws related to the environment, covering everything from swimming water standards to radioactivity and municipal waste incineration. For example, directive EU/2015/2193 focuses on air pollution emissions from medium-sized combustion sources and sets source-level pollution emissions concentration standards for SO_2, large particles (dust) and NO_x. These limit values are specified in Annex II of the directive.[9]

Directive 2010/75/EU sets standards for industrial pollution emissions and has specific requirements for so-called "integrated pollution control and prevention." Directive 2009/126/EC updates a 1994 directive on requirements for petrol (gasoline) vapor recovery during refueling of motor vehicles at service stations, and so on and so on. There are hundreds of such directives.

Now let us think about the ways all these standards and requirements are set. Are they based on efficiency and defined at the pollution reduction levels where the marginal benefits of abatement equal marginal costs? Probably not. These ways of looking at the pollution control problem no doubt enter the decision process, but it is simply not reasonable to suppose that efficiency dominates the calculations. Factors other than efficiency – protection of human health, risk reduction and politics – are likely substantially more important.

This realization that pollution reduction goals and objectives are fundamentally determined by what we might think of as "non-economic" criteria has important implications for our work on the economics of pollution control. Basically, it means that pollution control goals and objectives are set at least partially independently of the instruments used to achieve them.[10]

Incorporating this pollution control policy reality into our analysis takes us into new territory. What can economics possibly contribute when politics of one sort or another drive the determination of pollution control objectives? In the remaining chapters, we will discuss several contributions, but one of the most important is the idea of *cost-effective pollution control policy*, which means that policies to achieve given, politically determined, pollution reduction goals are achieved at lowest total social cost. This idea has led to potentially important – and in some countries fraught – policies, such as environmental taxes and cap-and-trade, which are discussed in the next four chapters. For now, let us consider the economics of cost-effective pollution control, which is an important rationale for using these instruments.

14.5 Cost-Effective Pollution Control

As we are focusing on the costs of pollution control, let us delve into the abatement costs in Fig. 14.3. Marginal abatement cost functions are typically "upward-sloping," meaning that they are increasing in the level of pollution reduction. Let us first make sure that we understand the relationship between marginal and total abatement costs. Suppose that Table 14.2 presents the abatement costs of one BOD polluter as a function of abatement. We see that total abatement costs (TACs) are increasing in the level of abatement, because pollution control is costly. In accord with Fig. 14.3, *marginal* abatement costs (MACs) are also increasing in the pollution control level.

Notice that MACs are just the change in total costs divided by the change in abatement. If Δ means "change in," MACs = $(TAC_{ending} - TAC_{beginning})/(BOD_{ending} - BOD_{beginning}) = \Delta TAC/\Delta BOD$. For example, from 0 to 1 ton reduced, the TAC goes up by €1.00. Going from 1 to 2 tons of abatement increases the TAC from €1.00 to €7.00 or by €6.00, etc. Increasing MACs may be considered the typical case, but it is not the only possible one. In Table 14.2, reducing the fifth ton of BOD is not more costly than the fourth – just an example.

Now suppose there are two BOD polluters in an industry in India that is subject to government pollution regulations. Each polluter is currently emitting 10 tons of BOD into Indian rivers. The Central Pollution Control Board (CPCB) in the Ministry of Environment, Forest and Climate Change would like to reduce total BOD emissions from the current total of 20 tons (10 + 10) to 12 tons, which is a reduction of 8 tons. Box 14.3 shows the abatement costs of these two firms.

Before reading on, please analyze the data in Box 14.3 and answer the three questions posed in the box.

Table 14.2 Biological Oxygen Demand (BOD) Abatement Costs of a Typical Firm

BOD Abatement (Tons)	Total Abatement Costs (TACs) (€)	Marginal Abatement Costs (MACs) (€)
0	0	Undefined
1	1	1
2	7	6
3	14	7
4	23	9
5	32	9
6	42	10

Box 14.3 Marginal Abatement Costs of Two BOD Polluters

	Costs of Reducing the Indicated Ton in Hundreds of Indian Rupees (₹100)	
	Polluter 1	Polluter 2
1st ton of BOD	1000	3000
2nd ton of BOD	2000	4000
3rd ton of BOD	3000	5000
4th ton of BOD	4000	6000
5th ton of BOD	5000	7000
6th ton of BOD	6000	8000
7th ton of BOD	7000	9000
8th ton of BOD	8000	10,000

Questions

1. Which polluter can reduce BOD pollution more cost-effectively?
2. What is the total abatement cost if the CPCB simply requires each polluter to reduce pollution by four tons (i.e., each polluter reduces by the same amount)?
3. What is the total abatement cost if the CPCB divides up the eight tons of abatement at *lowest total cost*?

The first question asks which polluter can reduce emissions at lower cost. Polluter 1 clearly has this advantage, because at every level of reduction the MAC is lower. If we were to interpolate between the points in Box 14.3 (i.e., connect the dots) and plot each polluter's MACs on one graph, it might look something like Fig. 14.4.

Question 2 asks about the total cost if the abatement burden were equally divided across the two polluters. As we need to reduce total emissions by eight tons, that would mean each polluter reduces their pollution by four tons. The total cost of that policy is just the sum of the MACs for each polluter, which is just ₹1,000,000 + ₹1,800,000 = ₹2,800,000 (please check that you know how to get this answer).

What if instead of dividing up the total reduction equally, the CPCB were to assign the responsibility to minimize total costs (Question 3)? For now, let's leave aside how that might work from a policy standpoint, as we will discuss those issues in subsequent chapters. Based on looking at the graph in Fig. 14.4, one might be tempted to conclude that Polluter 1 should reduce all eight tons, because its MAC is always less than Polluter 2 at all levels of abatement. But if we go back to Box 14.3 and always choose the cheapest ton, we see that this is not the case. The first cheapest ton is from Polluter 1 (₹100,000), as is the second cheapest ton (₹200,000), but the third is the same for each polluter (₹300,000), because we have moved up Polluter 1's upward-sloping MAC function. So, let's get the third ton from Polluter 1 and the fourth from Polluter 2. Once again, the next lowest MACs are equal for the two polluters, this time at ₹400,000, which gives us six tons reduced. We just need two more tons, which I will leave to you.

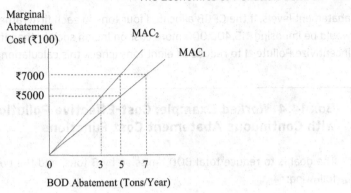

Figure 14.4 Smoothed and Interpolated Marginal Abatement Costs of the Polluters in Box 14.3

We find that to reduce eight tons at least cost will imply that Polluter 1 reduces its BOD emissions by five tons and Polluter 2 abates three tons. The total cost of this policy approach is just the sum of the MACs or ₹1,500,000 (Polluter 1) + ₹1,200,000 (Polluter 2) = ₹2,700,000. This total is less than the equal allocation policy, and indeed any other approach we might try. Explicitly focusing on achieving pollution reduction goals at least cost therefore saves money.

Notice, though, that the distribution of costs is different than when abatement is divided equally. Polluter 2 will be really happy with the cost-effective policy, because when we answered Question 2, they had to pay ₹1,700,000, whereas now they only have to pay ₹1,200,000. Not surprisingly, shifting one ton of BOD emissions reduction from Polluter 2 to Polluter 1 (the lower-cost reducer) costs Polluter 1 more money. This is something that will need to be addressed, but for now let us just note the issue.

Please notice the mathematics of cost-effective abatement. Starting with the lowest-cost pollution reducer, we climbed up its MAC function until it was no longer the lowest-cost reducer. We then switched to the higher-cost polluter and oscillated back and forth, always choosing the low-cost option. Though the MACs started out far apart, at the optimum, the MACs were equal. This finding tells us that one way to think about cost-effectiveness is that the MACs should be equal, or as close together as possible if they cannot be exactly equal (e.g., if our abatement goal in Box 14.3 was seven tons rather than eight). Equation 14.2 expresses this finding formally for any number of polluters.

Equation 14.2 Condition for Cost-Effective Pollution Abatement

MAC1 = MAC2 = MAC3 = MAC4 = ... MACk

A very important corollary to this finding is that the gains from paying attention to cost-effectiveness are larger the bigger are the differences in MACs across polluters. In our Box 14.3 example, the gains from paying attention to cost-effectiveness were only ₹100,000, but imagine if Polluter 2 had costs that were an order of magnitude (i.e., 10x) higher at all abatement levels. If the CPCB allocated four tons to each firm rather than cost-effectively, it would be imposing ₹15,400,000 more cost on Indian society than if it figured out a policy to incentivize Polluter 1 to reduce all eight tons (check this calculation yourself).

Box 14.4 Worked Example: Cost-Effective Pollution Abatement with Continuous Abatement Cost Functions

The goal is to reduce total BOD emission by 8 tons, and the polluters' MACs are the following:

$MAC_1 = 1.50*A_1$ and $MAC_2 = 2.5*A_2$. A_1 is the abatement of Polluter 1 and A_2 is the abatement of Polluter 2 (in tons of BOD).

Given these functions, the environmental objective and a cost-effective policy, how much should each polluter abate?

Step 1: Realize that $A_1 + A_2 = 8$ or equivalently that $A_2 = 8 - A_1$.
Step 2: Cost-effectiveness requires that $MAC_1 = MAC_2$ or that $1.50*A_1 = 2.5*A_2$.
Step 3: Substitute in $A_2 = 8 - A_1$, meaning $1.50*A_1 = 2.5*(8 - A_1) => 1.50*A_1 = 20 - 2.5*A_1$.
Step 4: Add $2.5*A_1$ to both sides => $4.0*A_1 = 20$ and $A_1 = 5$ tons.
Step 5: Recall that $A_2 = 8 - A_1$, meaning that $A_2 = 3$.

Suppose the two BOD polluters in Box 14.3 had continuous rather than discrete (i.e., with "jumps") MAC functions. Box 14.4 presents a worked example in which we can solve these continuous equations for the cost-effective abatement allocation. I chose the functions to give the same allocation, but please review the methodology and you will get a chance to practice in the practice problems at the end of this chapter. You will also get the chance to try your hand at a three-polluter problem, so you are convinced that this method works for any number of polluters.

14.6 Do Marginal Abatement Costs Really Differ?

We've shown that allocating abatement effort to polluters with low abatement costs can reduce total societal costs (but will alter the *distribution* of costs). How big are the gains from allocating abatement cost-effectively? In other words, do the abatement costs of polluters differ enough to make thinking about cost-effectiveness worth the trouble? The answer seems to be a resounding "yes!"

We can think of differences in costs in terms of differing technologies, industries, countries or vintages (i.e., how old polluters' equipment is). All are relevant to costs and have their own estimation challenges, which we will sidestep here. For example, some industries have inherently lower abatement costs than other industries. New production technologies tend to be easier to make low-pollution than retrofitted technology and old pollution control equipment tends to be worse than new tech. There is therefore generally a ladder of sorts

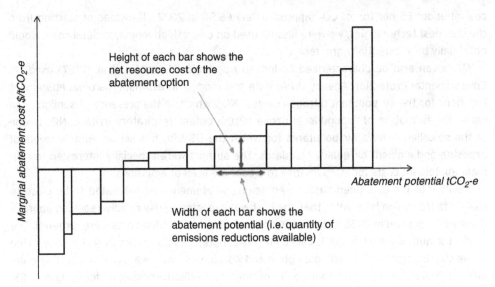

Figure 14.5 Stylized Example of a Marginal Abatement Cost Function for CO₂e Reductions

Source: Government of New Zealand, Ministry of the Environment (2020).

that we can climb from low-cost, low-impact pollution abatement technology up to very high-cost, comprehensive approaches. The basic idea is given in Fig. 14.5, which was produced by the Ministry for the Environment of the Government of New Zealand. On the horizontal axis is the abatement potential of various technologies, sectors, regions etc. in terms of tons of CO_2 equivalent (tCO_2-e) reduced. As discussed in Chapter 2, tCO_2e represents a conversion of non-CO_2 greenhouse gases to CO_2 "equivalents" using estimated global warming potentials. On the vertical axis is the "average" across all polluters of MACs in $/t$CO_2$e.

An influential 2009 report by the management consulting firm McKinsey & Co. attempted to estimate the costs of reducing greenhouse gas emissions sufficiently to limit the global average surface temperature increase to 2 degrees Celsius by 2030. The authors ordered abatement technologies from lowest to highest cost. Starting from the lowest-cost technologies, they estimated that the total abatement potential was about 38 gigatons (billions of tons), which was roughly equivalent to the emissions from world energy production and use.[11]

The lowest-cost technology was estimated to be switching from incandescent to LED lighting at ≈ –€95 per ton of CO₂ₑ reduced (–€125 per ton in 2022). The various carbon capture and storage (CCS) methods, whereby through chemical or physical processes CO_2 is stored underground or via other methods, were estimated to be the highest-cost (about €55 per ton of CO₂ₑ reduced; €72/ton in 2022). They also estimated that about half the abatement potential could be achieved at low cost or even *negative* cost. That is, some measures not only reduced greenhouse gas emissions, but they also saved us money! A variety of energy efficiency investments are in that so-called "win-win" category. Land use modifications, such as degraded land restoration, were estimated to be low-cost and other technologies, such as CCS and solar photovoltaic cells (which since 2009 have become cheaper than coal!), were estimated to be high-cost technologies.

Looking at abatement costs from the technology ladder standpoint, the McKinsey & Co. report suggests that we can reduce or offset our energy emissions by half at a marginal cost of about €5 per ton of CO_2 (approximately €6.50 in 2022). If instead of starting from the cheapest technologies, we were less focused on cost-effectiveness, societal costs would potentially be substantially greater.

MACs can also be characterized by industry or region. Loughlin et al. (2017) used US Environmental Protection Agency (EPA) data and models to estimate marginal abatement functions for the air pollutant nitrogen oxides (NO_x) which, in the presence of sunlight, can cause the formation of ground-level ozone (O_3), a potent respiratory irritant. NO_x is one of the so-called *criteria* air pollutants for which the US EPA has set acceptable levels of exposure and ambient air quality standards. The authors were especially interested to estimate abatement costs for 2035 relative to a baseline level of emissions.

Relative to the estimated baseline emissions, abatement was estimated to be close to free up to 1/2 million tons. After that, the cost was expected to rise rapidly, reaching approximately $60 per ton in 2035 if the maximum abatement of 2 million tons were targeted. The model the authors used, which is called MARKAL, optimizes the choice of technology – kind of like starting from the lowest-cost option and working its way toward higher-cost technologies. Of course, with policies that don't consider cost-effectiveness (e.g., focusing on CCS, because government officials think it's cool), the technology package will not be optimized, and costs will be higher.

One way to make abatement systematically more costly would be to target high-cost economic sectors for abatement and let others off the hook. The authors broke the economy down into electricity generating, industry, commercial (e.g., stores, restaurants etc.), residential, transport and natural resources, which was mainly fossil fuel extraction. Their analysis showed that most of the possible reductions were from the electric sector and, to a lesser extent, industry and transport. The estimated potential for commercial, residential and especially natural resources to contribute to NO_x abatement was very limited, because marginal costs rose very quickly. Imagine if the US EPA crafted policies that treated all sectors equally, perhaps in the interest of "fairness." They would potentially be subjecting the country to high costs, shifting the MAC function upward.

The Ministry for the Environment (2020) conducted a CO_2e abatement analysis for New Zealand relative to a projected 2030 baseline emissions level. They also broke down costs within sectors (e.g., car transport, rail, shipping etc.). They found that measures in some sub-sectors had negative CO_2e reduction costs and others (e.g., in the heating sector) had costs of about NZ$500 per ton (about $US320 in 2022). Basically, everywhere we look in New Zealand – and indeed around the world – whether within sectors, across sectors, within countries or across countries, there are enormous differences in the costs of abating the same pollutant. These costs depend both on who is doing the abatement and on how it is done.

14.7 Types of Instruments for Pollution Control

Though there are many ways governments control pollution, they can generally be divided into *standards* and *incentive-based* instruments. Standards specify what exactly pollution sources need to do to comply with regulations and this class of approaches forms the vast majority of environmental regulations around the world.

Technology standards specify the particular technologies that polluters must employ. As we have seen, the European Union specifies technologies through its Integrated Pollution Prevention and Control Directive. Directive 2009/126/EC specifies the requirements for petrol (gasoline) vapor recovery during motor vehicle refueling. The same is true in the US and around the world.

Standards can also focus on materials used. Lead is a heavy metal that is extremely toxic to humans and other species, and especially affects the brain function of children. The 1990 reauthorization of the US Clean Air Act made the use of lead in gasoline illegal. The law therefore specified that no lead could be used in petrol. This was after over ten years of phasing out lead using incentive-based mechanisms. There are also national standards in the US for the sulfur content in diesel fuels. Sulfur in fuels can form sulfur dioxide, which, among other effects, can cause acid rain, which lowers the pH (i.e., potential hydrogen) of anything the rain hits, such as rivers, lakes, monuments, swimming pools etc., making them more acidic.

Technology standards are especially important for regulating *mobile sources* of pollution, such as cars and trucks. In the 1950s, Arie Haagen-Smit, professor at the California Institute of Technology and founding chair of the California Air Resources Board, discovered that ground-level ozone (so-called "smog") was caused by nitrogen oxides (NO_x) and hydrocarbons emitted by vehicles in the presence of sunlight. Shortly after, the first version of a device to reduce smog was patented by Eugene Houdry. This *catalytic* converter was improved and applied specifically to motor vehicles by other researchers. Indeed, if you look under a car you are likely to see two squarish metal boxes that are part of the exhaust system. The one closer to the back reduces noise from the engine and in the US is called a muffler. Toward the front is the catalytic converter, which passes the exhaust through a honeycomblike structure that contains compounds that alter the chemical structure of the exhaust, and makes it less harmful.

Catalytic converters are on virtually all cars in the US, because starting with the 1975 model year they were required on all new cars. Catalytic converters have continued to be improved and are still required today. This important technology standard to help mitigate tailpipe emissions is also required in the European Union.

Technology standards are pretty rigid policies, because they offer polluters – who presumably know their businesses and technologies the best – little or no choice about how to reduce pollution. That technology standards offer fewer opportunities for private sector innovation is often a complaint of economists. If the goal is less pollution, why not, the logic goes, offer flexibility to polluters and thereby increase cost-effectiveness?

Technology standards are sometimes called *command-and-control* policies, a perhaps overly pejorative term in many situations. Mobile sources, which are subject to perhaps hundreds of technology standards, are controlled by millions of people and may be subject to very limited or no inspections. At the same time, pollution emissions are closely related to owners' driving and maintenance behaviors. Even choice of tires can dramatically affect performance! Requiring technologies therefore offers baseline protection for the environment, even as emissions depend on behaviors that are not readily observable by regulators.

Perhaps the most common class of standards, though, is the *performance standard*. Sticking with mobile sources for a moment, perhaps you live in an area where in order to renew vehicle licenses, vehicles must first be inspected. The West Coast states of the US all have requirements that to be relicensed vehicles must meet emissions standards. In Oregon, passenger vehicles are tested using onboard computer diagnostic equipment at licensed testing facilities throughout the state. Even old vehicles are tested. Light-duty cars and trucks that were made in 1981 or later must have hydrocarbon concentrations less than 220 parts per million (ppm) and carbon monoxide concentrations of 1% or less.[12]

California also tests vehicles at registration and re-registration, but the system is a bit unique in that it regulates the average emissions of new cars sold in the state. For example, starting in 2016, all manufacturers of new passenger vehicles under 3751 pounds (about 1700 kg) had to demonstrate that average CO_2e emissions per mile (1.61 km) were less than or equal to 205 grams.[13]

Earlier in this chapter, we mentioned the EU Medium Source Combustion Directive (EU/2015/2193). Table 14.3 presents the limit values for existing sources (new sources have tighter limits). Note that these standards are different than the ones we've seen so far, because they are concentration standards expressed as mg/m^3 and are differentiated by the type of fuels plants use.

There are thousands of such performance standards of perhaps hundreds of different types around the world for air, water and solid waste pollution. They all have in common that they specify that emissions or concentrations compared with some metric (e.g., year, day, miles travelled, MW produced, air volume) cannot be above regulated levels, otherwise polluters are penalized (e.g., they cannot relicense their vehicles or they can be fined). Importantly, performance standards offer polluters more flexibility than technology standards because they do not specify **how** polluters reach limits. This flexibility can be important even for mildly complicated sources, such as cars and trucks, because there are so many ways to reduce emissions. If we are right in our behavioral model that people pay attention to opportunity costs, offering polluters the chance to choose their abatement method should increase cost-effectiveness.

Incentive-based instruments are perhaps the most interesting pollution policy from an economic standpoint, because they rely on incentives rather than government dictates. They therefore offer the most flexibility and the greatest potential for cost-effectiveness. Moreover, because polluters choose methods and even **levels** of abatement, there is wide

Table 14.3 Emission Limit Values (mg/m^3) for Existing Medium-Combustion Plants (Thermal Input between 1 and 5 MW)

Pollutant	Solid Biomass	Other Solid Fuels	Gas Oil	Liquid Fuels Other than Gas Oil	Natural Gas	Gaseous Fuels Other than Natural Gas
SO_2	200*	1100	-	350	-	200*
NO_x	650	650	200	650	250	250
Dust	50	50	-	50	-	-

*There are some exceptions for particular types of plants.

Source: EU/2015/2193.

scope for human innovation. In the coming chapters we will discuss and analyze some of the most important incentive-based pollution control instruments. We therefore table this topic for now.

14.8 Chapter Conclusions

This chapter suggests that it makes most sense to focus pollution analysis and policy not on polluting goods and services, but on pollution itself or **abating** pollution from baseline levels. We can analyze the problem either way, because abatement = {uncontrolled/ baseline pollution} - {actual pollution}. We dug into the question of whether pollution control would best be addressed by the private sector, with governments mainly focusing on securing property rights. While it is certainly possible to develop examples where the so-called Coase Theorem holds, we can also think of many real-world complications that argue for government regulation. Unfortunately, the information gaps that make it unlikely the private sector can alone achieve optimal pollution levels also apply to government regulators. Cost-effectiveness in achieving perhaps politically determined abatement goals at least cost, rather than optimality, is therefore an especially important part of the economics of pollution control.

Though cost-effectiveness is certainly about getting polluters who can reduce at lowest cost to abate, cost-effective pollution policy becomes rather complex when (as is likely) polluters have marginal abatement costs that are increasing in the level of pollution control (decreasing in pollution). This reality means that marginal costs are likely dependent on how much pollution control firms, industries and indeed whole polluting communities have already done. Polluters who could initially control pollution at low cost work their way up their marginal abatement cost functions until previously "high-cost" firms become "low-cost." The mathematical condition for cost-effectiveness is therefore that marginal abatement costs are as close together as possible. With continuous cost functions, we can solve for the cost-effective pollution abatement allocation, and if we have discontinuous, step-type functions we would just choose the lowest-cost tons from marginal abatement cost tables until our goal was achieved.

From a policy standpoint, the utility of focusing on cost-effectiveness hugely depends on if there are differences in pollution abatement costs across polluters. We investigated marginal abatement costs and found that whether we focused on technologies, industries or firms, there we generally large differences in abatement costs. Cost-effective pollution control policies are therefore potentially very important, particularly for achieving expensive goals, such as limiting the increase in the average surface temperature of the Earth to 1.7 degrees Celsius.

The reality is that a combination of pollution control instruments are needed to save the planet, but it is worth noting that some instruments are inherently more cost-effective. Technology standards, though often the best choice, offer limited opportunities for innovation and pretty much box in polluters. Performance standards specify pollution levels but allow polluters to figure out how to reduce pollution, which offers more opportunity for cost-effectiveness. Incentive-based instruments, such as cap-and-trade and pollution taxes, allow the most room for achieving pollution abatement goals at least cost. As we will see in the coming chapters, though, they can also be politically fraught. It may therefore be the case that cost-effectiveness, which is second-best compared with optimal pollution, is in some cases only partly achievable.

Issues for Discussion

1. Please present one real-world example where the Coase Theorem might actually work.
2. Review the European Union directives in the Environmental Acquis. To what degree does this body of environmental regulations actually focus on cost-effectiveness in addition to getting the environmental job done?[14]
3. The chapter discusses several studies that estimated abatement costs by industry and technology and found significant differences. To what degree do you think cost-effectiveness can really be the basis of policy? How would you incentivize polluters who can abate pollution at low cost to actually do it?
4. Air pollution from *stationary source* polluters, such as electric power plants, is regulated using a variety of performance standards. The most common standards focus on emissions per unit of time (e.g., moment, day and year) or emissions per cubic meter of airflow out of chimneys (e.g., as in the European Union Medium Source Combustion Directive). Please list two advantages of each metric vis-à-vis the other one. Please also explain why they are indeed advantages.
5. Critics were quick to jump on McKinsey & Company (2009) for oversimplifying the challenge of climate change mitigation and perhaps underestimating the costs of land-based abatement.[15] Please discuss one circumstance in which such technology ladder approaches to estimating abatement costs would be appropriate. Also discuss one situation in which the McKinsey & Company (2009) approach would not be so useful.

Practice Problems

Please use the description and equations below for Questions 1-4.
Two plants, Plant X and Plant Y, are emitting NO_x into the air. *Emissions are currently uncontrolled and no abatement has occurred.* The two plant owners maximize profits. Because the air circulates throughout the region, the precise location where this pollutant is emitted is not relevant to the level of damages observed. NO_x is therefore a uniformly mixed pollutant. The Oregon Department of Environmental Quality decides to require reductions of 100 tons of NO_x per day.

Abatement costs can be approximated by continuous marginal cost functions. Suppose that Plant X has MACs of $MAC_x = 10 + 1.5x$, with TACs equaling $TAC_x = 10x + 0.75x^2$, where x is the level of NO_x abatement in tons of Plant X. *Plant X is currently polluting 150 tons.* Plant Y has MACs of $MAC_y = 60 + y$ and TACs of $TAC_y = 60y + 0.5y^2$, where y is the level of NO_x abatement in tons of Plant Y. *Plant Y currently pollutes 125 tons.*

1. Suppose regulators know the MACs of each plant. Using Eq. 14.2, what is the cost-effective *pollution abatement* of each plant if total abatement must equal 100? What is the industry *total abatement cost* to achieve the goal?

Hint: Recall that cost-effective pollution control implies that marginal costs are equal, and the pollution reduction goal is achieved. Use Box 14.3 as a guide for your calculations.

2. Suppose that instead of focusing on cost-effectiveness, the regulators require Plant X to reduce its emissions by 20 tons and Plant Y to reduce by the remainder needed to achieve the limit. What are the MACs of each plant after reducing emissions? Is this a cost-effective solution? Why or why not? What is the total cost of this policy? How does it compare to the total cost in Question 1?

Using the information above, answer the questions below in terms of NO_x *emissions* (not abatement). You will need to recast the MAC and TAC functions as marginal benefit and total benefit functions before solving. Please do not in any way directly use your answers from Questions 1 and 2.

Hint: Recall that abatement = uncontrolled/baseline pollution emissions - actual emissions. Let unknown actual emissions be represented by placeholders, such as E_x and E_y. Substitute these expressions in for x and y and solve, keeping careful track of the negative signs.

3. Suppose regulators know the marginal pollution benefits of each plant. What is the cost-effective *emissions level* of each plant if total abatement must equal 100?

4. Suppose that instead of focusing on cost-effectiveness, the regulators simply require Plant X to reduce its emissions by 20 tons and Plant Y to reduce by the remainder needed to achieve the limit. What are the *marginal pollution benefits* of each plant after abating? Is this a cost-effective allocation *of pollution*? Why or why not?

5. **Challenge Problem**: Add a Plant Z to the mix and answer Question 1. Here is the relevant information for Plant Z. Plant Z has MACs of $MAC_z = 2z$ and TACs of $TAC_z = 4z^{0.5}$, where Z is the level of NO_x abatement in tons of Plant Z. *Plant Z currently pollutes 100 tons.*

Suppose regulators know the MACs of each plant. What is the cost-effective pollution abatement of each plant if total abatement must equal 100? What is the industry total abatement cost to achieve the goal?

Hint: Solve the problem the same way you did Problem 1, but you will need to do more substitutions to make sure that you end up equalizing all MAC expressions. All expressions you work with will need to be independent of each other. For example, you can substitute the results of equalizing MAC_y and MAC_z in terms of y into another equality relating MAC_x and MAC_z. Don't forget that total abatement needs to equal 100 tons! If you know matrix algebra, you can also use that technique to solve this system of equations.

Please use the table to answer Questions 6 and 7.

Suppose the table shows the *total* abatement costs of two electric power plants emitting the pollutant SO_2 in the US State of Ohio as a function of emissions per month. **Both firms maximize profits, are currently uncontrolled and are emitting 20 tons each.** Both firms' TACs are therefore currently zero.

Emissions of SO_2 (Tons/Month)	Total Abatement Cost	
	Firm A	Firm B
20	0.00	0.00
19	1.00	2.10
18	3.10	6.70
17	6.40	16.10
16	11.00	35.40
15	17.00	67.90
14	24.60	122.80
13	34.00	205.70
12	45.50	322.60
11	59.40	479.50
10	75.90	684.40
9	95.20	949.30
8	117.50	1282.20
7	143.00	1689.10
6	171.90	2176.10
5	204.40	2753.10
4	240.70	3430.30
3	281.20	4217.50
2	326.10	5124.70
1	375.80	6161.90
0	430.70	7349.10

The Ohio Environmental Protection Agency, which is charged with environmental regulation in Ohio, would like to reduce emissions over time. In 2024, it would like to reduce emissions from the current 40 tons to 30 tons and by 2025 it would like to reduce emissions to 14 tons. Suppose the abatement functions given in the table are constant over time.

6. Using the method discussed with respect to Box 14.3, calculate the MACs at each level of emissions. Using this information and the criterion that MACs must be equal to achieve cost-effectiveness, for each of the abatement goals, please calculate the level of reductions each firm makes if reductions are to be made *at lowest total cost*. Please report the MACs. What are the total costs of achieving the goals in 2024 and 2025?

7. Using the method discussed with respect to Box 14.3, suppose instead that the reductions to achieve the goals are split equally across the two firms. How much is saved in total in each period by using the cost-effective allocation of responsibility?

Notes

1 https://education.nationalgeographic.org/resource/pollution
2 As we see in Section 14.3, for a number of reasons optimal pollution policy is likely to elude us. A so-called *second-best* policy is to achieve a given goal at lowest cost. Such cost-effective policies are an especially important part of pollution control. This term and the theory around it are attributable to Lipsey and Lancaster (1956-1957).
3 That CO_2 builds up in the atmosphere means it is a so-called "stock" pollutant that the environment cannot really process within a reasonable timeframe. As we know from Chapter 2, CO_2

molecules hang around in the air for hundreds of years before being sequestered by the planet. Stock pollutants are an extreme example of a pollutant with a positively sloped marginal cost function, but it would be completely reasonable for non-stock pollutants to also have upward-sloping marginal costs.

4 For more information see the US Department of Energy heat pump website at https://www.energy.gov/energysaver/heat-pump-systems

5 Private sector lobbying and political infighting can certainly slow down decisions. For example, it took the US basically 20 years, with the Inflation Reduction Act of 2022, to make meaningful legislative progress at the federal level on climate change. For example, the US Senate never ratified the 1997 Kyoto Protocol to the UN Framework Convention on Climate Change, which was ratified by 192 countries and regional blocs.

6 Please see https://en.wikipedia.org/wiki/Ministry_of_Natural_Resources_and_Environment_(Thailand) for more information on the structure of government environmental protection in Thailand.

7 e.g., see https://www.who.int/news-room/fact-sheets/detail/ambient-(outdoor)-air-quality-and-health for more info.

8 For a helpful discussion of the importance of political feasibility, please see Cullenward and Victor (2021).

9 The full text of EU/2015/2193 is available at https://eur-lex.europa.eu/legal-content/EN/TXT/PDF/?uri=CELEX:32015L2193&rid=4#:~:text=This%20Directive%20lays%20down%20rules,the%20environment%20from%20such%20emissions.

10 Let us also acknowledge that objectives can be intertwined with instruments. Examples include the requirements for BACT and RACT technologies.

11 https://www.iea.org/news/global-co2-emissions-rebounded-to-their-highest-level-in-history-in-2021

12 https://www.oregon.gov/deq/filterdocs/SSI-Standards.pdf

13 https://ww2.arb.ca.gov/sites/default/files/barcu/regact/2012/leviiighg2012/lev1.pdf

14 Summaries of the EU environmental directives can be found at https://eur-lex.europa.eu/summary/chapter/20.html

15 e.g., see Dyer and Counsell (2010).

Further Reading and References

Barbanell, M. 2022. "A Brief Summary of the Climate and Energy Provisions of the Inflation Reduction Act of 2022." World Resources Institute. Retrieved from https://www.wri.org/update/brief-summary-climate-and-energy-provisions-inflation-reduction-act-2022 May 21, 2024.

Coase, R. 1960. "The Problem of Social Cost." *Journal of Law and Economics* 3: 837-877.

Congressional Research Service (CRS). 2022. *Clean Air Act: A Summary of the Act and its Major Requirements*. CRS Report RL30853. Downloaded from https://crsreports.congress.gov/product/pdf/RL/RL30853 November 30, 2022.

Cullenward, D. and D.G. Victor. 2021. *Making Climate Policy Work*. Polity Press: Cambridge, UK.

Dyer, N. and S. Counsell. 2010. "McREDD: How McKinsey 'Cost-Curves' are Distorting REDD." *Rainforest Foundation UK Climate and Forests Policy Brief*. Downloaded from https://www.redd-monitor.org/wp-content/uploads/2010/11/McRedd-English.pdf February 6, 2023.

Hayes, A.W. 2005. "The Precautionary Principle." *Archives of Industrial Hygiene and Toxicology* 56 (2): 161-166.

Larsen, J., B. King, H. Kolus, N. Dasaro, G. Bower and W. Jones. 2022. "A Turning Point for US Climate Progress: Assessing the Climate and Clean Energy Provisions in the Inflation Reduction Act." Energy & Climate Report, August 12. Retrieved from https://rhg.com/research/climate-clean-energy-inflation-reduction-act/ May 15, 2024.

Lipsey, R.G. and K. Lancaster. 1956-1957. "The General Theory of Second Best." *Review of Economic Studies* 24 (1): 11-32.

Loughlin, D.H., A.J. Macpherson, K.R. Kaufman and B.N. Keaveny. 2017. "Marginal Abatement Cost Curve for Nitrogen Oxides Incorporating Controls, Renewable Electricity, Energy Efficiency, and Fuel Switching." *Journal of the Air & Waste Management Association* 67 (10): 1115-1125.

McKinsey & Company. 2009. "Pathways to a Low-Carbon Economy: Version 2 of the Global Greenhouse Gas Abatement Cost Curve." Retrieved from https://www.mckinsey.com/capabilities/sustainability/our-insights/pathways-to-a-low-carbon-economy December 10, 2022.

Ministry for the Environment. 2020. *Marginal Abatement Cost Curves Analysis for New Zealand: Potential Greenhouse Gas Mitigation Options and Their Costs*. Ministry for the Environment: Wellington. Downloaded from https://environment.govt.nz/assets/Publications/Files/marginal-abatement-cost-curves-analysis_0.pdf December 1, 2022.

15 Green Pricing: Theory and Practice of Taxes on Measured Pollution

15.1 Recap and Introduction to the Chapter

Waste sink services are some of the most important and conflicting private ecosystem services that assets such as air, rivers and land offer humans. Without some type of collective action intervention to deal with moral hazard, our behavioral models and experience suggest there will be too much pollution. Though other groups in society can potentially be very important, when it comes to limiting pollution, governments are by far our most important collective action institutions.

In the last chapter, we thought through some of the information limitations governments deal with as they try to limit pollution to an efficient level. The marginal benefits of pollution control, which accrue to societies as a whole, can perhaps at best be roughly estimated because the science is likely in-progress and most ecosystem services affected by pollution do not have markets that tell us values. But even the marginal costs of pollution control can be difficult to know because of polluters' moral hazard incentives. After all, why should polluters go to the trouble to focus in on abatement costs before it is necessary? Once they estimate them, why should they give government regulators that information? Our behavioral models, which predict that one's own benefits are very important for decision-making, tell us to expect polluters to keep such valuable information to themselves or, at minimum, to err on the side of overestimating marginal abatement costs.

Governments therefore generally need to regulate polluters with imperfect information. Abatement costs can differ substantially across polluters, technologies and industries, implying governments typically cannot ignore cost differences, even if they don't accurately

DOI: 10.4324/9781003308225-18

334 Environmental Economics and Ecosystem Services

know them. Cost-effective policies that offer polluters flexibility, with the goal to get as much pollution abatement as possible for the amount spent, are therefore likely very important. Indeed, cost-effectiveness may be critical to generating the political will to pursue strong environmental policies.

One of the most important classes of cost-effective policies is "green" environmental pricing, which uses fees, charges and taxes to reduce environmental moral hazard. We briefly encountered such policies in Chapter 9, when we modeled environmental moral hazard as an externality. The externality lens naturally leads to the conclusion that price instruments, such as environmental taxes, can be very useful for steering economies in greener directions and "internalizing" the environmental costs polluters generate for themselves and others.

We now extend our analysis of pollution control in two important ways. First, we do our best to focus on the pollution, which is what governments are hopefully trying to control, rather than on production of goods and services as in Chapter 9. Second, we take account of governments' information difficulties and evaluate the potential for green pricing to achieve environmental objectives and improve cost-effectiveness. We then examine the international experience with pricing of pollution and the evidence on its potential to reduce emissions.

15.2 Green Pricing: Theory, Opportunities and Challenges

Green pricing tries to help those who degrade the environment feel the pain they cause the world and thereby reduce environmental moral hazard. In many, if not most, countries around the world, emphasizing sink ecosystem services is the #1 way humans put common pool ecosystem services at risk. It is therefore perhaps not surprising that taxing pollution is the most important focus of green pricing in the world today. We therefore focus our attention in this chapter on that topic.

Green pricing is a more general term than pollution taxation or even environmental taxation. For example, the price premia typically found on organic produce, ethically sourced meat or sustainably grown lumber are all types of green pricing. But this is not what we are talking about here. Our interest is in the use of government tax policies to steer economies in greener directions.

Let us first acknowledge that the purpose of a "tax" is generally to raise revenues for governments, but the purpose we have in mind here is pollution control, which is something quite different. There is therefore an argument for using the term pollution "charge" or "fee," because polluters are using the public ecosystem services of clean air, water and land; governments are essentially just charging for those ecosystem services. This distinction is not made so much nowadays. For example, we typically use the term carbon *tax* rather than carbon charge. We will therefore use the tax terminology in the rest of this chapter and the remainder of the book.

Let us envision pollution tax possibilities as a ladder, with the bottom rung being taxing pollution itself, which is presumably the issue we want to address. Taxing pollution is our bottom rung on the ladder and starting point, because it is the most focused and offers us the widest set of options for reducing emissions. But what if, for some reason, taxing pollution is not possible? Well, we might climb up one rung, moving a bit away from taxing specific pollutants, toward a higher level of generality by taxing polluting production inputs

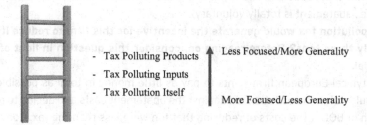

- Tax Polluting Products
- Tax Polluting Inputs
- Tax Pollution Itself

Less Focused/More Generality

More Focused/Less Generality

Figure 15.1 The Pollution Tax Policy Ladder
Source: Freepikcompany.com

based on their quality. As we will discuss in the next section, such an approach can potentially work well for air pollution, such as carbon pollution, which depends heavily on fuel efficiency and choice of fuels.

Finally, if it is not possible to tax pollution or polluting inputs, it may be necessary to move even further away from pollution, to the third rung on the ladder, and tax polluting products. As we will discuss in the final section of this chapter, such policies include taxes on tires, single-use paper and plastic bags, cans of oil-based paint (which will ultimately have to be disposed of as hazardous waste) and access to congested urban zones. Polluting product taxes are not as focused as other taxes, but they still incentivize less pollution.

We will come to the international experience with pollution taxes in a minute but let us first use the models we developed in Chapter 14 to consider how polluters would respond to a tax focused on pollution. Our behavioral model is once again a simple one – polluters try to make as much money as possible. In this case, that means spending as little as possible on abatement and pollution taxes. Suppose we consider our typical European firm's biological oxygen demand (BOD) abatement cost schedule discussed in Table 14.2, which is reproduced here with slightly different values as Table 15.1.

To start, let us suppose rather unrealistically that the executive part of the European Union, the European Commission, knows everything in Table 15.1. They would like this typical firm to reduce its emissions of BOD by three tons and plan to use a single pollution tax rate

Table 15.1 Biological Oxygen Demand (BOD) Abatement Costs of a Typical Firm

BOD Abatement (Tons)	Total Abatement Costs (TACs) (€)	Marginal Abatement Costs (MACs) (€)
0	0	Undefined
1	1	1
2	4	3
3	9	5
4	16	7
5	25	9
6	36	11
7	49	13

to incentivize that reduction. The firm can choose to abate as much or as little as it wants to reduce (i.e., abatement is totally voluntary).

What pollution tax would generate the incentive for this firm to reduce its emissions by exactly three tons? Before reading on, consider this question in light of our behavioral model.

If our typical European firm wants to pay as little money in total as possible, we expect that it would weigh the pollution tax against the abatement costs and decide to reduce a particular ton of BOD if the costs of reducing that ton were less than the tax. Conversely, if the marginal abatement cost (MAC) exceeded the tax, they would instead pay the tax.

So, what pollution tax would incentivize the firm to abate one ton? As the first ton costs €1 to abate, the tax must be bigger than €1, but less than €3, otherwise the firm would save money by reducing two tons. The firm abates two tons if the tax is set anywhere between €3 and €5. The firm is predicted to reduce its BOD emissions by three tons if the pollution tax were set at anything between €5.01 and €6.99. The total cost of abating three tons, as read from the second column, is €9 or €1 + €3 + €5.

Before reading on, please confirm that if the pollution tax were set, for example, at €6 it would indeed be cheaper to abate three tons than do anything else.

Our behavioral model predicts that, faced with a pollution tax of €6 per ton of BOD, the firm would voluntarily reduce its pollution by three tons. This prediction assumes that polluters abate if the tax is above the MAC, but do not reduce emissions if the tax is less than the MAC.

Challenge Yourself

Using Table 15.1, and assuming a pollution tax of €6 per ton of BOD, can you calculate the total abatement cost and total pollution charges *saved* by reducing BOD emissions by 3 tons rather than zero?

Suppose we set the tax at €5 per ton. What happens then? Well, for the very last bit of abatement the tax is exactly equal to the MAC. For just that last unit of pollution, our behavioral model suggests the firm would be indifferent between abating and emitting. Let us assume that being indifferent, firms always choose to abate that last ton (or part of a ton if the MAC is continuous). This assumption implies that, faced with a pollution tax, the profit-maximizing private equilibrium of the model would be for any firm (call it firm "i") to abate until the MAC exactly equals the pollution tax. This profit-maximizing equilibrium condition in a pollution tax model is given in Eq. 15.1.

Equation 15.1

Any Polluter "i" Abates Such that MAC_i = Pollution Tax

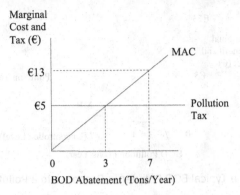

Figure 15.2 Abatement of a BOD Polluter in Response to a Pollution Tax of €5 and €13/Ton

Let us connect the seven points in Table 15.1 and approximate this stepwise discontinuous relationship between the amount of abatement and MACs with a linear function of the type used in previous chapters. This approach gives us something like Fig. 15.2.

As we saw in the previous chapter, any pollution abatement problem can be specified as a pollution-allowed challenge, because the pollution observed is just uncontrolled or baseline pollution minus abatement.

Before reading on, please take a minute to sketch a clearly labeled graph similar to Fig. 15.2 that phrases the policy problem in terms of calibrating the pollution tax to *allow* four tons of pollution. Suppose that the uncontrolled/baseline emissions of the polluter are seven tons, and it is possible to reduce emissions to zero.

Figure 15.3 starts from a situation of no tax, which is predicted to cause the polluter to emit seven tons, which is the uncontrolled/baseline level of BOD. As the tax rises, emissions are predicted to fall, with a tax of €5 incentivizing the firm to emit only four tons. We can now clearly see how adding a pollution tax reduces emissions.

Challenge Yourself

How much does Fig. 15.3 suggest the typical firm pays in pollution taxes? How much money does it save by not totally eliminating its pollution emissions?

Now suppose that, as is most likely, the European Commission does not know the firm's abatement costs. It therefore does some investigation and decides that a pollution tax of €9 would incentivize the firm to reduce its BOD emissions by three tons. Well, as is hopefully clear from Table 15.1 and Fig. 15.2, the regulator would not be correct. Notice that the cost of reducing the fifth ton is €9, so the firm would be just indifferent between abating five and four tons. Once again, let us assume now and going forward that if a pollution tax or other financial incentive is the same as an MAC value, the firm abates that ton, because the firm is

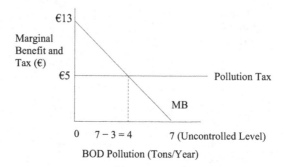

Figure 15.3 Pollution of a Typical BOD Polluter in Response to a Pollution Tax of €5/Ton

truly indifferent. So, in this case the tax of €9 would be expected to cause the firm to reduce five tons, not three tons as the European Commission predicted would happen.

Taxing pollution will reduce emissions, but unless regulators have very good estimates of polluter MACs, using pollution taxes alone they cannot be sure to hit specific abatement targets. This is an important limitation of pollution taxes, suggesting that they may need to be used in conjunction with other instruments, such as emissions or technology standards. For example, regulators might specify maximum levels of BOD emissions for each firm per hour, day and year and add a pollution tax on top of those emissions standards to give polluters incentives to reduce emissions as much as they find cost-effective.

Challenge Yourself

Is adding an emissions or technology standard to pollution tax policy consistent with worries about over- or underestimating MACs?

In the European Union and around the world, there are thousands of BOD polluters, and the European Commission does not really know the abatement cost functions of any of them. How would a pollution tax work with many firms? Answer: The same way. To illustrate the incentives, let us add a second firm to Table 15.1. This situation is shown in Table 15.2, with Firm 1 being the polluter in Table 15.1 and Firm 2 being our new and also higher-cost firm. Please notice that at all abatement levels the MAC of Firm 2 is greater than that of Firm 1.

If the two polluters have free choice to abate their BOD emissions or pay a €5 tax on any tons they decide not to abate, how much will the two firms reduce emissions? From our previous discussion we know that, based on our behavioral model, which assumes profit maximization, Firm 1 is predicted to reduce its emissions by 3 tons. Firm 2, which has higher MACs at all levels of BOD abatement, will only reduce its emissions by 2 tons for a total of 5 tons = 3 tons + 2 tons. That is, the lower-cost firm will reduce its emissions by more than the higher-cost polluter, meaning it will carry more of the total burden of reducing emissions.

Table 15.2 Biological Oxygen Demand (BOD) Marginal Abatement Costs of Two Firms

BOD Abatement (Tons)	Firm 1		Firm 2	
	Total Abatement Costs (TACs) (€)	Marginal Abatement Costs (MACs) (€)	Total Abatement Costs (TACs) (€)	Marginal Abatement Costs (MACs) (€)
0	0	Undefined	0	Undefined
1	1	1	2	2
2	4	3	7	5
3	9	5	15	8
4	16	7	26	11
5	25	9	40	14
6	36	11	57	17
7	49	13	77	20

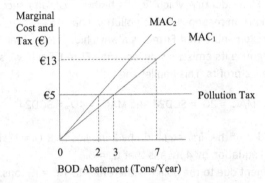

Figure 15.4 Response of Two BOD Polluters to a Pollution Tax of €5/Ton

If you are thinking that lower abatement cost firms carrying more of the burden than high-cost polluters sounds like cost-effectiveness, you would be right. The total abatement cost with this allocation of abatement is €7 + €9 = €16. Notice that there is no more cost-effective way to reduce total emissions by 5 tons. For example, if Firm 1 reduced by 2 and Firm 2 reduced by 3 it would cost €4 + €15 = €19, which is more than €16. **Please confirm for yourself that a €5 pollution tax indeed creates incentives to reduce 5 tons at the lowest possible cost.**

In Chapter 14, we learned that a cost-effective allocation of pollution abatement effort implies that MACs end up equal or as close together as possible. Notice that in Table 15.2 the abatement costs started out far apart (€1 and €2 for the first ton respectively, i.e., 100% difference). As Firm 1 reduced its emissions, though, its cost to reduce additional tons increased, closing the gap with Firm 2. That is, the MACs converged and ended up equal at €5 for the final tons abated. Pollution taxes therefore create incentives for cost-effective allocations of abatement responsibility. This situation is shown with linear MAC functions in Fig. 15.4.

Figure 15.4 and Table 15.2 show that a second polluter, who tries to make as much profit as possible, will also choose its abatement such that its MAC equals the pollution tax. The MACs of both polluters therefore end up equal to each other and the tax rate, which is the condition for cost-effectiveness. Indeed, no matter how many firms pollute, if our behavioral model holds, we can be confident that the outcome will be cost-effective. The rub, though, is that regulators cannot be so confident about how much total abatement will be incentivized, because they likely will not know the abatement costs.

Box 15.1 Worked Example: Equilibrium Abatement with Pollution Taxes

Suppose there were only two CO_2 polluters in the Canadian Province of British Columbia (not even close to true), which has long had a carbon tax. The provincial government sets the tax at \$CD24 per ton. The MAC of Firm 1 is $MAC_1 = 2Q_1$ and for Firm 2, $MAC_2 = 4Q_2$, where the "Qs" indicate the firms' abatement of CO_2 in tons per year.

Your tasks are: First, identify which firm is higher-cost; and second, solve for the cost-effective equilibrium response to the pollution charge.

To solve this problem, note that Firm 2 is always higher-cost than Firm 1. We therefore expect it will reduce its emissions by less than Firm 1. Each will set its MAC equal to the tax to maximize profits. This implies the following:

$$MAC_1 = 2Q_1 = \$CD24 \text{ and } MAC_2 = 4Q_2 = \$CD24$$

Dividing both sides of the first equation by 2 implies that $Q_1 = 12$ and dividing both sides of the second equation by 4 means that $Q_2 = 6$.

The total abatement due to the policy is therefore $12 + 6 = 18$ tons.

At those abatement levels, what is the MAC of each firm? Answer: \$CD24. The allocation of abatement effort is cost-effective.

If we specify the curves in Fig. 15.4 as mathematical functions, we can even solve for the abatement each polluter will voluntarily choose. Box 15.1 presents a worked example for two firms.[1]

Challenge Yourself

Can you add two more CO_2 polluters to Box 15.1 and solve for the total abatement if a carbon tax is set at \$CD24?

But what would polluters think about all this cost-effectiveness logic? Suppose that each of the two firms in Table 15.2 is emitting 7 tons of BOD and therefore could abate anywhere from none to all of that. At a tax of €5/ton, Firm 1 voluntarily abates 3 tons and Firm 1 abates 2. Total abatement cost is €16, which is the lowest possible.

Let us not forget, though, that because of how this instrument works, both firms will have to pay €5 on every ton they **do** emit. Firm 1 will therefore need to pay €5*4 tons = €20 and Firm 2 will pay €5*5 = €25. In this example, therefore, there will be a transfer of money from the private to the public sector of €20 + €25 = €45, which is vastly more than the reduction in abatement cost from using this so-called "cost-effective" policy!

But other than the polluters having to pay more money, is such a transfer a problem? Let us break the problem down a bit. If governments effectively utilize pollution tax revenues, such a transfer would not be a problem, because those revenues generate at least as much value in government hands as they do in the private sector. But from a firm's perspective, pollution taxes cost them more money. If there are going to be more stringent environmental regulations, polluters may therefore lobby for exemptions or perhaps simpler – and from a social standpoint less cost-effective – policies that specify exactly what polluters need to do to be in compliance. *Pollution taxes are therefore found to be cost-effective only from an overall social perspective, particularly if governments use revenues at least as effectively as the private sector. Pollution taxes are not cost-effective from a purely private sector perspective.*

Challenge Yourself

Do you think the revenue-generating feature of pollution tax policy is a positive or negative aspect of such policies?

Please consider what other benefits pollution taxes offer in addition to cost-effectiveness, which may make them worth adopting. Please think carefully about this question before reading on.

Regarding other potential benefits of pollution taxes, those who think polluters should be "punished" for polluting the Earth's air, water and land will get their wish. Polluters pay explicit, easily identifiable costs per ton of pollution, which makes clear that they are being held accountable and are forced by policy to pay their way.

Second, subject to the level of pollution tax chosen by regulators, polluters fully feel the effects of their actions and incorporate them into their decision-making; again, subject to whatever tax level regulators choose, pollution taxes therefore truly "internalize" the externalities discussed in Chapter 9.

Third (and related to the second point), pollution taxes offer incentives for polluters to innovate and change technologies to reduce their tax burdens. If tax rates are high enough, firms will have all the incentives they need to do the needful to reduce their pollution emissions. These steps may include internal waste minimization and pollution prevention audits, evaluating alternative production methods and pollution control equipment, or hiring environmental consultants and engineers who would be able to help them find cost-effective technological approaches.

Fourth, because pollution taxes target pollution itself, all abatement possibilities are incentivized. This stands in sharp contrast to our product tax policy presented in Chapter 9, because if products are taxed, the only thing those taxes incentivize is reduced production and consumption. As we discussed in that chapter, no incentives are created for adopting better pollution control technologies, alternative inputs, different fuels etc. Such cruder pollution control instruments presume that the only way to reduce emissions is by reducing output, which is often not correct.

Finally, because pollution taxes imply that polluters pay for the right to pollute, they generate government revenues. Such revenues can be used to support environmental protection and environmental education or may be used to offset other taxes, such as those on incomes, profits and purchases. Pollution taxes can therefore be part of a "green tax reform," which taxes pollution rather than positive things like your income.

Now, please consider the following question: What are potential downsides of using pollution taxes as a core part of pollution control policy? Please think about this question before reading on.

What are the downsides associated with pollution taxes? First, we should not forget that pollution taxes are at best only roughly calibrated to the marginal benefits of abatement. They can only really help with cost-effectiveness and perhaps offer some of the additional benefits just mentioned. Pollution taxes in general do not achieve efficiency, because as we discussed in Chapter 14, the abatement goal is the result of political processes rather than a true balancing of marginal benefits and costs.

Challenge Yourself

If a regulator sets its tax rate too high to achieve its desired goal, what does that say about its estimate of firms' MACs? Too high? Too low? Correct?

Second, because regulators likely have only a rough estimate of MACs, it is unlikely that pollution taxes alone can incentivize precise levels of abatement. As in our example in Table 15.1, regulators thought that a tax of €9 would cause our typical European firm to reduce BOD emissions by 3 tons, when in fact the firm voluntarily chose 5 tons. The regulator therefore overshot its target by almost 2/3. In fact, overshooting pollution control targets is likely, because not only are regulators unlikely to know firms' MACs, but firms have incentives to claim it is too expensive for them to abate; such lobbying, if firms know they can abate cheaply, is just one example of environmental moral hazard.

Third, to tax based on measured pollution emissions, pollution sources must be identifiable. You will recall from Chapter 14 that a pollution source is the point where pollution meets the environment. In some cases, such as so-called "non-point" pollution sources, it may be difficult to pin particular emissions on specific sources. Agriculture is a great example, because when agricultural chemicals, such as nitrogen fertilizers, are put on fields, some are absorbed by plants (which is, of course, the purpose) and some run off into the environment and eventually waterways. This runoff can cause eutrophication and the environmental dead zones that are of such concern in the planetary boundaries literature that we considered in Chapter 1. The amount of runoff is not known with certainty and, in any case, pollutants from different polluters get mixed up, making it difficult to identify the source of particular emissions.

Fourth, to collect taxes based on pollution emissions, regulators need to know how much pollution is emitted into the environment. In other words, pollution must be measurable at reasonable cost, so regulators can monitor emissions and assess the correct amount of taxes. A number of air pollutants are measurable using a variety of direct-measurement techniques in the environment at large (so-called *ambient* pollution monitoring) and sometimes also at the source. In the US, for example, the six "criteria" air pollutants subject to federal control can be monitored.[2]

Many water pollutants are also measurable at the source and the US Environmental Protection Agency has developed a variety of compliance monitoring protocols for facilities such as wastewater treatment plants, pre-treatment facilities and industrial sites. Pollution monitoring is a huge topic that we need not fully get into here, but let us just say it can be very complex, technical and costly. Measurement difficulties are therefore sometimes important barriers to implementing pollution taxes, which rely especially heavily on direct pollution monitoring.[3]

Challenge Yourself

Do you think that pollution taxes are a useful pollution policy instrument? If so, why do you think they have not been especially widely implemented?

Fifth, pollution taxes are instruments that offer polluters complete flexibility, including with respect to choosing their pollution emissions levels. Hazardous (sometimes called "toxic") pollutants that may have effects at even low concentrations are not, however, the place for allowing flexibility. Examples of officially designated toxic air pollutants in the US include lead (also a criteria pollutant), asbestos, benzene, chromium and mercury. Often the relationships between hazardous chemicals and emissions are sufficiently complex that regulatory rules are quite prescriptive and tailored to specific industries. The US, for example, has specific toxics rules for glass manufacturing, hydrochloric acid production, spandex production, plywood manufacturing, phosphate fertilizers and many, many, many more. These rules are made specifically because flexibility of the type offered by pollution taxes is too dangerous and therefore not appropriate.[4]

Finally, pollution taxes generate revenues, which represent additional costs for polluters and a transfer of funds from the private to the public sector. In many settings, despite the potential cost-effectiveness, incentive and fairness properties of pollution taxes, such transfers can make pollution taxes unpopular.

In sum, pollution taxes on monitored emissions are sort of the gold standard for green pollution pricing, because when pollution is taxed, all abatement options are incentivized. Polluters can reduce output, change fuels, change inputs, use pollution control equipment, etc., etc. etc. It is a good option, if it is feasible, but often implementation costs, practical issues and political difficulties seem to keep us from implementing pollution taxes. Perhaps these complications explain why, despite their theoretical appeal, we do not observe terribly widespread use of pollution taxes on directly measured emissions.[5]

15.3 Pollution Taxes on Measured Emissions

There are a lot of taxes around the world that target pollution of various types, but not very many that are based on measured pollution. Exceptions are largely in China and the former communist countries of Russia and Central/Eastern Europe. Most of these systems were put in place in the 1980s and 1990s as part of efforts to address the types of serious pollution problems associated with economic planning gone wild, which we studied in Chapter 12.

I and my colleague Bruce Larson edited a book that mainly focused on pollution taxes in Central and Eastern Europe and the former Soviet Union, which until the 2000s had by far the most extensive pollution tax systems in the world. As we noted in the introductory chapter of that edited book, economic reforms moved forward in the 1990s and output plummeted by anywhere from 18% to 60% between 1989 and 1994. By 1994 and into the 2000s, "all countries to varying degrees include[d] pollution charges [taxes] levied on estimated or, *much less frequently, on directly measured pollution emissions*" (Bluffstone and Larson, 1997, p. 4; emphasis added). Several contributors to the book found that marginal abatement costs were low and/or that low pollution tax rates had substantial effects on pollution emissions.

Later in the book, Vincent and Farrow (1997) surveyed pollution tax systems across the region and found that most countries taxed over 100 pollutants, few of which were measured. They found that emissions were instead estimated using methodologies, often developed during the Soviet period. One of the many environmental policy reforms that took place during the 1990s and 2000s was to limit pollution taxes that were designed to affect polluting behavior to measurable pollutants.

Interestingly, often revenues from pollution taxes in Central and Eastern Europe and the former Soviet Union went to environmental funds that supported environmental investments by the private and public sectors. For example, in Lithuania, where I worked for three years, 70% of the pollution taxes went to municipal environmental funds and the rest to the government general budget. Poland had the largest pollution tax system in the region, generating about $500 million per year (about $1 billion in $US 2023), with all revenues going to national or sub-national environmental funds (Fiedor and Anderson, 1997).

Few of those systems survive today, as most countries joined the European Union and did not choose to maintain their pollution taxes. One exception is the Czech Republic, which since 1967 (when it was part of Czechoslovakia) has had what they call "fees" on air and water pollution. These fees are closely linked to environmental permits, which allow facilities to operate, and are only charged for pollutants that are easily measurable at stationary sources. Water pollution is charged per kilogram of pollution only when pollution exceeds permitted limits.

Air pollution fees focus on pollutants with significant impacts on human health and are assessed at the regional level. Rates per ton of pollution are published well in advance so polluters can prepare, and are calibrated to motivate polluters to reduce emissions. Past rates are given in Table 15.3. These rates have certainly been high enough to be significant for many polluters.

Table 15.3 Czech Republic Air Pollution Fee Rates per Ton of Pollutant (2023 $US)

Pollutant	2017	2018	2019	2020	2021+
PM_{10}	$231	$316	$401	$485	$625
SO_2	$77	$105	$134	$162	$208
NO_x	$62	$83	$160	$127	$166

PM_{10} – particulate matter (i.e., dust) 10 microns or larger in diameter. SO_2 – sulfur dioxide. NO_x – nitrous oxides. Currency conversion from Czech Crowns to $US using Oanda.com. Inflation calculated using consumer price index calculator from the US Federal Reserve Bank of Minneapolis.

Source: European Union Publications Office (2021).

Since 1979, China has had a system of pollution taxes on both air and water pollution, which was called a pollution "levy." In 2015, for example, the government collected about $2.5 billion in levy revenues from almost 300,000 businesses (Zhang, 2017), making it the biggest pollution tax system in the world. In 2018, a new law went into effect, which replaced the levy system with what is now called an "Environmental Protection Tax." The main goals of the new system are to improve enforcement and increase incentives for stationary source polluters to reduce their emissions. As has been discussed at several points in this book, many areas in China have had very serious air and water pollution, with potentially significant effects on the environment and human health. The tax law seeks to use the power of incentives to improve environmental outcomes.

Box 15.2 The Nitrogen Oxide (NO_x) Tax in Sweden

In the early 1990s, Sweden introduced a tax on nitrogen oxide (NO_x) emissions by power plants. The goal of the tax was and is to reduce NO_x emissions, which cause soil acidification and nitrogen pollution of waterways, contributing to eutrophication. NO_x is formed in the atmosphere and therefore it is not possible to know emissions based on, for example, the fuels used; emissions must be directly measured.

To curb emissions, a high tax of SEK 40 was levied per kilogram (about $4000 per ton in 1992!), which was developed based on estimates of marginal abatement costs. Only about 200 large combustion plants had to pay the tax, which was calibrated to reduce emissions by 5000–7000 tons, which at the time was a 30% reduction. All sources were required to install continuous NO_x emissions monitoring equipment.

The Swedish NO_x tax is especially innovative, because of what they do with the tax revenues. In Sweden, all revenues from the NO_x tax except administrative costs of about 0.7% were and are returned to polluters themselves *in proportion to their production of energy*. The tax therefore directly incentivizes power production efficiency in terms of NO_x emissions. Sources with high NO_x emissions relative to their production are net payers to the system, whereas sources who are highly efficient receive more money back than they pay in. Amazingly, because of how it recycles its revenues, this pollution tax system creates a new revenue stream for highly efficient firms!

Source: OECD (2013).

As is the case in the Czech Republic and elsewhere, the Environmental Protection Tax is linked to the environmental permitting system, which specifies pollution source limits. The tax covers 71 water pollutants, plus general sewage, and applies to 44 air pollutants. Notably, the tax system excludes CO_2, which is addressed separately. Compared with the >40-year-old pollution levy system, tax rates are higher, provinces may adjust rates based on local conditions and there are more possibilities for polluters to reduce their tax costs if they pollute less than permitted limits.

Air and water pollutants are combined into what are called "pollution equivalent values" in kilograms, which are linked to toxicities and typical volumes emitted. For example, both SO_2 and NO_x have pollution equivalent values of 0.95 kg, but particulates, which are a particular problem in many parts of China, are worth 4 kg.[6] As of 2018, provinces could choose taxes on air pollution anywhere between $0.16 and $1.60 per kg of pollution equivalent (i.e., $160–$1600 per metric ton) and the water tax could range from $0.20 to $2.00 per kg.[7]

Before reading on, please evaluate if the basic Environmental Protection Tax rates in China are higher or lower than in the Czech Republic.

Most provinces have the minimum tax of $0.16 per pollution equivalent, but more heavily polluted areas can pay over four times that amount. Polluters pay 50% less if they are 50% below their limits and 25% lower taxes if they are 30% below their limits (Zhang, 2017).

There have already been a number of seemingly careful evaluations of China's Environmental Protection Tax, especially focusing on air pollution. In general, results suggest that the tax significantly reduces urban air pollution, with key pollutants declining by 25%–30% after the tax reform (Guo et al., 2022). At the regional or provincial level, which also includes rural areas, Li (2023) found beneficial effects of the Environmental Protection Tax on $PM_{2.5}$, SO_2 and NO_x concentrations in the atmosphere, presumably via reducing emissions. The author concluded that a US$0.14 per pollution equivalent increase in the provincial-level tax (1 Chinese yuan) reduced $PM_{2.5}$ concentrations by 3.8% and SO_2 by 18.6%.

Also at the provincial level, Lin et al. (2023) analyzed the effect of the Environmental Protection Tax on total SO_2 air pollution emissions. They estimated large declines in emissions for what they called "tax rate raising" provinces with higher than basic tax rates (Hebei, Jiangsu, Shandong, Henan, Hunan, Sichuan, Chongqing, Guizhou, Hainan, Guangxi, Shanxi, and Beijing). They also estimated that improvements were caused at least partly by the technological innovations of polluters. The conclusion that the tax is causing firms to adopt greener technologies is echoed by Deng et al. (2023), who used firm-level data from heavily polluting companies and found that the tax caused most types of companies to adopt additional environmental protection measures.

Figure 15.5 shows the trend in the standardized index of provincial-level SO_2 emissions analyzed in Lin et al. (2023).

What stands out to you about this graph, noting that the Environmental Protection Tax Law was announced in 2017 and implemented in 2018? Can you attribute the large declines observed between 2012 and 2019 solely to higher environmental protection taxes? There is no right answer here, but what seems most likely?

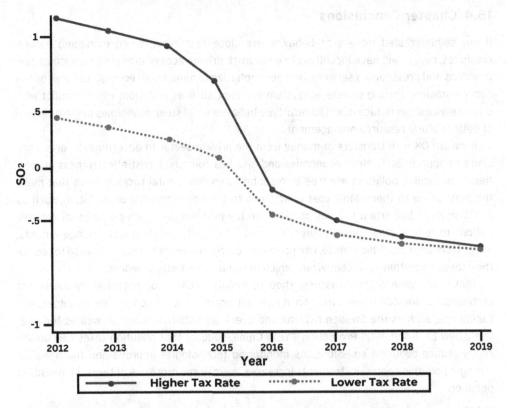

Figure 15.5 Trend in Standardized SO_2 Emissions for Provinces with High and Low Tax Rates in 2018
Source: Revised and approximated for clarity by the author based on Lin et al. (2023).

The Swedish NO_x tax discussed in Box 15.2 is notable because it is quite high, levies pollution taxes on measured emissions and because of its very innovative revenue recycling system. The system distributes all taxes except administrative costs back to polluters, but does so not based on pollution, which would be counterproductive, but on efficiency of electricity production in terms of NO_x emissions (i.e., NO_x/kwh). The evidence suggests that the policy, which has been in place for more than 30 years, reduced emissions by 35% within 2 years (OECD, 2013) and allowed electricity production to continue increasing. The NO_x tax very quickly created broad innovation effects, reducing emissions of NO_x per gigawatt (i.e., 1 billion) hour of power produced from 557 to 300 kg within 5 years, with the most polluting plants improving the most (Höglund-Isaksson and Sterner, 2009).

In sum, therefore, there is some interesting international experience with taxes on measured pollution, but it is pretty limited. The evidence on effectiveness that exists suggests that low tax rates may have big effects, perhaps due to lower-than-expected marginal abatement costs, among other possible reasons, including focusing firms' attention on pollution abatement. They may also simply prefer to invest in pollution control rather than pay taxes.

15.4 Chapter Conclusions

If our self-interested models of behavior are close to correct, when managing natural resources, people will have incentives to emphasize private ecosystem services, such as sink functions and provisioning services, and de-emphasize common pool ecosystem services like climate stability. Pricing private ecosystem services, such as pollution, which conflict with other services, can reduce moral hazard–type behaviors and steer economies in the direction of better natural resource management.

If we are OK with transfers of money from the private sector to governments, green pricing can appropriately steer economies and also has some nice cost-effectiveness properties. In particular, polluters are free to respond to environmental taxes in ways that make the most sense to them. Most cost-effective is to tax environmental degradation, such as pollution itself, because it focuses precisely on the problem and gives polluters all possible options to reduce emissions. Polluters can reduce output, switch inputs, change outputs, adopt technologies, reduce waste, use pollution control equipment, yada yada yada to reduce their taxes. Everything is incentivized, which maximizes cost-effectiveness.

There are some practical issues, though, mainly because of potential measurement problems, so we don't see taxes on measured pollution as much as we might expect. Exceptions, such as the Sweden NO_x tax and Czech air pollution taxes, as well as the relatively new Environmental Protection Tax in China, suggest that pollution taxes can significantly reduce pollution emissions. As monitoring technologies improve and the need for stronger pollution control instruments increases, we may see more use of taxes on measured pollution.

Issues for Discussion

1. Do some research. Are pollution taxes used where you live? If yes, what kind?
2. Is adding an emissions or technology standard to pollution tax policy consistent with worries about over- or underestimating MACs?
3. If a regulator sets its tax rate too high to achieve its desired goal, what does that say about its estimate of firms' MACs? Too high? Too low? Correct?

Practice Problems

1. Add two more CO_2 polluters to Box 15.1 using your own abatement cost functions and solve for the total abatement if a carbon tax is set at \$CD24. How much will CO_2 decline? Is the solution cost-effective? How do you know?
2. Suppose there are three main stationary source NO_x polluters in a jurisdiction in South Africa. This jurisdiction wants to reduce NO_x emissions by 12,000 tons. Each of the three polluting firms analyzes its abatement opportunities and they find the following abatement functions denominated in Rand, which is the South African currency:

   $MAC_1 = 1000 + 6A_1$.
   $MAC_2 = 4000 + 15A_2$.
   $MAC_3 = 3000 + 4A_3$.

The jurisdiction in South Africa levies a tax on measured NO_x emissions of RND 19,000 per ton, which is approximately $1000 per ton.

With this tax rate, would the jurisdiction achieve its goal? Would the result be cost-effective? Please show your work and explain your answers.

Notes

1   Regulators are unlikely to have more than a rough estimate of polluters' MACs, but let us leave that complication aside for purposes of the worked example in Box 15.1.
2   These US criteria air pollutants are sulfur dioxide, nitrogen oxides, particulate matter, ozone, lead and carbon monoxide. They have this name, because criteria for ambient concentration standards are based on the latest science. See https://www.epa.gov/criteria-air-pollutants for more information. Please also note that in the US, implementation of plans to achieve ambient standards is done at the state level through so-called "state implementation plans."
3   For more information on the US federal government approach to water pollution monitoring and inspections, please see https://www.epa.gov/compliance/compliance-inspection-manual-national-pollutant-discharge-elimination-system. See https://www.epa.gov/compliance/clean-air-act-caa-compliance-monitoring#neshap for details on air monitoring.
4   To go down this particular US EPA pollution regulation rabbit hole, please see https://www.epa.gov/stationary-sources-air-pollution/national-emission-standards-hazardous-air-pollutants-neshap-8. Please note that government revenues can still be raised off such emissions, but this goal is distinct from the objective to incentivize cost-effective emissions reductions.
5   See Fullerton et al. (2010) for additional information on the pros and cons of a variety of environmental taxes.
6   To take a look at an unofficial English translation of the law, please see https://www.chinalawtranslate.com/en/2016-environmental-protection-tax-law-of-the-p-r-c/. In addition to air and water pollution, the law also covers solid waste and noise pollution.
7   This is based on the Chinese yuan to $US exchange rate in October 2023.

Further Reading and References

Bluffstone, R. 1999. "Are the Costs of Pollution Abatement Lower in Central and Eastern Europe? Evidence from Lithuania." *Environment and Development Economics* 4 (4): 449–470.

Bluffstone, R. and B.A. Larson, eds. 1997. *Controlling Pollution in Transition Economies: Theories and Methods*. Edward Elgar: London.

Dasgupta, S., M. Huq, D. Wheeler and C. Zhang. 2001. "Water Pollution Abatement by Chinese Industry: Cost Estimates and Policy Implications." *Applied Economics* 33 (4): 547–557.

Deng, J., J. Yang, Z. Liu and Q. Tan. 2023. "Environmental Protection Tax and Green Innovation of Heavily Polluting Enterprises: A Quasi-Natural Experiment Based on the Implementation of China's Environmental Protection Tax Law." *PLoS One* 18 (6): e0286253.

European Union Publications Office. 2021. *Ensuring that Polluters Pay: Czech Republic*. Luxembourg. Retrieved from https://op.europa.eu/en/publication-detail/-/publication/b55b443a-9dc8-11ec-83e1-01aa75ed71a1/language-en October 2024.

Fiedor, B. and G.D. Anderson. 1997. "Environmental Charges in Poland." In R. Bluffstone and B.A. Larson, eds. *Controlling Pollution in Transition Economies: Theories and Methods*. Edward Elgar: London.

Fullerton, D., A. Leicester and S. Smith. 2010. "Environmental Taxes." In Institute for Fiscal Studies (IFS), ed. *Dimensions of Tax Design*. Oxford University Press: Oxford.

Guo, B., Y. Wang, Y. Feng, C. Liang, L. Tang, Z. Yao and F. Hu. 2022. "The Effects of Environmental Tax Reform on Urban Air Pollution: A Quasi-Natural Experiment Based on the Environmental Protection Tax Law." *Frontiers in Public Health* 10: 967524.

Höglund-Isaksson, L. and T. Sterner. 2009. *Innovation Effects of the Swedish NO_x Charge*. Paper presented at the OECD Global Forum on Eco-Innovation, November 4–5, 2009. Retrieved from https://pure.iiasa.ac.at/id/eprint/14600/

Li, Z. 2023. "Can Environmental Tax Curb Air Pollution: Evidence from China." *Applied Economics Letters* 30 (11): 1449–1452.

Lin, Y., L. Liao, C. Yu and Q. Yang. 2023. "Re-Examining the Governance Effect of China's Environmental Protection Tax." *Environmental Science and Pollution Research* 30 (22): 62325–62340.

OECD. 2013. *The Swedish Tax on Nitrogen Oxide Emissions: Lessons in Environmental Policy Reform.* OECD Environmental Policy Paper #2. Retrieved from https://www.oecd.org/en/publications/the-swedish-tax-on-nitrogen-oxide-emissions_5k3tpspfqgzt-en.html October 2024.

Vincent, J. and S. Farrow. 1997. "Pollution Charges in Central and Eastern Europe." In R. Bluffstone and B.A. Larson, eds. *Controlling Pollution in Transition Economies: Theories and Methods.* Edward Elgar: London.

Zhang, L. 2017. "China: New Law Replacing Pollution Discharge Fee with Environmental Protection Tax." Retrieved from the US Library of Congress, https://www.loc.gov/item/global-legal-monitor/2017-02-08/china-new-law-replacing-pollution-discharge-fee-with-environmental-protection-tax/

16 Green Pricing: Carbon Taxes, Green Product Taxes and Green Tax Reform

16.1 Recap and Introduction to the Chapter

In the previous chapter, we learned about the pollution tax policy ladder and discussed the first rung, which taxes measured pollution. This is the most specific and targeted pollution tax subject. It also aligns best with our environmental goal, which is typically to reduce pollution emissions, and incentivizes a wide spectrum of options. Abatement options incentivized in principle could include end-of-pipe emissions controls, production process alterations, input changes, better efficiency and changes to product lines, among others.

Unfortunately, in many cases taxing measured pollution is infeasible. Emissions monitoring can be a highly technical and expensive enterprise and for some pollutants measurement and attribution may be difficult. Historically, Central and Eastern Europe and the former Soviet Union tried to tax measured pollution, but in actuality tax liabilities were mainly based on pollution estimates. Currently, the Czech Republic, Sweden, China, and perhaps a few other countries, have taxes on measured pollution.

Contemporary pollution tax systems mainly move up the policy ladder, away from actual pollution. The first and perhaps most important set of pollution taxes focuses on polluting inputs. Taxing polluting inputs reduces the menu of abatement options that are incentivized - for example, end-of-pipe controls are not incentivized - but in some cases restricted incentives may be less important. Furthermore, there may be important gains in terms of implementation feasibility.

This chapter analyzes taxes on polluting inputs, with a primary focus on carbon taxes, which are some of the most important instruments for addressing climate change. We will also briefly consider cases where taxes on polluting products - important real-life versions

DOI: 10.4324/9781003308225-19

of our externality taxes from Chapter 9 – make sense. Finally, we also evaluate the case and possibility for so-called green tax reform in which tax systems shift toward taxing bad things like pollution, dirty inputs and polluting products, and away from taxation on our purchases, property and incomes.

16.2 Pollution Taxes Based on Quality of Production Inputs: Carbon Taxes

Most taxes on pollution are not actually based on measured pollution. Instead, generally in the interest of feasibility, regulators move up the pollution tax policy ladder. The first rung up on that ladder toward more generality is to tax based on the quantity and quality of polluting inputs, rather than pollution itself.

Taxes on polluting inputs, such as the carbon content in fuels, operate very much like the pollution taxes already discussed. Indeed, our analysis in Chapter 15 (see Figs. 15.1-15.3) would basically be the same, except the tax rate on the vertical axis and the quantity reduced on the horizontal axis would be based on the amount of the pollutant (e.g., CO_2 in tons) contained in a particular input, such as a fossil fuel. Regulators must *presume* that the best way to reduce pollution emissions, such as CO_2, is to a) shift away from carbon-intensive fuels toward carbon-free or low-carbon fuels or b) reduce fossil fuel use. These are the only shifts such taxes incentivize, which is a more limited set of choices than for taxes on measured pollution.

Challenge Yourself

If instead of taxing *pollution*, regulators tax *inputs and materials* they believe to be linked to pollution, what options for reducing tax costs do such approaches incentivize?

If monitoring emissions is impossible, difficult or expensive, however, and the two presumptions are pretty much correct, taxes on polluting inputs may be a good second-best option.[1] For the specific example of CO_2 emissions from fossil fuels, the two presumptions are probably reasonable, because, as we will discuss in a minute, the carbon content in fossil fuels is well-known and rather easily measured.

You will recall that most greenhouse gas emissions come from the burning of fossil fuels to produce heat and usable energy for human purposes. Carbon taxes have been adopted in many jurisdictions around the world to control greenhouse gas emissions. There are two key questions implementers of carbon taxes need to answer. The first question is what is the tax rate per ton of carbon, measured in terms of CO_2e using the global warming potential measurements discussed in Chapter 2. The second key question to answer is, "What exactly are we taxing and how is it measured?" Taxes need to be assessed according to the standard units that are used (e.g., liters, tons, cubic meters etc.), and it turns out there are a couple of calculation complications we need to address before achieving that result. We therefore start there and come back to the question of tax rates.[2]

First, fuels come in different forms. Some are gaseous, such as so-called "natural gas," which contains anywhere from 70% to 90% methane, a potent greenhouse gas (Lacroix et al., 2020). Others, such as diesel, home heating oil, kerosene, gasoline (i.e., petrol) and jet fuel, are liquids and are therefore measured in gallons and liters. Coals (anthracite, bituminous, lignite etc.), municipal solid waste and tire-derived fuels are solid and are therefore measured in "long" (i.e., metric) or "short" (US and UK) tons. A key first step is therefore to normalize all these fuels based on their energy contents, which is itself confusing, because there are several energy units. These include British thermal units (Btu), joules and calories, kilowatt hours (for electricity) and therms, which are used for heating. In addition, sometimes energy content is expressed in terms of fuels themselves (e.g., tons of oil equivalent, tons coal equivalent or even cubic meters of natural gas).

This can all be rather confusing, but measures are easily convertible to each other. For example, according to the American Physical Society, the joule is the basic metric of energy and 1 calorie equals 4.187 joules. The Btu is the English analog of the calorie and 1 Btu = 1055 joules. The kilowatt-hour is the main metric for electricity production and consumption (a kilowatt = 1000 watts) and 1 kWh = 3.6 million joules (APS, undated). The US Energy Information Agency in the US Department of Energy uses Btu as its basic energy metric, so that is what we will use in this chapter.

Equation 16.1 Carbon Tax Calculation Method

$$\text{Carbon Tax Charged} = \text{Tax Rate} * \text{Amount of Fuel Sold} * \frac{\text{Total Btu}}{\text{Amount of Fuel Sold}} * \frac{\text{Total CO2}}{\text{Total Btu}}$$

After choosing one of these metrics, the next step in answering the question of "What exactly are we taxing?" is to use scientific information on the CO_2e content per Btu and the number of Btus per standard unit of volume (liters) or mass (tons) to calculate the CO_2 content of each fuel in the normal units in which fuels are sold. Once we make this calculation, regulators can apply the chosen tax rate per ton of CO_2 to sales of the fuels (phew!). Equation 16.1 shows how such calculations work, and Table 16.1 gives the official US Government CO_2e values of each fuel per million Btu and per standard unit in the marketplace. To keep things a bit simpler, I only present the metric units.

We see from the table that, among carbon-based fuels, anthracite coal is the least carbon-efficient fuel per million Btus of energy (104 kg of CO_2/million Btu). Municipal waste is the most efficient, at about 50 kg of CO_2/million Btu, which is a bit better than natural gas. Perhaps rather than using natural gas as a fuel to "bridge" our transition to renewable energy as has been occurring in the US, we should be adding more "waste-to-energy" plants so we can reduce solid waste in landfills as we enjoy lower-carbon power production!

Table 16.1 CO_2 Content of Various Fuels

Select Fuels	Kg of CO_2/Million Btu	Kg of CO_2/Standard Unit
Home and Business Fuels		
Propane	62.88	1.52 kg/liter
Fuel Oil (also for electricity)	74.14	2.69 kg/liter
Kerosene	73.19	2.61 kg/liter
Natural Gas (also for electricity)	52.91	1.94 kg/cubic meter
Transport Fuels		
Gasoline (petrol)	70.66	2.32 kg/liter
Jet Fuel	72.23	2.59 kg/liter
Diesel	74.14	2.69 kg/liter
Coals Mainly for Electricity Production by Type		
Anthracite	103.69	2868 kg/metric ton
Bituminous	93.24	2392 kg/metric ton
Lignite	98.27	1404 kg/metric ton
Other Fuels Mainly for Electricity Production by Type		
Municipal Solid Waste	49.89	781 kg/metric ton
Tire-Derived Fuels	85.97	2653 kg/metric ton

Source: US Energy Information Agency (undated).

Now let us move on to carbon tax rates and revenues around the world. According to the World Bank Carbon Pricing Dashboard,[3] as of 2023 there were a total of 73 carbon pricing initiatives covering about 23% of global greenhouse gas emissions (11.66 Gt CO_2e), and varying emissions percentages within legal jurisdictions. Thirty-nine systems are national in scope and 33 are at sub-national (e.g., province or state) levels. About half of these structures are *cap-and-trade systems* (sometimes called "emissions trading schemes" after the EU system), which we will discuss in Chapter 18. China and the US, which are the two largest greenhouse gas emitters, have cap-and-trade systems at the national and/or sub-national levels.

It is worth noting that there are currently 193 countries who are UN members and 196 parties who signed the 2015 Paris Agreement on climate change, which we discussed in Chapter 2. Only 20% of countries price the right to put carbon emissions into the air, otherwise that ecosystem service is given away for free. OECD (2019) analyzed 44 OECD and select partner countries.[4] They found that 85% of energy-related CO_2 emissions occur in power plants, factories, buildings, trains, ships etc. – anything but road transportation – but taxes price less than 20% of those emissions. Carbon pricing in the OECD and its six partner countries is highly focused on road transport, which often means that petrol (gasoline) and diesel taxes are often significant.

A total of 37 global CO_2 pricing initiatives are carbon tax programs, representing about 6% of annual global emissions (2.76 Gt CO_2e). Ten of those systems are at the sub-national level, meaning 27 are national in scope, plus several more are under development or being considered. In a number of cases, especially in Europe, carbon taxes are combined with cap-and-trade.

Finland was the first country to adopt a national carbon tax in 1990 and was quickly followed by Sweden, Norway and Poland. As of 2023, Uruguay had the highest carbon tax rate in the world at $156/ton CO_2e, followed by Switzerland and Liechtenstein at $131/ton and Sweden at $126/ton. A total of 14 countries have carbon taxes greater than $20 per ton. According to the World Bank Carbon Pricing Dashboard, among those who have carbon taxes there are 9 countries with carbon taxes below $10/ton and over 150 have no carbon taxes (or cap-and-trade system) at all. Figure 16.1 presents what we believe to be a census of carbon tax rates around the world.

Before reading on, please take a moment to consider whether the rates in Fig. 16.1 seem significant to you. How would you judge whether $10, $30 or $150 per ton is significant? What data would you need to make such an assessment?

As pointed out by Kojima and Asakawa (2021), OECD (2019) and many others, carbon taxes around the world, where they exist, tend to be low compared with the social cost of carbon (SCC), which is the estimated value of marginal damages from a ton of CO_2 (remember the SCC from Chapter 9?) emitted at any point in time, and the estimated CO_2 abatement costs. Though the current official US Government SCC estimate is $51/ton (Asdourian and Wessel, 2023), as part of a revised rule proposed in September 2022, based on recent scientific evidence, US EPA (2022) estimates the SCC to be anywhere between $120 and $340 per ton, depending on the discount rate used.[5] If these marginal damage estimates are even close to correct, it means there is substantial room for carbon taxes to be instituted and increased to internalize carbon pollution externalities.

There is, of course, more to carbon taxes than just rates. Table 16.2 presents data on carbon tax coverage, in terms of the percentage of total greenhouse gas emissions subject to taxation, including the four Canadian provinces where carbon taxes exist. As shown in the table, carbon tax coverage varies dramatically across the world. For example, Japan, Ukraine and South Africa, which have quite low tax rates, include the vast majority of emissions. The small country of Liechtenstein has wide coverage and a very high tax rate ($131/ton). Many countries cover around 40% of emissions, including the Scandinavian countries (Denmark, Finland, Norway, Sweden), and some, including Uruguay, the highest carbon tax rate country, cover less than 20% of emissions. Canada as a whole makes 30% of its emissions subject to its meaningful $48 per ton carbon tax. The four provinces that have their own carbon tax laws in most cases include substantially higher percentages of total emissions in their tax schemes than the country as a whole. Western hemisphere carbon tax pioneer British Columbia covers 70% of emissions, by far the most in Canada.[6]

To what degree do these carbon taxes actually reduce emissions? After all, many rates are quite low. On the other hand, even if rates are well below the SCC, perhaps they can still incentivize reduced carbon emissions. Firms may try to make as much money as possible, but attention is limited and people may make decisions based on assumptions, past events or easy rules of thumb that avoid the need for deeper analysis. Low carbon taxes could therefore potentially spur polluters to focus on their energy choices, perhaps having an outsized effect on emissions compared with what profit maximization alone might suggest. As mentioned in the previous chapter, for example, evidence from the pollution tax systems in the former European planned economies suggests that a little pollution tax can go a long way toward changing polluting behavior.[7] Maybe something similar is true for carbon taxes? Or not!

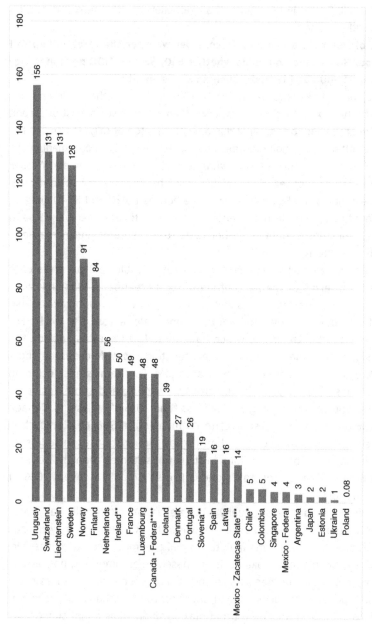

Figure 16.1 Carbon Tax Rates around the World per Ton CO_2e

Most countries have some rate differentiation by fuel or greenhouse gas.

*   Based on values quoted in Kojima and Asakawa (2021).
**   Average. Rates differ somewhat by fuel.
***   Four other Mexican states have adopted carbon taxes, but rates and coverage are unavailable.
**** Four provinces have their own systems, but are required to meet the federal minimum price.
Rates are from World Bank (undated). Retrieved October 23, 2023.

Table 16.2 Carbon Tax Coverage and Year Instituted

Country	Percentage of Greenhouse Gases Covered	Year Tax Instituted
Argentina	20%	2018
Canada - Federal	30%	2019
Canada - British Columbia	70%	2008
Canada - Newfoundland/Labrador	47%	2019
Canada - Prince Edward Island	56%	2019
Canada - New Brunswick	39%	2020
Chile	29%	2017
Colombia	23%	2017
Denmark	40%	1992
Estonia	6%	2000
Finland	36%	1990
France	35%	2014
Iceland	55%	2010
Ireland	40%	2010
Japan	75%	2012
Latvia	5%	2004
Liechtenstein	81%	2008
Luxembourg	65%	2021
Mexico - Federal	44%	2014
Netherlands	52%	2021
Norway	63%	1991
Poland	4%	1991
Portugal	40%	2015
Singapore	80%	2019
South Africa	80%	2019
Sweden	40%	1991
Switzerland	33%	2008
Ukraine	71%	2011
Uruguay	11%	2011

It is important to realize that how much particular carbon tax levels reduce emissions is an empirical issue that has to do with numbers and estimation rather than theory and ideas. One can claim, for example, that Japan's carbon tax of $2.00 per ton is too little to have any effect, but that would just be a claim. We need to test such a hypothesis using data and a key way to measure the effect of carbon taxes on emissions is to use the elasticity measures we discussed in Chapters 7, 9 and 10. You will recall that elasticities measure effects in percentage terms, which are comparable regardless of units. These percentages are then formed into a ratio, with the dependent variable on the top and the independent variable on the bottom. To actually estimate the elasticity of carbon emissions with respect to carbon tax rates, we would use econometric statistical methods, such as those overviewed in Chapter 4.

What have researchers found? Basically, the evidence suggests that, at the current rate levels observed in the world, carbon taxes reduce carbon emissions, at least in the transportation sector. Effects are, however, not nearly one-to-one. Wolde-Rufael and Mulat-Weldemeskel (2021) found in a sample of upper middle-income countries that, depending on the tax measure, carbon-focused energy taxes reduced CO_2 emissions by 0.11%–0.14% for a 1% increase in the tax rate. This means they estimated elasticities of CO_2 emissions with respect to carbon taxes of –0.11/1 = –0.11 to –0.14/1 = –0.14.

Andersson (2019) studied the Swedish carbon tax, which as of 2023 was very high by world standards ($126/ton) and covered 40% of CO_2 emissions, with a heavy focus on transportation. Calculating the elasticity, the author found that the elasticity of transport emissions with respect to the tax rate was –1.57 in 2005.

Before reading on, if the elasticity is –1.57, which changes more, the tax rate or CO_2 emissions? By how much in percentage terms?

Though previous research found that the British Columbia tax reduced CO_2 emissions between 5% and 15%,[8] Pretis (2022) used several methods and found that the tax only reduced emissions in the transportation sector, which is the largest source of emissions, but the elasticity was very low at –0.002. Importantly, he found that the tax did not affect overall emissions or emissions from non-transport sectors.

Two 2021 studies of the first-in-the-world carbon tax in Finland analyzed CO_2 emissions *per capita* using a well-respected empirical analysis method called "synthetic control," which tries to estimate what would have happened if a policy (in this case a carbon tax) had not been implemented. Elbaum (2021) found that when the carbon tax was increased from about $1.75 per ton in 1990 to around $23.39 in 2005 (a whopping 1236% increase), CO_2 emissions dropped 48%, for an elasticity of about –0.04 (please see if you can replicate this calculation). Mideska (2021) found an elasticity of about –0.09 over the same period, which is not a bad correspondence with Elbaum (2021).

As shown in Fig. 16.2, though, the author estimated that the elasticity declined over time – and as the carbon tax rose – from an elasticity of about –0.4 (i.e., every 1% increase in the carbon tax reduced CO_2 emissions per capita by 0.4%) to something more like the

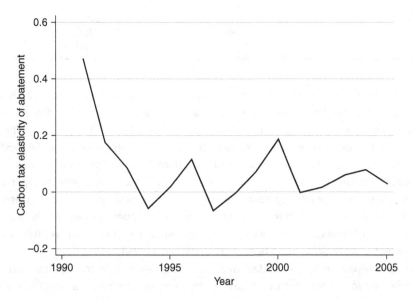

Figure 16.2 Elasticity of CO_2 Emissions to Changing Carbon Taxes in Finland
Source: Mideska (2021).

Elbaum (2021) estimate of 0.03 by 2005. This type of result suggests that the economy responded very strongly to even the initially very low 1990 carbon tax, which is consistent with Finland "using up" cheap abatement opportunities before moving on to more expensive options. Maybe the biggest effects of carbon taxes are therefore felt when such programs are instituted, even if tax rates are low?

Moving on to thinking about carbon tax revenues, how much money do governments actually collect and what do they do with those resources? Table 16.3 presents estimates of carbon tax revenues from the World Bank Carbon Pricing Dashboard. Uses of revenues are extracted from a variety of believed credible sources (e.g., government, UN-affiliated agencies and peer-reviewed publications).

World Bank (2019) offers a very useful summary of the use of carbon tax revenues around the world. The authors found that as of the 2017/2018 fiscal years carbon taxes generated

Table 16.3 Carbon Tax Revenues Collected and Use of those Revenues

Country	2023 Revenues ($US Millions)	Uses of Carbon Tax Revenues
Argentina	$167	Multiple beneficiaries
Canada - Federal	$5451	100% returned to provincial/territorial governments
Chile	$171	Unavailable
Colombia	$92	80% for protection of ecosystems. 20% for financing illicit crop substitution program.
Denmark	$493	Environmental protection and industry compensation
Estonia	$2	Unavailable
Finland	$1707	50% government budget. 50% to offset other taxes
France	$9103	2/3 government budget. 1/3 climate investments
Iceland	$52	Unavailable
Ireland	$709	Energy efficiency, social protection, CO_2 reduction investment
Japan	$1648	Climate investments and to offset other taxes
Latvia	$8	Unavailable
Liechtenstein	$5	Unavailable
Luxembourg	$282	Unavailable
Mexico - Federal	$239	General government budget
Netherlands	$0	Only applicable when the EU emissions trading scheme price is below a minimum threshold. Not the case in 2023.
Norway	$1800	General government budget and development assistance
Poland	$6	Unavailable
Portugal	$487	Same as other excise taxes
Singapore	$149	Unavailable
South Africa	$95	Unavailable
Sweden	$2125	General government budget
Switzerland	$1629	1/3 climate investments. 2/3 redistributed to households and businesses
Ukraine	Not available	30% to reduce CO_2 emissions. 70% to government budget
Uruguay	$271	Unspecified portion to fund reducing greenhouse gas emissions

Source: All revenues except for France are from the World Bank Carbon Pricing Dashboard. Data for France are from 2017/2018 and are taken from World Bank (2019). Uses of revenues are from a variety of believed credible sources and may not be current as of 2024.

about $24 billion in revenues, almost 1/3 of which were collected in France. A majority of revenues (59%) went into general government budgets where they could be spent on all the things discussed in Chapter 13 that governments generally do. Just over 23% of revenues went toward climate investments, 14% were used to cut/offset other taxes, such as income taxes, and 5% financed development projects in lower-income countries.

Only in Finland and British Columbia Province in Canada did carbon taxes make up more than 2% of government revenues, so carbon taxes represent only a very small portion of total taxes. This finding is perhaps not surprising, because as was just discussed, carbon taxes incentivize lower emissions. Declining emissions means a lower tax base, making them problematic ways to finance schools, roads, militaries, social services etc.

16.3 Environmental Taxes on Products, Transport and Green Tax Reform

The term "environmental taxation" is much more general than pollution or carbon taxes and can encompass anything from taxes on energy, transportation and natural resource use to volumetric garbage pricing or taxes on specific products, such as tires and plastic bags. Sometimes, such taxes are refundable if one reuses or recycles the product, as is the case for glass soda bottles that must be returned to the seller for reuse, or deposits on aluminum cans. These are vast topics about which many books have been written that we are unfortunately not able to cover here. The common element across environmental taxes is the emphasis on taxing behaviors that deplete common pool ecosystem services and perhaps raising revenues for governments in the process.[9]

Road pricing to reduce pollution and congestion is an especially interesting form of environmental taxation. Traffic jams are very costly for people and the planet, producing greenhouse gases, local air pollutants, accidents, lost time, frayed patience and road rage. Driving in the world's largest cities is the most harmful, because traffic is greater, congestion is higher and people inhale the most exhaust fumes. Congestion and pollution externalities alone can be US$25 per kilometer, with 2016 estimates for higher-income countries at anywhere from $500 to $1300 per capita per year due to urban congestion. Fortunately, because of high population densities and more transportation options, people in cities are likely to be responsive to transport pricing. Price elasticities in urban areas are estimated to be -0.3 to -0.5 (i.e., 0.3%–0.5% change in driving for a 1% change in the cost of driving) compared with 0.1 in rural areas (Creutzig et al., 2020).

Box 16.1 London Central Business District Congestion Pricing

London charges a fixed £15 charge for vehicles traveling within an inner-ring road between 7:00 a.m. and 6:30 p.m. Monday through Friday. The congestion charge, combined with improvements in public transit financed through revenues from the charge, led to 15% less traffic in central London, with no significant increase in the use of roads outside the area. Congestion was reduced by 30% and bus delays fell by 1/3.

Drivers pay on daily, weekly, monthly or annual bases by a variety of methods. A network of fixed and mobile cameras records the license plates of vehicles traveling within the zone, which are matched against a database of those who have paid the charge. Taxis, motorcycles, mopeds and vehicles registered to disabled users are exempt from the congestion charge. Residents of central London are eligible for a 90% discount.

Sources: US DOT (2008) and Transport for London (undated).

Congestion pricing, what the US Department of Transportation calls "zone-based" pricing, is a method of road pricing in which drivers pay to enter some or all of a city. New York is in the process of developing a congestion pricing program, which is expected to be in place in 2024, but the key current examples are in London, Stockholm and Singapore (US DOT, 2008).

Singapore, which is a city-state in Southeast Asia, was the pioneer in congestion pricing, with peak-traffic pricing that goes back to 1975. After introducing green pricing for access to the central city during the morning rush hour period, traffic fell by 50% and average speeds doubled (Christainsen, 2006). Initially, transportation workers checked payments manually, but by 1998, automatic monitoring was introduced, greatly increasing the efficiency of the system. A similar system was instituted in Stockholm in 2006, immediately causing a 22% decline in vehicle trips, 14% less tailpipe emissions, fewer traffic accidents and major shifts to public transit (US DOT, 2008).

Before reading on, please guess the answers to the following two questions: a) What is the average environmental taxation as a percentage of GDP in OECD countries? and b) Which country has the highest percentage in the world?

The OECD compiles information from around the world on overall environmental taxation.[10] Taxes in the average OECD country make up about 34% of GDP, though some countries run more through governments (e.g., Norway at 44% and Austria with 43%) and some less (e.g., Mexico at 17% and Chile with 24%) (OECD, 2023a). Adding up everything, OECD (2023b) finds that the average OECD country uses only about 2% of GDP for environmental taxes, but the OECD as a whole is much lower at 1.5% of GDP.

Before reading on, please think for a second about the difference between these calculations. Why does it make sense that they would differ?

The bottom line is that governments in the average OECD country get about 6% of their revenues from environmental taxes, meaning 94% comes from other sources. Most non-OECD countries, especially those with low incomes per capita, get very little or no revenue from environmental taxes. The absolute top in the world is non-OECD member Croatia with about 4%, followed by Greece at about 3.8% of GDP, and then Slovenia, with 3.5% of GDP taken as environmental taxes. There is then a rather steep drop-off, with several countries collecting environmental taxes to the tune of 3% of GDP, including Denmark, Italy and the Netherlands. In general, though, environmental taxes are a minor fiscal player, which is not necessarily bad, if their goal is to reduce environmentally harmful behaviors. One thing is for sure. These data do not suggest that a large-scale green tax reform, which focuses on

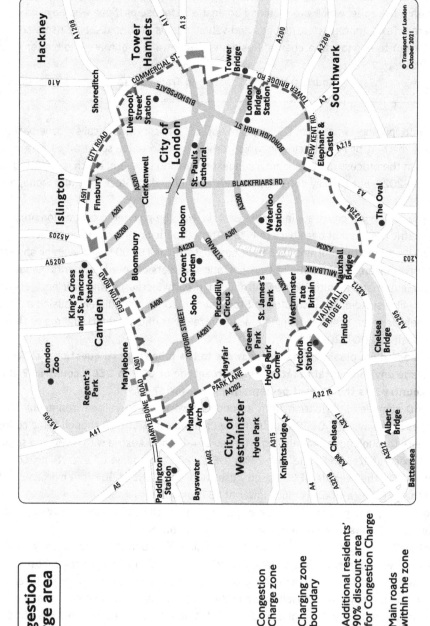

Figure 16.3 Central London
Sources: Map data ©2023 Google and Transport for London (undated).

taxing environmentally harmful behaviors rather than sales, incomes, wealth and profits, is anywhere on the horizon.

16.4 Chapter Conclusions

The most important pollution taxes currently in use move one rung up the pollution tax ladder away from pollution itself, to target polluting inputs, such as quantities of fossil fuels burned differentiated by carbon content. Though not discussed in the chapter, other air pollutants, such as sulfur dioxide, which causes acid rain and respiratory irritation, can also be taxed based on fuel quantity and quality. The evidence on carbon taxes suggests that when taxes are first introduced and there are lots of abatement options (e.g., as was analyzed in Finland), as well as when taxes get very high (e.g., in Sweden), such pricing policies can dramatically reduce emissions.

The third rung on the pollution tax ladder is to tax polluting products, basically the externality scenario we discussed in Chapter 9, where the only option incentivized is to use fewer polluting products. Examples include taxes on plastic bags, aluminum cans, tires and access to city downtowns, as well as volume-based garbage disposal fees. Though the evidence suggests that green taxes can be powerful waste-reducing and pollution-controlling instruments, and despite the attractiveness of taxing bad things like pollution and sparing good things like your income, nowhere in the world is there even a hint of a major green tax reform. It therefore seems unlikely that green taxes will soon take over the world's tax systems.

Issues for Discussion

1. How would you judge whether $10, $30 or $150 per ton CO_2 taxes are significant? What data would you need to make such an assessment?
2. If garbage is collected where you live, how is garbage disposal priced? What is actually priced and how high are the costs? Does the system incentivize less waste disposal?

Practice Problems

1. Suppose that Poland currently uses 1000 tons of anthracite coal to produce electrical power. Using the conversion factors given in Table 16.1 and other information in Section 16.1, please calculate the following:
   a. How much CO_2 is Poland emitting?
   b. How many metric tons of municipal solid waste would it need to burn to generate the same amount of CO_2 emissions?
   c. If Poland were to switch over to municipal solid waste fuel and keep its CO_2 emissions constant, how much more power (in million Btus) would it be able to produce?
   d. If Poland were to switch over to municipal solid waste fuel and keep its power production constant, how much less CO_2 would it produce?

2. Suppose two countries have carbon taxes. Country A covers 45% of its emissions at a rate of $25 per ton and Country B covers 20% of its emissions with a carbon tax rate of $50 per ton of CO_2 emitted into the atmosphere. If both countries have a total of 1000 tons of CO_2 emissions, which country earns more carbon tax revenue? Please calculate the difference in revenues.

3. Suppose in Finland the elasticity of CO_2 emissions with respect to tax rates ranged from -0.40 to -0.03 as estimated by Mideska (2021). The table summarizes the estimation results over time, with the 1989 CO_2 tax representing the baseline costs as a carbon tax equivalent.

Year	Elasticity	CO_2 Tax per Ton Emitted	CO_2 Emissions
1989	Not Applicable	$1.00	1000
1990	-0.40	$1.75	
1995	-0.10	$7.00	
2000	-0.07	$15.00	
2005	-0.03	$23.40	

Using the data in the table, please predict Finland's CO_2 emissions for the end of each year 1990, 1995, 2000 and 2005 after the carbon tax has had a year to influence emissions.

Hint: Remember that elasticity of CO_2 emissions with respect to the carbon tax is defined as %Δ Emissions/%Δ Carbon Tax, where %Δ indicates the "percentage change." Please see Chapters 7, 9 and 10 for how to manipulate elasticities and calculate %Δ.

4. **Challenge Problem**: Suppose that Finland starts 1990 using 170 kg of anthracite coal and 356 kg of lignite coal for electricity production. Please use the fuel conversion factors provided in Table 16.1 to answer the questions below.

   a. How many million Btus is Finland using for electricity in 1990?
   b. Using the CO_2 emissions estimates you calculated in Practice Problem 3, use the fuel conversion factors provided in Table 16.1 to suggest how the mix of electricity-producing fossil fuels might change as emissions fall. Suppose that policymakers try to keep electricity production roughly constant as they incentivize fuel shifts.
      Hint: there are many correct answers. Please find one combination of fuels for each year that 1) achieves the CO_2 goal and 2) gets as close as possible (if it is possible!) to keeping energy use in million Btu constant.
   c. Given the information in Table 16.1 and your calculations in Practice Problem 3, is it possible to actually increase electricity production (in million Btus) from 1990 levels by changing energy technologies even as the tax rate rises?
   d. Given the information in Table 16.1 and your calculations in Practice Problem 3, is it possible to keep electricity production (in million Btus) at 1990 levels as the tax rate rises or are energy efficiency investments also needed?

Notes

1 Actually, independent of feasibility, it would be the third-best tax, because a first-best tax would achieve efficiency (i.e., set tax so MB = MC) and the second-best would achieve cost-effectiveness based on measured pollution.

2 The US EPA offers a handy calculator that tells how much CO_2 common fuels contain and compares those amounts with other fuels and common uses of fuels. Please see https://www.epa.gov/energy/greenhouse-gas-equivalencies-calculator

3 Please see https://carbonpricingdashboard.worldbank.org/ for more information.

4 Recall from earlier chapters that the Organisation for Economic Co-operation and Development (OECD) is a non-UN organization made up of 38 mainly high-income countries.

5 Please see US EPA (2022) for a summary and analysis of the proposed SCC to be used for official US Government cost-benefit and other studies.

6 To learn some details and nuances of the Canadian carbon tax system, including on the enabling legislation and incentive payments, please see Government of Canada (undated).

7 In addition to chapters in Bluffstone and Larson (1997), which cover a variety of countries, please also see Dasgupta et al. (2001) for China and Bluffstone (1999), which focuses on Lithuania, among others.

8 e.g., see Metcalf (2019) and Murray and Rivers (2015).

9 For much, much more information on environmental taxation policy instruments, please see Sterner and Coria (2011).

10 OECD (2023b) defines environmental taxes in the following way: "Environmental taxes are environmentally related tax revenues. The characteristics of such taxes include revenue, tax base, tax rates and exemptions. They are applied on the following environmental domains: energy products (including vehicle fuels); motor vehicles and transport services; measured or estimated emissions to air and water, ozone depleting substances, certain non-point sources of water pollution, waste management and noise, as well as management of water, land, soil, forests, biodiversity, wildlife and fish stocks. This indicator is measured as a percentage of GDP and of tax revenue."

Further Reading and References

American Physical Society (APS). Undated. *Energy Units*. Retrieved from https://zwellhome.com/wp-content/uploads/2021/11/Energy-Units.pdf October 2024.

Andersson, J.J. 2019. "Carbon Taxes and CO_2 Emissions: Sweden as a Case Study." *American Economic Journal: Economic Policy* 11 (4): 1-30.

Asdourian, E. and D. Wessel. 2023. "Commentary: What is the Social Cost of Carbon?" The Hutchins Center Explains Blog. Retrieved from https://www.brookings.edu/articles/what-is-the-social-cost-of-carbon/#:~:text=An October 2024.

Bluffstone, R. 1999. "Are the Costs of Pollution Abatement Lower in Central and Eastern Europe? Evidence from Lithuania." *Environment and Development Economics* 4 (4): 449-470.

Bluffstone, R. and B.A. Larson, eds. 1997. *Controlling Pollution in Transition Economies: Theories and Methods*. Edward Elgar: London.

Christainsen, G. 2006. "Road Pricing in Singapore after 30 Years." *The Cato Journal* 26 (1): 71-88.

Creutzig, F., A. Javaid, N. Koch, B. Knopf, G. Mattioli and O. Edenhofer. 2020. "Adjust Urban and Rural Road Pricing for Fair Mobility." *Nature Climate Change* 10: 591-594.

Dasgupta, S., M. Huq, D. Wheeler and C. Zhang. 2001. "Water Pollution Abatement by Chinese Industry: Cost Estimates and Policy Implications." *Applied Economics* 33 (4): 547-557.

Elbaum, J.-D. 2021. *The Effect of a Carbon Tax on Per Capita Carbon Dioxide Emissions: Evidence from Finland*. IRENE Working Paper 21-05, Institut de recherches economique, University of Neuchâtel.

Government of Canada. Undated. "How Carbon Pricing Works." Retrieved from https://www.canada.ca/en/environment-climate-change/services/climate-change/pricing-pollution-how-it-will-work/putting-price-on-carbon-pollution.html October 2024.

Kojima, S. and K. Asakawa. 2021. "Expectations for Carbon Pricing in Japan in the Global Climate Policy Context." In T. Arimura and S. Matsumoto, eds. *Carbon Pricing in Japan*. Springer Nature: Singapore.

Lacroix, K., M.H. Goldberg, A. Gustafson, S.A. Rosenthal and A. Leiserowitz. 2020. "Should it Be Called 'Natural Gas' or 'Methane'?" New Haven, CT: Yale Program on Climate Change Communication. Retrieved from https://climatecommunication.yale.edu/publications/should-it-be-called-natural-gas-or-methane/#:~:text=Natural%20gas%20is%20composed%20of,major%20contributor%20to%20global%20warming October 2024.

Metcalf, G.E. 2019. "On the Economics of a Carbon Tax for the United States." Brookings Papers on Economic Activity, BPEA Conference Draft.

Mideska, T. 2021. *Pricing for a Cooler Planet: An Empirical Analysis of the Effect of Taxing Carbon.* CESifo Working Paper 9172 2021. The Munich Society for the Promotion of Economic Research.

Murray, B. and N. Rivers. 2015. "British Columbia's Revenue-Neutral Carbon Tax: A Review of the Latest 'Grand Experiment' in Environmental Policy." *Energy Policy* 86: 674–683.

OECD. 2019. *Taxing Energy Use 2019: Using Taxes for Climate Action.* OECD Publishing: Paris.

OECD. 2023a. *Revenue Statistics 2023: Tax Revenue Buoyancy in OECD Countries.* Retrieved from https://www.oecd.org/tax/tax-policy/revenue-statistics-highlights-brochure.pdf December 10, 2023.

OECD. 2023b. Environmental Tax (indicator). Retrieved from https://data.oecd.org/envpolicy/environmental-tax.htm November 2, 2023.

Pretis, F. 2022. "Does a Carbon Tax Reduce CO_2 Emissions? Evidence from British Columbia." *Environmental and Resource Economics* 83 (1): 115–144.

Sterner, T. and J. Coria. 2011. *Policy Instruments for Environmental and Natural Resource Management* (2nd ed.). Routledge: New York.

Transport for London. Undated. *Congestion Charge Zone.* Retrieved from https://tfl.gov.uk/modes/driving/congestion-charge/congestion-charge-zone October 2024.

US Department of Transportation (US DOT). 2008. *Congestion Pricing: A Primer: Overview.* Retrieved from https://ops.fhwa.dot.gov/publications/fhwahop08039/fhwahop08039.pdf October 2024.

US Energy Information Agency (US EIA). Undated. *Carbon Dioxide Emissions Coefficients.* Release Date September 7, 2023. Retrieved from https://www.eia.gov/environment/emissions/co2_vol_mass.php October 2024.

US EPA. 2022. "EPA External Review Draft of Report on the Social Cost of Greenhouse Gases: Estimates Incorporating Recent Scientific Advances." National Center for Environmental Economics, US EPA.

Wolde-Rufael, Y. and E. Mulat-Weldemeskel. 2021. "Do Environmental Taxes and Environmental Stringency Policies Reduce CO_2 Emissions? Evidence from 7 Emerging Economies." *Environmental Science and Pollution Research* 28 (18): 22392–22408.

World Bank. 2019. *Using Carbon Revenues.* Partnership for Market Readiness Technical Note 16. Retrieved from https://openknowledge.worldbank.org/server/api/core/bitstreams/376a5822-a3cd-561f-a707-8ac43a1a3812/content October 2024.

World Bank. Undated. *Carbon Pricing Dashboard.* Retrieved from https://carbonpricingdashboard.worldbank.org/ October 2024.

17 Green Markets: Non-Carbon Pollution Trading

What You Will Learn in this Chapter

- What are green markets and how do they work?
- About the pollution markets ladder
- The theory of green markets and why they can achieve cost-effectiveness
- Pollution markets in the US
- The evolution of cap-and-trade in the US

17.1 Recap and Introduction to the Chapter

As we saw in Chapters 15 and 16, incentive-based pollution control instruments, such as green taxes, are implemented in a variety of countries, often with important effects on pollution emissions. One of the most important applications of green pricing is carbon taxes, whereby governments charge polluters a fixed price for the right to emit CO_2 and other greenhouse gases into the atmosphere. Particularly if pollution – or a very good proxy for it – is directly taxed, pollution taxes offer polluters incentives for cost-effective pollution control.

The second major way to price pollution is via green markets,[1] in which rights to engage in environmentally degrading activities, such as pollution, are allocated and then polluters have opportunities to reallocate those rights via trade. Cap-and-trade is the most sophisticated of those systems, but other, simpler green market mechanisms also exist. As we will see in this chapter, green markets offer the potential for both environmental improvements and cost-effectiveness.

Green markets are being used as important tools particularly to control greenhouse gas emissions; exploring the potential for green markets to help us save the planet cost-effectively is therefore well worth our time. To better understand how such systems

DOI: 10.4324/9781003308225-20

can be expected to operate, we start our exploration by examining the theory of cap-and-trade. We then examine the US experience with cap-and-trade in pollutants other than CO_2.

17.2 The Theory of Green Markets

We discussed green taxes in the previous chapter, so you are already familiar with a number of key green market elements. Like green taxes, green markets are systems that try to use the self-interested aspects of human behavior to curb environmental moral hazard and reduce excessive depletion of common pool ecosystem services. Though green markets can focus on a variety of natural resources and their ecosystem services (e.g. wetlands and their services), the most important, sophisticated and complex applications of green markets focus on controlling pollution. For this reason, pollution markets are our focus in this chapter.

Like pollution taxes, green markets focusing on pollution have their own ladder, with the rungs defining differing levels of market complexity. There are also a variety of names, depending on where we are on the ladder, and which aspects of systems advocates, regulators and researchers want to emphasize. You may have heard the terms bubbling, offsets, banking, joint implementation and netting. These are all programs and/or aspects of pollution markets. Cap-and-trade itself goes by several names, including emissions trading, pollution trading and marketable permits. All of these systems are cap-and-trade.

Within the panoply of green markets, let us first consider cap-and-trade, which is used to control greenhouse gas emissions in the European Union, China, South Korea, the US State of California and the Canadian Province of Quebec. We can then look at simpler forms of pollution markets, such as offsets, where there is no market per se, but polluters can make market-type deals with other polluters. Bubbling systems are even simpler and allow polluting facilities that have many sources to be regulated as a single source.

As the name suggests, cap-and-trade systems have two main components. The first part is the pollution cap, which is set exogenously (i.e., outside our framework of analysis) as we discussed in Chapter 14, based on scientific information, political realities and believed pollution abatement costs. Under a cap-and-trade system, unlike with stand-alone pollution taxes, total emissions (often annual in a jurisdiction) are capped at a fixed level below current emissions, so environmental improvements will occur.

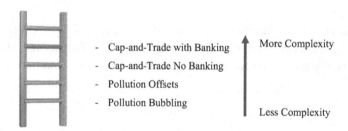

- Cap-and-Trade with Banking   More Complexity
- Cap-and-Trade No Banking
- Pollution Offsets
- Pollution Bubbling   Less Complexity

Figure 17.1 Pollution Markets Ladder

Challenge Yourself

Do you think giving polluters the *right* to pollute and even to sell those rights is good, ethical public policy?

The aggregate cap must be met and so achieving a targeted maximum emissions level is truly baked into cap-and-trade systems. Let's home in on the word *aggregate*, because the cap is across a whole regulated jurisdiction. To use cap-and-trade, the location of pollution sources relative to those affected by pollution must be basically irrelevant. Think of the situation as one where pollution is released into the air or water and immediately gets fully distributed and mixed into the environment. The implication is that there is little or no possibility for pollution to build up in a particular location and pollution concentrations are spread relatively equally. A pollutant that, by its nature, mixes in this way is known as a *uniformly mixed pollutant*.

Air pollution is more likely to be uniformly mixed than water pollution, because the atmosphere is so large and wind mixes pollutants. CO_2 pollution is close to an ideal example of a uniformly mixed pollutant; though other local pollutants may be emitted along with carbon pollution, whether a CO_2 molecule is released in Buenos Aires, Tokyo or Berlin matters little, because each one uniformly mixes in the atmosphere and causes climate change.

Major advantages of cap-and-trade systems compared with pollution taxes are that regulators do not allow more than the cap to be emitted, the environmental quality should improve compared with the status quo and we know from the start how much emissions will decline. Those who like cost-effectiveness, but are most concerned with environmental quality, will therefore appreciate cap-and-trade for getting the environmental job done.

The second aspect of cap-and-trade systems is trading. Once pollution caps are set, there must be one or more mechanisms for polluters to exchange rights to pollute. Rights go by various names, including tradable permits, marketable permits and pollution allowances, but in carbon markets, "allowance" is the most common terminology.

To be able to sell or otherwise exchange pollution, such as CO_2, polluters must have secure *rights* to pollute. This feature is different than pollution taxes, where governments retain pollution rights and companies need to buy them if they want to emit. There is therefore an important aspect of the property rights policy inspired by the Coase Theorem, which we discussed in Chapter 14, at work in cap-and-trade systems. We leave aside for now how polluters get their rights, but we will come to that in a minute. For now, we just note that the sum of all individual rights must add up to the total pollution cap and those rights can be traded.

In analyzing cap-and-trade, let us once again assume our simple behavioral model in which polluters try to make as much money as possible. Suppose further that the government somehow allocates secure rights, which guarantee that if firms want to pollute, they can emit up to the amounts of rights they own. Firms therefore truly have rights to pollute, though there are no guarantees they will be enough to meet their needs. Assume governments are trying to reduce the total amount of a pollutant, so it makes sense that the pollution allocated to at least some polluters will be less than the amount to which they are accustomed, otherwise cap-and-trade would not reduce total emissions.

Third, recall that governments are unlikely to have the full picture of pollution abatement options available to polluters. They therefore will not know whether allocations are just right, too tight or too loose. In the language of Chapter 14, governments probably do not know polluters' marginal abatement costs and polluters are unlikely to fully reveal that information. As is the case for basically all pollution control instruments, governments are flying at least partially blind when it comes to the abatement burdens they are imposing on members of the polluting community. It is also not clear that all polluters will even *want* to use all the pollution rights they have.

In sum, in our model governments know the following:

a)   What they want to get done (i.e., how much they would like to reduce total emissions).
b)   Polluters' histories of pollution emissions, which could offer some policy guidance.
c)   That additional pollution abatement possibilities probably exist.
d)   There is likely a wide range of available abatement options and costs will vary across pollutant sources.

Suppose the government of Chile wants to limit emissions of SO_2 into the atmosphere in a jurisdiction to ten tons, which is below the current level. This emissions goal is set using non-economic criteria, such as the best science and political realities, but with an eye toward polluters' abatement costs. Assume for simplicity, and so we can graph things, that there are only two sources subject to regulation, Source A and Source B.

The government has decided to give away rights to the ten tons at no cost based partly on historical firm emissions. This is one possible way to allocate pollution rights. Before instituting the cap-and-trade regulatory system, both sources are emitting at their uncontrolled/baseline emissions levels.[2] Of the ten-ton cap, Source A is given eight tons and Source B receives two tons. The two polluters are free to buy and sell the rights to emit SO_2 that they have received. Each polluter is welcome to make any deal with the other pollution source, and we assume they try to maximize their profits.

Before reading on, please think about how these polluters will decide whether to buy or sell rights to emit tons of SO_2. How will they determine how much they want to buy or sell? How will they know what they would be willing to pay or accept? What are the factors we expect them to consider, if they want to make as much money as possible?

Let's take this problem from the perspective of Source A, which starts off with eight tons that it can emit, if it would like to do it, or it can sell some or try to buy more rights. The three possible decisions are therefore to stand pat, buy or sell SO_2. If it wants to make as much money as possible, our behavioral model predicts that the key factors influencing whether Source A will buy or sell are the (opportunity) costs and benefits. If Source A buys rights, it will give up the money it had to pay to get those emissions but will avoid abating and will be able to earn whatever profits it gets from the higher emission level. If Source A sells, it gets the proceeds from those sales, but will have to abate and/or change its production level or process, potentially reducing profits. Our behavioral model therefore suggests that polluters will weigh permit prices, abatement costs and profitability of production as they decide whether to buy, sell or stick with the amount of permits they have.

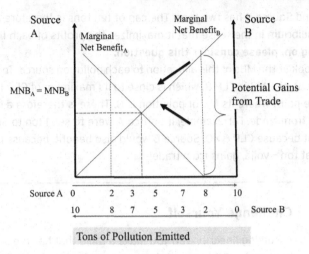

Figure 17.2 Marginal Benefits of SO_2 Air Pollution for Two Sources in Chile: A Two-Firm Static Cap-and-Trade Model

It is critical to remember that not only does the government not perfectly know how costly it will be for polluters to reduce emissions, but firms don't have this information about each other. The implication of this situation is that just like in a normal market those considering buying will utilize information about the market (e.g., past prices and sales) and their own firm information. Firms leaning toward selling their pollution rights to emit SO_2 are expected to do the same thing. They will then decide together if there is a deal to be made.

Figure 17.2 shows a graphical example of two Chilean firms interacting in a single period, which makes it a *static* rather than *dynamic* cap-and-trade model. The graph includes two vertical axes, with Chilean pesos (CLP) as the units. Suppose currently both firms' SO_2 emissions are uncontrolled.

On the left side is the marginal benefit (MB) of pollution to Source A and the right axis is the MB of Source B. The emissions of Source A are on the top of the horizontal axis, running from left to right, starting at 0 tons and going to the maximum possible, which is 10 tons (i.e., the pollution cap). Source B is on the bottom of the horizontal axis, running from right to left from 0 to 10 tons.

The vertical intercept for Source A is CLP 8 (i.e., at 0 tons the MB = 8) and for Source B the vertical intercept is CLP 10. The MB of pollution to Source A reaches 0 at 8 tons and, for Source B, suppose the MB is zero at 10 tons.

Before reading on, what does it mean for the MB of pollution to reach CLP 0? What are the uncontrolled/baseline emissions levels of each source, if they maximize profits? Before implementing cap-and-trade, what are the total pollution emissions of SO_2 of the two Chilean polluters? Would both pollution sources actually want to use 10 tons of SO_2? How do you know?

Suppose now a system of cap-and-trade is instituted. Let us consider the profit-maximizing levels of pollution of the two sources, if they are allowed to trade. Recall that Source A starts with eight tons and Source B has two tons. The cap of ten tons is therefore achieved. Is this allocation an equilibrium in the sense that it maximizes the profits of each firm?

Before reading on, please consider this question.

Let us take a look at the MBs of this allocation to each pollution source. To Source B, which received only 2 tons, the MB is CLP 8, which is close to its maximum of CLP 10, but at 8 tons, Source A is at the point where its MB of pollution is 0. There is therefore a huge gap, which could offer gains from trade. For example, if Source A were to sell 1 ton to Source B for CLP 6, it would benefit because CLP 6 > 0. Source B would also benefit, because CLP 6 < 8, which is its MB from that ton – voila, gains from trade!

Challenge Yourself

Assuming linearity, can you make a table that has the same informa-tion as Fig. 17.2 and gives the same cap-and-trade equilibrium?

But are there actually more gains to be had? If Source A has 7 tons, the MB as read off the left axis of the graph is about CLP 2, while the MB of Source B is approximately CLP 7. Actually, both firms could again perhaps benefit from a deal based on a price of CLP 6 per ton for both tons. Suppose these 2 tons were sold. The gap between the MB curves is smaller, but there is still a gap, meaning there are still gains from trade to be exploited. We would expect deals could be struck for 2 more tons, which would use up all possible gains from trade. This maximum occurs when the MBs of both sources are CLP 4.

You might say that for that last small part of a ton transacted, the opportunity cost to Source B is the same as its MB and you would be correct. Source B would pay no more than CLP 4 and Source A would accept no less than CLP 4. Let us assume, as we did for green pricing, that if a firm is truly indifferent between trading and not trading, it chooses to trade. Just an assumption.

Challenge Yourself

Assume both MB functions in Fig. 17.2 are linear. Can you find their equations? Please confirm that the equilibrium I calculated is correct.

Accepting this assumption, the predicted equilibrium of this two-polluter cap-and-trade model is therefore that Source A, which received an 8-ton allotment from the government, will voluntarily choose to emit 4 tons, and will sell the rights to 4 tons to Source B. Source B received only 2 tons from the government, but voluntarily bought 4 more tons from Source A, which allowed them to emit a total of 6 tons. Critically, and basically by construction, 4 + 6 = 10 and the pollution cap is exactly met.

Source A started out with uncontrolled emissions of eight tons, which we know because that is the emissions level at which its MB reached zero. Under the cap-and-trade system, it received an eight-ton allotment from the government, so it would have been possible for Source A to continue on as if no regulation had happened. Instead, it voluntarily decided to emit only four tons, so it abated four tons. Source B had uncontrolled/baseline emissions of ten tons but was only allocated two tons. It therefore chose to buy four tons from Source A, ended up emitting six tons and abated four tons. Everything here is voluntary and driven by own-incentives, leading to meeting the pollution cap.

Also importantly, notice that in the profit-maximizing equilibrium the two firms end up with MBs of pollution (i.e., the marginal abatement costs and/or lost profits avoided) that are equal to each other. Our profit-maximizing model therefore predicts that, as was the case with green taxes, cap-and-trade yields a cost-effective allocation of pollution and, because chosen pollution = uncontrolled/baseline pollution – chosen abatement, also cost-effective abatement. This finding is shown in Eq. 17.1.

Equation 17.1 Predicted Equilibrium in a Static Cap-and-Trade Model

Polluters Choose to Pollute Such that $MB_i = MB_j$ for All Firms i and j.

So far, we have thought about two firms, but now let us open this problem up to the many hundreds or thousands of pollution sources that interact in a big cap-and-trade market, kind of like the product markets we discussed back in Chapter 3. Though in principle all sources could be buyers and sellers, given the cap-and-trade market conditions, some will not have enough pollution rights (like Source B) and others will have too many (like Source A).

If there were many, many buyers and sellers in this market for a completely homogeneous product, such as the right to emit SO_2 or another pollutant, we know from Chapter 3 that the equilibrium price will be where MB = MC or, equivalently, where demand = supply. In a cap-and-trade system, the predicted price would therefore be the one equal to the MBs of pollution of all sources, which we know from Eq. 17.1 in equilibrium are themselves equal.[3] The predicted equilibrium price condition therefore supports the cost-effective allocation of pollution rights and is given in Eq. 17.2.

Equation 17.2 Equilibrium Price in a Static Cap-and-Trade Model with Many Polluters

Price = $MB_i = MB_j$ for All Firms that We Might Label "i" and "j."

Let us remind ourselves what goes into the MB of pollution. It includes the profits that can be earned from production and the abatement costs that are avoided. If MBs from pollution across polluters are equal, this means the total benefits are at a maximum and any *losses from not polluting are minimized.* We therefore find that cap-and-trade systems that work smoothly enough so all gains from trade can be exploited will allocate pollution rights and abatement effort cost-effectively. Notice, though, that in contrast to pollution taxes, all money changing hands stays within the private sector; if rights to pollute are given away for free (hardly a given), there would be no transfer from the private sector to the government.

In sum, our model of cap-and-trade, which is based on our assumptions that polluters try to make/save as much money as possible and trading is low-cost or no-cost, predicts the following outcomes:

1. Pollution caps will be exactly achieved.
2. Regardless of how rights are initially distributed, polluters who maximize profits trade until the market achieves the allocation of pollution rights whereby the total benefits of pollution are at a maximum. This also means that the costs of pollution abatement are minimized.
3. With many polluters participating in a cap-and-trade market, the equilibrium price is predicted to be equal to the MBs of pollution of all firms. The equilibrium price therefore supports the cost-effective allocation of pollution and abatement effort.
4. As long as rights to pollution are given away for free, there are no resources transferred from firms to governments.

Box 17.1 Worked Example: Equilibrium in a Two-Firm Cap-and-Trade System

The policy goal is to allow a total of 9 tons of CO_2 emissions per year and allocate those tons across two polluters using a cap-and-trade policy. The polluters' MBs of pollution are the following: $MB_1 = 10 - 0.50*Q_1$ and $MB_2 = 5.5 - Q_2$. Q_1 is the CO_2 emissions of Polluter 1 and Q_2 is the same for Polluter 2 (in tons of CO_2/year).

Given these functions, the policy objective and supposing that the firms will trade until all gains are exhausted, what is the predicted equilibrium?

Step 1: Realize that $Q_1 + Q_2 = 9$ or equivalently that $Q_2 = 9 - Q_1$.

Step 2: The polluters are predicted to trade until their MBs are equal, which is the cost-effective allocation of pollution. That means $10 - 0.50*Q_1 = 5.5 - Q_2$.

Step 3: Substitute in $Q_2 = 9 - Q_1$, meaning $10 - 0.50*Q_1 = 5.5 - (9 - Q_1)$. Subtract 5.5, add 9 and add $0.50*Q_1$ to both sides of the equality, implying $13.5 = 1.5Q_1$.

Step 4: Divide both sides of the equality by 1.5, which means $Q_1 = 9$.

Step 5: Recall that $Q_2 = 9 - Q_1$, meaning that $Q_2 = 9 - 9 = 0$.

Result: In this problem, the predicted equilibrium is that Polluter 1 will buy up all pollution allowances, either directly from Polluter 2 or at auction by bidding more than Polluter 2.

Rationale: Polluter 1 values pollution much more than Polluter 2 in terms of the initial tons, represented by the vertical intercepts of the equations (i.e., 10 > 5.5). Polluter 1's MB falls much slower than Polluter 2, as CO_2 emissions increase (0.50 < 1.0). This feature is represented by the slopes of the equations.

Let us consider how things would change if pollution rights were sold by the government, rather than given away for free. As we will learn in the coming sections, the main way pollution allowances are sold is via auctions.[4] If there are many buyers and sellers in auctions, we would have exactly the conditions for well-functioning markets we discussed in Chapter 3. We would therefore expect that the equilibrium price would be where MB = MC, and pollution allowance markets would end up at the equilibrium in Eq. 17.2. As the main cap-and-trade markets around the world transact allowances representing hundreds of millions of metric tons of pollution each year, the condition that markets need many buyers and many sellers is typically met. Furthermore, one of the key advantages of auctions is that transaction costs can be kept quite low, implying there will be limited barriers to participating.

There are three key theoretical advantages of auctioning allowances rather than giving them away for free. First, auctions are markets created by governments and those who want to trade allowances do not need to figure out how, where and with whom to buy and sell. Second, auctions provide immediate and solid information on prices. One problem with giving allowances away for free and then allowing them to be reallocated through trade is it may not be obvious what rights to pollute are worth. This price uncertainty, which can be expected to be limited for auctions, may represent an important transaction cost.

Finally, the purpose of the "trade" part of cap-and-trade is to move allowances quickly into the hands of polluters who will most benefit from having them. As we discussed in Chapters 3 and 4, among others, willingness to pay has important linkages with benefits and helps us understand value from the buyer side. Auction winners are therefore likely to be the polluters who are most profitable and/or have the most difficulty abating pollution. Auctioning therefore transforms a two-stage process for getting allowances into the hands of those who value them most into a one-stage process where allocations are made directly to those with the highest willingness to pay.

The theoretical advantages of auctioning pollution allowances are focused on efficiently and effectively getting the rights to pollute to those who value them the most, therefore reducing costs. The downsides are more on the distributional and political sides of the equation. As was the case with pollution taxes, allowance auctions involve transfers of resources from the private sector to public sectors. Existing polluters who behave in accord with our behavioral model can therefore be expected to prefer – and lobby for – distributing some or all pollution allowances to existing polluters at zero or low cost.

17.3 The US Experience with non-Carbon Pollution Trading

Though a lot of the contemporary action on cap-and-trade, especially related to carbon, is taking place in Europe and Asia, much of the world's fundamental understanding of pollution-related green markets comes from the US. The Clean Air Act of 1970 established the basic structure for how the US would try to improve air quality and reduce pollution's impact on common pool ecosystem services, such as human respiratory health, soil pH stability and dust-free surfaces. It charged the recently created US Environmental Protection Agency (EPA) with implementing the law.[5] As we discussed in Chapter 14, the Clean Air Act established National Air Quality Standards and required the US states to develop air quality management plans to bring areas that were out of compliance with those standards into attainment status. It also required that the US EPA develop performance standards for large stationary sources, tailpipe standards for vehicles and systems for managing toxic pollutants.

As you might imagine, these changes caused quite an uproar within the polluting community, creating pressure to find cost-effective solutions. Flexible compliance mechanisms therefore started to make their way into the nascent US air protection system. In 1974, for example, the US EPA created rules that approved polluting facility *bubbling*. This rule allows facilities, such as factories and power plants, that are located in one place with many smokestacks, to be regulated as if they are one big pollution source.

The *netting* program then extended this practice to cross firm boundaries, allowing firms to pay other firms to do part of their compliance for them. Schmalensee and Stavins (2019) note that by the middle of the 1980s, there were more than 50 nets authorized by the US EPA, and many more were permitted by states, offering firms flexibility that saved an estimated $430 million.

Offsets were explicitly allowed in 1977, when the first amendments to the Clean Air Act were enacted. Offsets created a system of earned *pollution credits* in which reducing emissions below permitted levels created headroom that could be used elsewhere in the same firm or sold. Critically, offsets created flexibility by allowing polluters wanting to site new sources in regions that were out of compliance with National Air Quality Standards, such as Los Angeles, California and Houston, Texas, to offset the new emissions by reducing pollution from existing sources in the same region. These sources could be within the same firm or other firms.

The 1977 amendments also allowed firms to *bank* any emission credits they generated for later use or sale. The banking program allowed flexibility not only within and across firms at one point in time, but extended that leeway across time. Banking was based on the belief that it was unlikely that emissions credits would create toxic "hot spots" in the future.

All these programs – bubbling, netting, offsets and banking – required administrative approval, which made trades cumbersome. Nevertheless, the early air pollution trading programs are believed to have saved the polluting community somewhere between $5 and $12 billion (Schmalensee and Stavins, 2019).[6]

The US gasoline (petrol) lead phasedown program was the first to eliminate the requirement for prior trade approval. As we discussed in previous chapters, lead is toxic at low doses, especially for children. When it is burned in internal combustion engines it improves engine performance, but the element can also interfere with catalytic converters used in vehicles to reduce NO_x emissions.

Starting in the early 1970s, the US EPA began preparations to reduce lead in gasoline and focused its attention on gasoline refineries. Gasoline additives other than lead were available, but refineries would have to make costly adjustments to their production processes. Furthermore, it was believed that refineries could face different abatement costs. As has been the case throughout its history, when any sort of trading was introduced, the driver was concern over compliance costs.

The US lead trading program allowed refineries to trade rights to add lead to gasoline even as the average lead content in fuels moved toward zero. The program was short-lived – only lasting from 1982 to 1987 – and midway through added banking. Administrative requirements were minimal. At the end of each quarter, refineries were required to submit reports to the US EPA on gasoline production, amounts of lead used, banked lead rights used and remaining lead rights. The EPA could audit refineries it suspected of misrepresenting lead use (Hahn and Hester, 1989).

Through much of the 1980s, lead trading was active and if a driver burned gas that included lead, chances were significant that the right to that lead had been purchased. Lead rights were also actively banked, particularly by large refiners for use at the end of the program when the average lead content in gasoline was required to be at its lowest level. At its peak in 1987, about 55% of the lead used in gasoline in the US had been traded and an even higher percentage had been banked (Hahn and Hester, 1989).[7]

The application of trading and banking to the lead phasedown reduced the average lead content in gasoline faster than initially anticipated, resulted in savings of about 20% compared with a purely standards-based regulatory approach, and incentivized diffusion of cost-reducing technologies (Schmalensee and Stavins, 2019). The success of the program offered a strong proof-of-concept that pollution trading could effectively get the environmental job done and save money in the process. It also highlighted the critical roles of low administrative costs and banking in generating gains from trade (Schmalensee and Stavins, 2017).

The federal-level peak of green markets, and where much of the initial learning about the functioning of cap-and-trade took place, was within the US EPA Acid Rain Program. The Acid Rain Program is a key part of the Clean Air Act amendments of 1990, which sought to reduce stationary source emissions of SO_2 by 50% below 1980 levels by 2005. The main source of SO_2 in the US was burning coal to produce electricity.

Firms were free to buy and sell what were called SO_2 *allowances* that were denominated by year. Transactions could take place either in anonymous markets using brokerage firms, or in business-to-business transactions. The overall emissions cap was ratcheted down over time until the goal was reached. If firms did not have enough allowances, they could buy them in the market. If they had more than they needed and did not want to save them for later (yes, banking was allowed), they could sell them (Schmalensee and Stavins, 2013).

Phase I (1995–1999) covered the 263 most-polluting electricity generation plants and Phase II included all fossil fuel-burning power plants over 25 megawatts of energy input, which was basically all power plants. The overall goal was to cap annual SO_2 emissions at 10 million short tons per year. Allowances were distributed for free based on past fuel use, with a small share auctioned by the US EPA, largely to provide the market with price guidance. All plants regulated under the Acid Rain Program were required to install continuous emissions monitoring technology, which automatically transmitted source-level emissions data to the US EPA. Any plant caught emitting SO_2 without an allowance could be fined $2000 per ton, which was well above the estimated marginal abatement costs (Schmalensee and Stavins, 2013; 2017; Aldy et al., 2022).

How did this system do and where is it today? The evidence suggests that the Acid Rain Program was very successful during its first decade (1995-2005). As shown in Fig. 17.3, by 2004 the goal had basically been reached. By 2010, total emissions were about half of the emissions cap. Like the lead phasedown program, this system overachieved its goal.

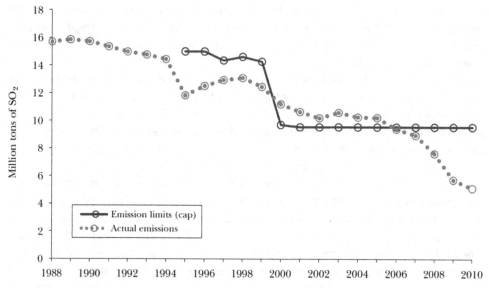

Figure 17.3 SO$_2$ Caps and Emissions, 1988–2010
Source: Schmalensee and Stavins (2013).

There was a lot of trading and banking in the Acid Rain Program. By 1998, over 20 million tons had changed hands and annual trading increased sixfold between 1995 and 1998. During much of its history, allowance prices were stable at roughly $175 per ton. Furthermore, *ex post* (i.e., after-the-fact) estimates suggested that allowing trading as part of the Acid Rain Program significantly reduced abatement costs. Estimates suggest that compliance costs were at least 15% lower and perhaps as much as 90% below a no-trading counterfactual situation. It also likely promoted technical change in the use of fuels and end-of-pipe technologies, such as flue gas desulfurization (Schmalensee and Stavins, 2013).

How is the cap-and-trade program doing now? Unfortunately, it is basically inoperative due to new regulations and court rulings. New regulations targeting PM$_{2.5}$ and SO$_2$ were more stringent than those under the Acid Rain Program, which effectively made the system non-binding. Furthermore, court rulings related to trading and banking across the US states found that without specific new legislation, the US EPA could not make rules that differentially affected people across state lines. This finding effectively restricted trading to within states. Though officially still in place,[8] little within-state trading takes place and the program is no longer a major player in air pollution control (Schmalensee and Stavins, 2013).

In addition to the US federal trading programs, the 1990 Clean Air Act amendments spurred two regional initiatives. As we discussed in Chapters 14 and 15, NO$_x$ pollution is a problem mainly because in the presence of sunlight and other air pollutants, it can form ground-level ozone, which is a potent respiratory irritant. Because of the potentially serious health effects, especially for children and older people, controlling NO$_x$ emissions is of significant importance, particularly in densely populated regions.

Starting in 1999 and continuing through 2008, 11 and eventually 19 northeastern US states plus the District of Columbia, where the US capital is located, developed and implemented the NO_x Budget Trading Program. The goal of the program was to reduce ozone concentrations during the summer season by at least 50% compared with 1990. NO_x allowances that added up to declining annual caps were distributed free to sources by the program via the US states. Trading was allowed across state lines and ultimately included 2500 large sources. The results were very positive. NO_x emissions fell by 3/4 between 1990 and 2006, with significant estimated cost savings compared with an alternative without trading (Schmalensee and Stavins, 2017; 2019; Aldy et al., 2022).

As was the case for the federal Acid Rain Program, though, regulatory changes and legal challenges ultimately dictated a relatively quick end to the NO_x Budget Trading Program. The interstate trading aspect proved to be especially problematic after the federal court ruling that interstate trading would require additional legislation.

The second regional non-carbon NO_x trading scheme emerging from the 1990 Clean Air Act amendments was the Regional Clean Air Incentives Market (RECLAIM). RECLAIM was started in 1993 by the South Coast Air Quality Management District, which manages pollution in Southern California, in the four counties near Los Angeles. This federal air quality management district had and continues to have among the worst ground-level ozone problems in the country.

RECLAIM sought to reduce NO_x emissions from approximately 350 stationary sources by 70% over ten years. The program also regulated SO_2 emissions from these same sources, though that program was substantially smaller than the NO_x program (Schmalensee and Stavins, 2019). No banking of allowances (what were called RECLAIM Trading Credits) was allowed and trading was prohibited between Pacific Ocean coast (upwind areas) and inland pollution sources (downwind areas). This prohibition was put in place because of differences in average and maximum ozone concentrations and marginal damages across the two regions (Aldy et al., 2022).

RECLAIM achieved its ten-year 70% NO_x reduction goal. As of October 2023, a total of almost 600,000 short tons of NO_x and 157,000 tons of SO_2 had been traded (SCAQMD, undated a). Emissions at RECLAIM facilities were estimated to be about 20% below comparable conventionally regulated facilities and except for the year 2000, when California experienced an electricity crisis and increased production was needed, RECLAIM had met its NO_x and SO_2 emissions reduction goals. In its audited yearly compliance report for 2021, the RECLAIM board of directors found that NO_x emissions were 22% below and SO_2 emissions were 17% lower than the credits allocated (SCAQMD, undated a). Like basically all other pollution trading systems, RECLAIM overachieved its goals.

Have you ever wondered what a ton of NO_x was worth? Perhaps not. Over much of RECLAIM's history, NO_x cost about $2000 per ton. Prices ramped up starting in compliance year 2020, though, averaging $8300 in 2020, over $30,000 per ton in 2021 and $47,000 in 2022 before dropping back to an average of $17,000 in 2023. In 2018, the RECLAIM board of directors announced that RECLAIM would close and regulated sources would transition to a conventional standards-based regulatory approach. As of 2023, this had not yet occurred, but almost 1/3 of the original 350+ RECLAIM sources had either closed or exited the program (SCAQMD, undated b).

17.4 Chapter Conclusions

This chapter explored green markets, with a focus on pollution trading. The world of green markets includes pollution control policies that construct markets for rights to pollute. These systems achieve environmental goals while also seeking to increase incentives for innovation and improve cost-effectiveness. Green markets are potentially appropriate instruments when pollutants are uniformly mixed or when systems are geographically restricted (e.g., within one facility). If many buyers and sellers face low transaction costs and look out for their own interests, cap-and-trade markets are predicted to allocate pollution abatement effort cost-effectively across polluters. This prediction implies that the MBs of pollution and marginal abatement costs are equal for all polluters.

The different types and levels of pollution trading can be thought of as a ladder, where higher rungs represent more complex systems. Cap-and-trade, particularly with banking, offers polluters lots of flexibility to abate at the pollution source, change production systems or ... buy allowances and pollute, all while achieving the designated pollution cap.

Cap-and-trade started in the US and the various systems have in general more than achieved their environmental and cost-effectiveness goals. Truly important national cap-and-trade experience was gleaned from SO_2 trading under the US Acid Rain Program, but this now continues only at the sub-national level. Other than for CO_2 and other greenhouse gases, which we will discuss in the next chapter, sub-national cap-and-trade has also been waning in the US. The decline of non-carbon cap-and-trade in the US partly was spurred by regulatory changes, but was also due to court rulings that prohibited cross-state trading under the Clean Air Act. Many US states simply do not have enough large polluters to support cap-and-trade systems located wholly within one US state.

Issues for Discussion

1. Does talking about the "marginal benefits" of pollution make sense to you?
2. Please present two reasons why it is useful to think in terms of the benefits of pollution to polluters and two reasons it is not helpful.
3. To what degree do you think cost-effectiveness should be a core consideration in pollution policy?
4. Are green markets ethical? Do they inappropriately reward polluters with rights they do not deserve?

Practice Problems

This description applies to Questions 1-4 below.

Two polluters emit CO_2 in a jurisdiction in Latin America. The MBs of pollution of these firms are the following:

Firm 1: $MB_1 = 40 - 2Q_1$ and Firm 2: $MB_2 = 60 - Q_2$, where $Q_i = CO_2$ emissions in tons/year and i can represent either Firm 1 or Firm 2. MB is net of any private costs associated with pollution (e.g., energy purchases, machinery costs etc.).

Suppose emissions are currently unregulated, but the environmental regulator would like to cap total CO_2 emissions at 50 tons.

1. If the polluters maximize profits and MB_1 and MB_2 include all benefits of pollution net of costs to generate those benefits, without any controls how many tons does our model predict each polluter will emit? Please explain your answer.

2. Suppose the two firms are regulated using a cap-and-trade regime. Suppose further – perhaps rather unrealistically, because there are only two firms – that they bargain with each other and fully exhaust all gains from trade. Using Box 17.1 as a guide, how much will each polluter emit and what will be their MB values?

   Hint #1: See Fig. 17.2 for guidance and remember that pollution allocations under well-functioning cap-and-trade regimes are predicted to be cost-effective.
   Hint #2: Please remember that the pollution cap must be respected.

3. **Challenge Problem**: Suppose a third firm with $MB_3 = 30 - 3Q_3$ is added to the cap-and-trade system. How much do you predict that each firm will pollute if the cap remains at 50 tons? Using the three MB functions, please explain in words why your predicted cost-effective allocation of abatement effort makes sense.

   Hint #1: You will need to be very careful with algebraic substitution and make sure all equations used are independent. The equilibrium solution is not in integer (i.e., whole number) form, so please round to the nearest whole number.
   Hint #2: Please see the hints for Practice Problem 2.

4. Assume we once again have only Firms 1 and 2. Suppose uncontrolled/baseline emissions of the two firms are as you calculated in Practice Problem 1 (you may want to recheck your calculations). Noting that uncontrolled/baseline emissions – actual emissions = abatement, please redo your calculations in Practice Problem 1 wholly in terms of abatement. How much do you predict that each firm will abate compared with the uncontrolled emissions level?

This description applies to Questions 5-8.

   Graph 1 gives the MB of biological oxygen demand (BOD) pollution for our typical European BOD polluter from Fig. 15.3 in Chapter 15, which is called Firm 1 in Table 15.2. You may want to quickly refer to that chapter. Suppose we add the second polluter called Firm 2 in Table 15.2, which has higher abatement costs than Firm 1, to Graph 1 and the result is Graph 2.

   Suppose the European Commission decides to limit total BOD emissions of the two firms to 7 tons. It implements a cap-and-trade system and gives 2 tons to Firm 1 and 5 tons to Firm 2 for free.

5. Please interpret the meaning of Graph 2, including the significance of the vertical and horizontal intercepts. In particular, please mention which firm gets more benefit from being able to pollute and how does that relate to abatement costs? How much are the uncontrolled emissions of each firm?

6. Is the European Commission's allocation of BOD pollution rights cost-effective? How do you know?

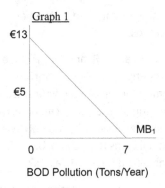

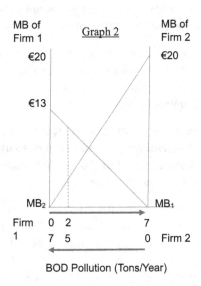

Figure 17.4 Marginal Benefits of Pollution to Firm 1 (Graph 1) and to Firm 1 and Firm 2 (Graph 2)

7. If the two firms at no cost can freely trade rights to pollution with each other, is the European Commission's allocation of BOD pollution rights an equilibrium? Why or why not? If it is not an equilibrium, please explain the mechanism by which the two firms would reach an equilibrium.

8. **Challenge Problem**: Assume the MB functions in Graph 2 are linear. Please convert the information in Graph 2 to equations and solve for the equilibrium allocation of BOD pollution if the European Commission only allows 7 tons of BOD to be emitted. Is the equilibrium cost-effective? How do you know?

Note: your result will not be whole numbers, but will include decimals.

Notes

1   I would like to acknowledge that the first time I encountered the term "green markets" was in the title of the book of the same name published by my former colleague Theodore Panayotou (Panayotou, 1993).

2   Recall from Chapter 14 that uncontrolled/baseline emissions = actual emissions + abatement.
3   To elaborate just a bit, in a cap-and-trade market the MC of the sellers is what they have given up to get the pollution permit payment (i.e., the benefit they would have received from the right to pollute). The MB for buyers is the additional profit they gained and the abatement costs avoided – on the margin – because they bought permits.
4   For general information on auction processes, please see Fine (undated).
5   Though much of the information in this subsection is widely known and available via the US EPA website and other sources, I would like to acknowledge a number of review articles that form the foundation of this subsection. These include Schmalensee and Stavins (2013; 2017; 2019) and Aldy et al. (2022), which presented much of the material I discuss, in concise and comprehensive ways. I draw freely from these authors' considerable expertise on the topic of air pollution trading and in some cases rely on the authors to accurately cite sources in their reviews.
6   Citing Hahn and Hester (1989).
7   Lead was completely eliminated from gasoline in the US in 1996 and Algeria was the last country to stop adding lead to gasoline in 2021 (NPR, 2021).
8   See the US EPA Acid Rain Program website at https://www.epa.gov/acidrain/acid-rain-program

Further Reading and References

Aldy, J., M. Auffhammer, M. Cropper, A. Fraas and R. Morgenstern. 2022. "Looking Back on 50 Years of the Clean Air Act." *Journal of Economic Literature* 60 (1): 179–232.

Fine, L.R. Undated. *Auctions*. Econlib. Retrieved from https://www.econlib.org/library/Enc/Auctions.html December 10, 2023.

Hahn, R. and G. Hester. 1989. "Marketable Permits: Lessons for Theory and Practice." *Ecology Law Quarterly* 16 (2): 361–406.

National Public Radio (NPR). 2021. "The World Has Finally Stopped Using Leaded Gasoline. Algeria Used The Last Stockpile." *All Things Considered*, August 30. Retrieved from https://www.npr.org/2021/08/30/1031429212/the-world-has-finally-stopped-using-leaded-gasoline-algeria-used-the-last-stockp October 2024.

Panayotou, T. 1993. *Green Markets: The Economics of Sustainable Development*. Island Press: Washington, DC.

Schmalensee, R. and R. Stavins. 2013. "The SO$_2$ Allowance Trading System: The Ironic History of a Grand Policy Experiment." *Journal of Economic Perspectives* 27 (1): 103–122.

Schmalensee, R. and R. Stavins. 2017. "Lessons Learned from Three Decades of Experience with Cap and Trade." *Review of Environmental Economics and Policy* 11 (1): 59–79.

Schmalensee, R. and R. Stavins. 2019. "Policy Evolution under the Clean Air Act." *Journal of Economic Perspectives* 33 (4): 27–50.

South Coast Air Quality Management District (SCAQMD). Undated a. *Historical Trade Registrations*. Retrieved from http://www.aqmd.gov/home/programs/business/about-reclaim/reclaim-trading-credits/historical-trade-registrations October 2024.

South Coast Air Quality Management District (SCAQMD). Undated b. *Twelve-Month NOx Rolling Average Reports*. Retrieved from https://www.aqmd.gov/home/programs/business/about-reclaim/twelve-month-rolling-average-price-of-nox-rtcs October 2024.

18 Green Markets: International Climate Policy and Carbon Markets

---

What You Will Learn in this Chapter

- About green markets and international climate change policy
- Voluntary versus compliance carbon markets
- Experience with carbon cap-and-trade around the world
- About the European Union Emissions Trading Scheme (EU ETS)

18.1 Recap and Introduction to the Chapter

As we discussed in the previous chapter, governments can use cap-and-trade and other types of green market instruments to reduce pollution emissions and improve cost-effectiveness. They contribute to achieving these goals by issuing rights to pollute and promoting markets to trade those rights. The US has had some useful experience with trading pollutants other than carbon, particularly in the 1990s and 2000s. Of special note were the pathbreaking lead phasedown and acid rain reduction programs, which were both able to more than achieve their environmental goals and save money in the process.

The contemporary world of green markets focuses mainly on controlling greenhouse gas emissions. As we discussed in Chapters 2 and 16, the international climate framework, which began with the 1992 UN Framework Convention on Climate Change (UN FCCC), now has basically 100% country membership. The UN FCCC included the possibility for carbon trading, and subsidiary agreements have only expanded the use of those instruments. We begin our discussion of carbon markets with reference to international climate policy, before moving on to the major contemporary carbon markets around the world. We then take an especially deep dive into the European Union Emissions Trading Scheme (EU ETS).

18.2 Green Markets and International Climate Policy

The use of green markets has been allowed and encouraged as part of the international climate policy architecture essentially from the beginning. The UN FCCC set goals, defined responsibilities, specified how the convention would work and required periodic meetings of

DOI: 10.4324/9781003308225-21

parties to the convention, called conferences of the parties (COP). In December 2023, the 28th COP was held in Dubai in the United Arab Emirates.

The UN FCCC established the broad climate change goals and placed the in-principle burden on higher-income countries and also European countries that at the time were transitioning from planned to market economies. These countries are all listed in Annex 1 to the Convention. Annex 2 lists the countries that are responsible for providing finance to lower-income countries and is made up of the high-income countries from Annex 1.

The UN FCCC explicitly allowed for flexible compliance, which essentially allowed offsets, but only between the Annex 1 countries that were identified as being responsible for addressing climate change. This mechanism for offsets within the group of Annex 1 countries is called *joint implementation* (JI). As is the case with all offsets, projects must be proposed, justified and approved by a review body, in this case the Joint Implementation Supervisory Committee of the UN FCCC[1] and the financing country government representative. At the end of the project, the parties funding JI projects must commission a "determination report," which lays out the results of the project. These results have to be approved before the GHG *credits*, which are known as *Emission Reduction Units* (ERUs), can be issued.

As with all things United Nations, if one digs into the right website hard enough, it is possible to get full information on activities and outcomes. JI projects are between institutions within countries, often firms and municipalities, and have run the gambit of GHG-reducing projects, including forest management, energy generation, energy efficiency, renewable energy, implementation of no-till agriculture and methane capture from landfills. Interestingly, most projects that were implemented between 2000 and 2015 were hosted by European countries that were in transition (e.g., Bulgaria, Ukraine, Russia, Estonia) and funded by higher-income European countries (e.g., Denmark, Germany, France). This feature of JI projects is perhaps not surprising, given what we learned in Chapter 12 about the enormous energy inefficiencies prevalent in the former planned economies; there were just a lot of opportunities for improvement to exploit and countries in Western Europe were well-placed to support GHG projects elsewhere in Europe.[2]

Box 18.1 UN FCCC Joint Implementation of a Wind Energy Project in Bulgaria

The Municipality of Kavarna on the Black Sea coast of Bulgaria donated land for the Kaliakra Wind Power Project (KWPP), which was designed to generate 35 MW of power annually. This electricity is enough to power roughly 8000 homes. The project cost was €65 million and was projected to reduce GHG emissions by 81,400 tons per year from 2008 to 2012, mainly by substituting wind power for fossil fuel-generated energy. The operational lifetime of the wind farm was 20 years.

The project host in Bulgaria was INOS-1, which is a Bulgarian energy company, and funding for the project came from Japan Carbon Finance, Ltd., which was formed to develop GHG reduction projects and claim credits for use during the Kyoto Protocol first commitment period. INOS-1 developed the project in partnership with Mitsubishi Heavy Industries in Japan.

The project determination report, which was prepared by the consulting company Jaco CDM, verified that the project reduced emissions by approximately 300,000 tons over the five-year crediting period, which was about 25% less than projected in the project design document.

Source: https://ji.unfccc.int/JIITLProject/DB/O3G4FV0BYW6RVN1 OP8PESF1BY7I8AX/details.

As we discussed in Chapter 2, the third UN FCCC conference of the parties was held in 1997 in Kyoto, Japan. At that meeting, the parties to the UN FCCC adopted an agreement under the convention called a *protocol*. The Kyoto Protocol set a goal to reduce GHG emissions by 5% below 1990 levels between 2008 and 2012, which was defined as the Kyoto Protocol first commitment period. It also required Annex 1 countries to submit emissions inventories (now annual) and set binding emissions reduction goals for most, but not all, of the UN FCCC Annex 1 countries. For example, the EU (at that time 15 countries) committed to reducing emissions by 8%, the US 7% and other countries agreed to smaller reductions (e.g., Japan 6%) or even emissions increases (e.g., Norway +1% and Iceland +10%).

Importantly, the parties to the UN FCCC agreed to extend trading from between Annex 1 countries to between Annex 1 and non-Annex 1 countries. There was no need for trading between non-Annex 1 countries, because those countries had no responsibilities. This second offset program supplemented JI and was known as the *Clean Development Mechanism* (CDM).

A total of 7841 CDM projects were registered between 2008 and 2020. Virtually all of them (7803) were completed by 2018 when the UN issued a report on the achievements of the CDM program. The report noted that the 7803 projects were implemented in 111 non-Annex 1 countries and were funded by institutions in 29 Annex 1 countries. A total of $304 billion was invested in climate and sustainable development projects that reduced or *sequestered* (i.e., captured, generally in trees) almost 2 billion metric tons of carbon emissions in the Global South. These investments included purchase and distribution of 1 million efficient cookstoves, 1 million megawatts (MW) of renewable power (enough to fully supply Ecuador, Morocco, Myanmar and Peru together!) and planting 152 million trees, among others (UN, 2018). The most recent projects were registered in 2020 and included a number of solar power installations in India, wind energy in Egypt, hydropower in Indonesia and a wind farm in Chile.[3]

Though signed by US President Bill Clinton, the US Senate did not ratify the Kyoto Protocol as required, and so the at-that-time largest GHG emitter did not take on binding Kyoto emissions reduction targets during the first commitment period or the second one, which was added in 2012 and covered the period 2013–2020.[4] Canada later withdrew from the Kyoto Protocol.

The world needed a follow-up to the Kyoto Protocol, which was agreed in December 2015 in Paris, France. The 2015 Paris Agreement continued the longstanding UN FCCC practice of allowing emissions offsets, which in the text of the Agreement is called *voluntary cooperation*. The key difference between the Paris Agreement and previous agreements is, not only is participation voluntary but everything about country commitments is country-generated.

Countries submit and agree to implement what we called in Chapter 2 *Nationally Determined Contributions* (NDCs). These commitments can be different in a variety of respects. Take a look at the NDC registry, which has all countries' most up-to-date commitments and discusses what each country has promised to do to reduce climate change.[5]

Challenge Yourself

Go to the Paris Agreement NDC registry at https://unfccc.int/NDC REG. Choose two countries you are interested in. How are their NDCs similar and different?

18.3 Voluntary Carbon Markets

There are two types of carbon markets operating around the world. *Compliance markets* are formal markets like the US EPA Acid Rain Program, which seek to provide a cost-effective system within which polluters can comply with laws to reduce carbon emissions. *Voluntary markets* are less formal and, as the name suggests, may or may not fulfill legal obligations for those who buy carbon credits. They are essentially a form of carbon offset and represent individual and institutional (mainly corporate) contributions to reducing climate change.

Voluntary market projects are often developed in lower-income countries and may be similar in character to CDM projects. Those who want to buy carbon credits contract with those who develop projects, to gain access to the credits. Carbon credits typically must be certified by an outside organization and several non-profits verify carbon offsets. These organizations, which may have slightly different methodologies for verifying the amount of carbon credits generated by projects, include The Gold Standard, Climate Action Reserve, American Carbon Registry and Verra.

In 2022, a total of 254 million metric tons of CO_2 credits were generated by 600 projects. This trading volume was less than half of what was transacted in 2021, but the metric tons of CO_2 credited in 2021 were about 150% of the previous high, which occurred in 2020. Credits in 2022 were about 25% above 2020. The average price per ton of CO_2 offset was $7.34, which was the highest price in the last 15 years and a total of $1.9 billion was transacted, about the same as in 2021.[6]

The voluntary carbon market focuses significantly more on forestry, land use and agriculture than JI and CDM, which mainly targeted alternative energy. Almost half of the 2022 voluntary carbon market volume was for such "nature-based climate solutions," but those projects made up a greater percentage of the revenues; carbon credits from projects involving land on average sold for twice the price of technology projects, such as renewable energy.

As we know from previous discussions, people in rural areas of low-income countries – the Global South – on average rely much more heavily on land for their livelihoods than those in urban areas or higher-income countries. Projects that manage land-based natural resources for carbon sequestration and reduced emissions may therefore be complementary to livelihood goals and even improve the conditions of households in the Global South.[7] For this reason, voluntary carbon market buyers, who are often trying to sequester carbon as they improve people's livelihoods, can be especially interested in social co-benefits from projects that generate carbon credits. Carbon credits that have certified non-carbon social benefits sell for almost double the price of those without social benefits. Similarly, projects that in 2022 were explicitly linked to the Sustainable Development Goals (SDGs) sold for twice the price of credits not linked to the SDGs.

18.4 Compliance Carbon Markets

Compliance markets are linked to legislation and government regulations, and help polluters meet their pollution reduction obligations. The main carbon compliance cap-and-trade markets are in the European Union, Republic of Korea, the US State of California, the Canadian Province of Quebec and New Zealand. The World Bank Carbon Pricing Dashboard shows a total of 36 operating or piloted cap-and-trade systems at national and sub-national levels.[8]

The California cap-and-trade system was launched in 2012 and initially sought to reduce total CO_2e emissions to 1990 levels by 2020. California achieved that goal four years early, in 2016, which is indicative of some of the incentives the system created (CARB, undated). The long-run goal is to reduce emissions by 80% compared with 1990. The 2022 cap was 308 million tons of CO_2e and covers 3/4 of total emissions. In 2023, the system generated over $4 billion in revenues.

Most emissions allowances are auctioned, and as of 2021 there was a minimum (i.e., "reserve") auction price of $17.71/ton, which increased yearly based on the US rate of infla-tion[9] plus 5%. The average price in 2023 was $30 per ton of CO_2e and polluters in so-called "energy intensive trade exposed" (EITE) sectors were eligible to receive initial allocations of allowances at no cost.

In 2014, the California system was linked with the cap-and-trade market in the Canadian Province of Quebec, meaning that regulated facilities in each place could trade with each other. The Quebec cap-and-trade system was launched in 2013 and covers about 3/4 of emissions in the province. In 2022 the system's cap was 54 million metric tons, like the California system it had a reserve auction price, and it auctioned most allowances.[10]

The Republic of Korea Emissions Trading Scheme (Korea ETS) was launched in 2015 and covers 74% of emissions under a cap of 589 million metric tons (2022). The Korea ETS includes six GHGs emitted by the industrial, power, domestic aviation, public and waste sectors, as well as buildings. Only 5% of this cap is allocated by auction and EITE economic sectors get their initial allocations 100% for free.

The New Zealand Emissions Trading Scheme (NZ ETS) was launched in 2008 and was originally envisioned as a system that would link with other Kyoto Protocol–focused trading regimes. When those international markets did not develop, in 2015 it became a domestic-only trading system and as of 2022 had a cap of 35 million metric tons. In its Paris Agreement NDC, New Zealand indicated its desire to meet its obligations by linking its system with other high-quality cap-and-trade regimes.

Though the NZ ETS is less than half the size of the systems of California and Quebec in terms of total emissions, it has wide coverage of GHGs and economic sectors, including power, waste management, transportation and forestry. Agriculture is also included in the NZ ETS and the government is working to bring agriculture fully under the cap. Allowances are both auctioned and given away for free, but starting in 2021 the NZ ETS began phasing out free allocation to industrial polluters at the rate of 1% per year through 2030, increasing by 2% per year through 2040 and 3% thereafter. Auction revenues in 2023 totaled approximately $1.3 billion.

The European Union is home to about 450 million people, making it the third most populous jurisdiction in the world. It established the EU ETS in 2003 and the cap-and-trade system was launched in 2005. As of 2023, it is the largest compliance-focused carbon trading system in the world. It has gone through four trading periods, corresponding to obligations associated with international climate agreements and in each subsequent period rules and caps were tightened.[11]

Trading period 1 (2005–2007) was a pilot period covering only CO_2 emissions from power producers and large industries. It also established the infrastructure for monitoring, reporting and verification. Countries prepared what were called *National Allocation Plans* (NAPs) on how they would use the portion of the cap they received from the European Commission and trading was only allowed within countries. The penalty for emitting CO_2 without allowances was set at €40 per ton. To avoid penalties, by April of each year polluters had to have allowances for each ton of CO_2. By 2007, 2.1 billion metric tons had been transacted. None could be banked.

The second trading period (2008–2012) corresponded to the first commitment period under the Kyoto Protocol, in which the EU promised to reduce its CO_2e emissions by 8%. Allowances were reduced by 6.5% compared with the first period and 10% were sold rather than given away. Allowances could be banked for the future. Countries again prepared and submitted NAPs, but this time were allowed to buy and sell 1.4 billion metric tons of CO_2e outside their countries. The penalty for emitting CO_2 without having enough allowances by April of the following was increased to €100 per ton.

The third trading period focused on the second Kyoto commitment period from 2013 to 2020. It further reduced allowances by 1.74% per year. One-third to one-half of all allowances were auctioned and an EU-wide cap replaced country-level caps. The aviation sector was added to the EU ETS. More GHGs and industries were included and 300 million metric tons in allowances were made available for renewable energy and carbon capture and storage projects.

Phase 4 started in 2021, when the European Climate Law was adopted (Regulation 2021/1119) and prescribed that the EU would be climate neutral by 2050. It also pledged to reduce GHG emissions by at least 55% compared with 1990 levels by December 2030, which was more ambitious than in previous legislation. The European Climate Law reaffirmed the EU ETS as the cornerstone of its cost-effective climate policy. Importantly, the law specified that "… it is possible to decouple economic growth from greenhouse gas emissions" and noted that EU GHG emissions fell by 24% between 1990 and 2019 even as the economy grew by 60%.

To achieve its emissions reduction goals, the EU ETS was once again revamped and expanded to cover more countries, gases and sectors, including maritime transport, and a higher percentage of total allowances was auctioned. According to the European Commission's 2021 report on the EU ETS (the latest as of December 2023),[12] the scheme covers about 40% of total emissions in all 27 EU member states, plus Iceland, Liechtenstein and Norway. It also links with the trading system in Switzerland, for a total of 31 countries. The EU ETS regulates almost 9000 electricity, heating and industrial plants, as well as approximately 400 aircraft operators. A total of about 1.57 billion metric tons of CO_2e allowances were issued to stationary sources in 2021 and the aviation sector was allowed to emit 28 million metric tons. Each year between 2021 and 2030, the EU-wide cap will decline by 2.2%.

Challenge Yourself

If the EU-wide CO_2e cap declines by 2.2% per year between 2022 and 2030, over those 9 years by what percentage will emissions have declined?

Table 18.1 presents stationary source and aviation allowances in circulation between 2013 and 2021. In those 9 years, the cap for stationary sources, which make up most regulated sources, fell by about 23% = (1,571,583,007 − 2,084,301,856)/2,084,301,856. Figure 18.1 presents the emissions caps over time and projects them until 2030 (which is easy to do, because the European Commission sets the caps!). It also compares select values to 2005, when the EU ETS was started.

Table 18.2 shows the total auctioned allowances between 2013 and 2021. Throughout the third trading period, power generators had to buy all allowances at auction. Allowance auctions generated significant revenues, particularly starting in 2019 when prices started to increase. In 2021, allowance auctions generated €31 billion, which was almost double the revenue in 2020, and all but €6 billion went back to member states. The EU requires that

Table 18.1 Stationary Source and Aviation Allowances in Circulation

Year	Stationary Sources	Aviation
2013	2,084,301,856	32,455,296
2014	2,046,037,610	41,866,834
2015	2,007,773,364	50,669,024
2016	1,969,509,118	38,879,316
2017	1,931,244,873	38,711,651
2018	1,892,980,627	38,909,585
2019	1,854,716,381	38,830,950
2020	1,816,452,135	42,803,537
2021	1,571,583,007	28,306,545

Source: European Commission (2021).

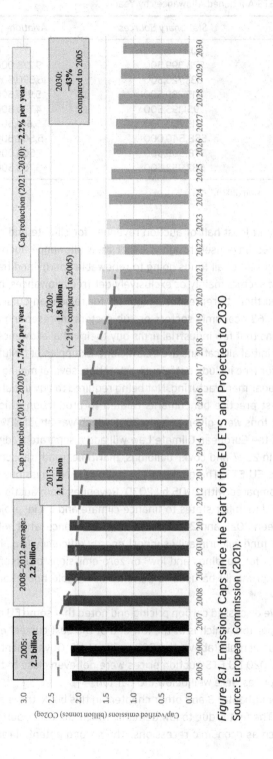

Figure 18.1 Emissions Caps since the Start of the EU ETS and Projected to 2030
Source: European Commission (2021).

Table 18.2 Total EU ETS Auctioned Allowances by Year

Year	Stationary Sources	Aviation
2013	808,146,500	0
2014	528,399,500	9,278,000
2015	632,725,500	16,390,500
2016	715,289,500	5,997,500
2017	951,195,500	4,730,500
2018	915,750,000	5,601,500
2019	588,540,000	5,502,500
2020	778,505,000	7,505,000
2021	582,952,500	3,785,500

Source: European Commission (2021).

member states use at least half of auction revenues for climate and energy projects. Since 2013, member states have used roughly 3/4 of their allowance auction revenues for such projects, with about half of all funds going to renewable energy and transportation.

Some economic sectors mainly or exclusively get free allowances, because the European Commission judged them to represent a high risk for relocation outside the EU.[13] For Phase 4, this list includes 63 economic sectors or sub-sectors representing almost all of the EU's industry. This means that few industrial firms buy their initial allowances at auction, but may still need to buy if initial allocations are insufficient. Examples of polluters that receive their initial allowances for free include petroleum extraction, several mining industries, paper, leather clothes and sugar manufacturing. Not being required to buy initial allocations at auction is linked to 54 best practice benchmarks related to production efficiency. Approximately 545 million metric tons were given away in 2021, which was about 43% of the total.

Looking ahead, the European Climate Law will add a separate trading scheme called ETS 2 that will launch in 2027 and cover buildings, maritime, road transport and small industry not covered by the EU ETS. The goal of ETS 2 is to reduce emissions from those economic sectors by 42% compared with 2005 by 2030. Revenues from auctioning ETS 2 allowances will mainly go to EU member states to finance climate and social projects, but an expected €86.7 billion between 2026 and 2032 will be used to finance what will be called the *Social Climate Fund*. This fund will primarily support energy efficiency (e.g., insulation), renewable energy investments for buildings and low- or zero-emissions transport solutions. Critically, the funds need to primarily be used to benefit "vulnerable households, micro-enterprises and transport users."[14]

In Chapter 16, we discussed carbon pricing and noted that some EU countries have carbon taxes above €100 per ton, which is at the low end of the best current estimates of the social cost of carbon, but actually above the 2023 official US Government value.[15] As shown in Fig. 18.2, through 2020 EU ETS auction prices were not even close to that level and in much of Phases 1 and 2 prices were well below €10 per ton.

As we know from Chapter 3 and other chapters in this book, there are three reasons that prices can be low. The first is due to low demand, which can occur during periods of low economic activity, such as economic recessions. The second potential reason for low prices is

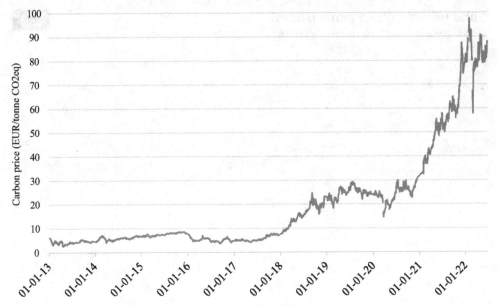

Figure 18.2 EU ETS Stationary Source Auction Prices (€ per Ton)
Source: European Commission (2021).

too much supply, which might happen if the carbon pollution cap was loose compared with the total amount of allowances needed by polluters. Third, of course, both low demand and high supply can occur at the same time, particularly in a constructed market like the EU ETS, where the number of allowances to be released is planned years in advance.

As shown in Fig. 18.2, the situation of low allowance prices started to change in 2019 when the European Commission launched the *Market Stabilization Reserve* (MSR), which is an account that holds back from auctions allowances that are judged to be "surplus" in any particular year. As we think about the meaning of a surplus in such a context, we must remember that allowances can be banked and therefore leftover allowances carry forward. Many allowances from the past could remain available for use well after they were first released, potentially driving down prices when they are sold.

Challenge Yourself

If allowance prices merely reflect supply and demand conditions, why worry about prices if emissions caps are met? In this regard, do you think the EU ETS MSR makes sense?

In May 2022, which was one month after the 2021 reconciliation period ended, the so-called *Total Number of Allowances in Circulation* (TNAC), which defines the surplus, was 1.45 billion. Since 2013, the total surplus had been between 1.4 and 2.1 billion allowances. Starting

Solar (photovoltaic) panel prices

This data is expressed in US dollars per Watt, adjusted for inflation.

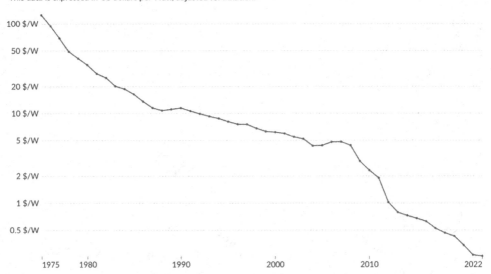

Data source: International Renewable Energy Agency (2023); Nemet (2009); Farmer and Lafond (2016)
Note: Data is expressed in constant 2022 US$ per Watt.
OurWorldInData.org/energy | CC BY

Figure 18.3 Solar Panel Prices per Watt Generated 1976-2022 (2022 $US)

in 2019, if the TNAC exceeded 833 million, 24% of the previous year's surplus was to be held back in the MSR from the next year's auction. Between 2019 and 2022, 1.4 billion allowances – basically one year's worth – were placed in the MSR, including 887 million that were unused during the EU ETS Phase 2. A total of 350 million allowances that were part of the 2022 and 2023 caps were placed in the MSR and at least for now have been withdrawn from the market.[16]

Box 18.2 Solar Photovoltaic Panel Prices Plunge over Time and Installed Capacity Soars

Prices of photovoltaic (PV) panels have declined dramatically over time. In 1992, the world average price per watt of power output was $9.92 in 2021 $US, making PV too expensive for widespread use. By 2012, the cost had dropped to $1.02 and as of 2022 the world average price was just $0.26 per watt.

This 75% real price decline during the last decade and 97% price drop over 30 years drove and was driven by an enormous increase in installed capacity. In 1992, only about 309 MW of solar PV capacity were installed across the globe. By 2012, there were 102,000 MW and as of 2022, capacity stood at 1.05 million MW. In the last ten years, installed capacity therefore increased over ninefold.

Solar (photovoltaic) panel prices vs. cumulative capacity

This represents the learning curve for solar panels. This data is expressed in US dollars per Watt, adjusted for inflation. Cumulative installed solar capacity is measured in megawatts.

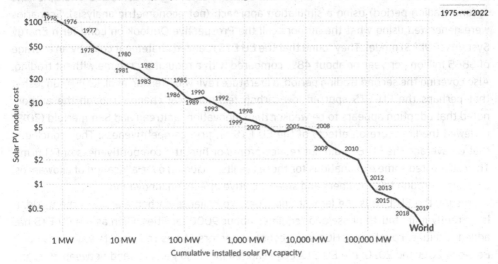

Data source: International Renewable Energy Agency (2023); Nemet (2009); Farmer and Lafond (2016)
Note: Data is expressed in constant 2022 US$ per Watt.
OurWorldInData.org/energy | CC BY

Figure 18.4 Solar Panel Prices as a Function of Installed Capacity
Source: Ritchie et al. (2023).

To what degree has the EU ETS contributed to the decline in carbon emissions in the European Union and saved money in the process? As we have discussed at many points in this book, correlation is not the same thing as causation. We therefore need to be mindful that just because the EU has implemented carbon trading – even with declining caps – does not mean it could not have happened anyway! We must also recall that the EU ETS only covers about 40% of EU CO_2e emissions and the low auction prices throughout much of the EU ETS's history should especially give us pause. Such low prices could very well imply a serious oversupply of allowances.

The literature suggests that the EU ETS indeed **has** contributed to carbon emissions reductions, but, perhaps not surprisingly, is not responsible for all of it. Using data at the level of the economic sector (e.g., energy, metals, minerals, chemicals etc.), Bayer and Aklin (2020) found that despite low auction prices, between 2008 and 2016 the system eliminated about 1.2 billion tons CO_2e or about 3.8% of the total emissions, which is within the range quoted in an early review by Laing et al. (2014). Colmer et al. (2020) used data on CO_2 emissions from 4000 French manufacturers between 1996 and 2012 to estimate the effect of the EU ETS on covered firms versus unregulated firms. They found the EU ETS reduced manufacturers' emissions by 8%-12% below what would have happened without the cap-and-trade system. This effect was about 1/4 of the total decline in emissions of the manufacturing sector.[17]

Evidence on the cost-effectiveness properties of the EU ETS is more limited than evidence on emissions reductions, but it seems the trading system may have reduced costs. Cludius et al. (2019) analyzed the power and manufacturing sectors between 2008 and 2012 (the second trading period) using a simulation approach (not econometric analysis). Scenarios were generated using what the authors call the Prospective Outlook on Long-Term Energy Systems (POLES) model. They found that the EU ETS reduced abatement costs by an average of $865 million per year or about 48% compared with a regulatory regime without trading. Also covering the second trading period, a literature review by Teixidó et al. (2019) suggested that perhaps the EU ETS spurred low-carbon technological change, though the authors noted that adoption appears to be weaker than innovation. Joltreau and Sommerfeld (2019) reviewed the literature on effects of the EU ETS on firm competitiveness. They concluded that on average the EU ETS did not reduce profits or hurt the competitiveness of EU firms. They suggested some explanations for those results, including over-allocation of allowances, cost pass-through to consumers and small, positive effects on innovation.

In sum, the EU ETS is the largest and oldest compliance carbon market in the world. It is currently in its fourth phase and regulates about 9000 facilities. Even as the EU ETS has added facilities, countries, GHGs and economic sectors, it has reduced its emissions caps. Between 2013 and 2020, the EU ETS CO_2e cap fell by 1.74% per year and between 2021 and 2030 it will decline by 2.2% per year, further reducing annual emissions. As of 2021, the cap was about 1.6 billion metric tons, which was about 25% less than in 2013. Over time, the EU introduced auctioned allowances, which now make up almost 60% of the available volume. Since 2019, it has used these auctions to help control surplus allowances and stabilize prices by auctioning fewer allowances when in any year surpluses are above 833 million tons.

18.5 Chapter Conclusions

This chapter discussed carbon trading, which has been explicitly encouraged under climate agreements going back to 1992. Offsets among institutions within UN FCCC Annex 1 countries (Joint Implementation) and with non-Annex 1 country institutions (Clean Development Mechanism) helped move the international climate agenda forward and perhaps reduced costs. The Clean Development Mechanism has also provided an important framework for the development of voluntary carbon markets. These markets now focus on both climate and social goals and include a substantial focus on forests and other land-based resources.

Major national compliance carbon cap-and-trade systems exist in New Zealand, Canada, South Korea and the European Union. Sub-national systems exist in several countries, including the US and Canada, and systems have been linked together, meaning that participants in these different jurisdictions can trade with each other. The EU ETS is the oldest, largest and probably the most important compliance carbon market on the planet. A lot of the experience with carbon trading has come from the EU ETS and the system has evolved significantly over time. Currently in its fourth phase, the EU ETS seeks to reduce its emissions by 55% compared with 1990 by 2030 and achieve carbon neutrality by 2050. Over time, the EU ETS has evolved and adapted as caps have been progressively tightened and emissions declined. Careful econometric analyses appear to suggest that the EU ETS had an important hand in the overall CO_2e emissions reductions of the EU without hurting competitiveness, but evidence on the degree to which the EU ETS reduced abatement costs compared with systems without trading is still very limited.

Issues for Discussion

1. Carbon taxes are an important tool for reducing carbon emissions, but in fact cap-and-trade may be more important. Can you suggest three possible reasons why pollution trading might be preferred to carbon taxes?

2. In 2023, cap-and-trade systems regulated about 18% of global GHG emissions and carbon taxes priced about 6% of emissions for a total of approximately 24%. Are these low or high numbers? Why or why not?

3. According to the Center for Global Development, the developing world – the Global South – was responsible for about 2/3 of the 2015 CO_2e emissions and is probably responsible for more today. Should institutions like the World Bank, which provide technical assistance to lower-income countries, promote instruments that price carbon in these countries? Please explain your reasoning.

4. If allowance prices merely reflect supply and demand conditions and if emissions caps are met, why worry about prices? In this regard, do you think the EU ETS MSR makes sense?

Practice Problems

1. If the EU-wide CO_2e cap declines by 2.2% per year between 2022 and 2030, over these 9 years by what percentage will emissions have declined?

2. Please compare the information in Tables 18.1 and 18.2 and calculate the percentage of all allowances that were auctioned in each year. By how much did the auction percentage change between 2013 and 2021? Is there a clear pattern? Please express the increase as 1) a percentage point increase and 2) a percentage change. Please see Chapter 12 for information on the difference and how to make the calculations.

Notes

1. This organization has since been disbanded.
2. Everything JI can be found at https://unfccc.int/process/the-kyoto-protocol/mechanisms/joint-imp lementation and you can search and review JI project documents at https://ji.unfccc.int/JI_Projects/ ProjectInfo.html. The peak of the JI program was in 2011, when institutions in 16 countries bought over 520 million tons of CO_2 credits as JI ERUs. See https://ji.unfccc.int/statistics/2015/ERU_Issua nce_2015_10_15_1200.pdf for details.
3. See https://cdm.unfccc.int/Projects/registered.html for details on all 7841 CDM projects.
4. There is nothing mysterious about these international climate agreements and the texts are available on the UN FCCC web page starting at https://unfccc.int/. For example, you can learn about the UN FCCC and download the text at https://unfccc.int/process-and-meetings/what-is-the-united-nations-framework-convention-on-climate-change#:~:text=The%20UNFCCC%20entered%20into%20fo rce,called%20Parties%20to%20the%20Convention or read about the Kyoto Protocol and its amendments at https://unfccc.int/kyoto_protocol#:~:text=During%20the%20second%20commitm ent%20period,is%20different%20from%20the%20first and the Paris Agreement at https://unfccc. int/process-and-meetings/the-paris-agreement
5. Available at https://unfccc.int/NDCREG

6 Material on the state of voluntary carbon markets is taken from Ecosystem Marketplace (2023).

7 For information on the issue and importance of so-called "land grabs," please see Borras et al. (2011).

8 The material on these cap-and-trade systems is drawn mainly from World Bank (undated).

9 In Chapter 12 we noted that inflation is a measure of the average economy-wide increase in prices.

10 The US State of Washington's cap-and-trade system was launched in January 2023 and was modeled on the California system. There are ongoing discussions related to linking the Washington system with those in California and Quebec.

11 The overview of the first three EU ETS trading periods is from https://climate.ec.europa.eu/eu-action/ eu-emissions-trading-system-eu-ets/development-eu-ets-2005-2020_en#evolution-of-the-europ ean-carbon-market. The amended version of the EU ETS directive that is in force as of December 2023 can be accessed at https://eur-lex.europa.eu/legal-content/EN/TXT/?uri=CELEX%3A02003L0 087-20230605

12 Unless otherwise noted, information on the EU ETS is from European Commission (2021), which is available at https://eur-lex.europa.eu/legal-content/EN/TXT/PDF/?uri=CELEX:52022DC0516

13 They call this phenomenon *carbon leakage*, because, if correct, the policy would reduce carbon emissions in the EU, but those emissions would simply move to countries outside the EU. Global emissions would not decline, because they "leaked" out of the EU-wide cap. The regulation around giving allowances for free and the list of economic sectors are available at https://eur-lex.europa.eu/ legal-content/EN/TXT/PDF/?uri=CELEX:32019D0708&from=EN

14 See https://climate.ec.europa.eu/eu-action/european-climate-law_en and https://climate.ec.europa. eu/eu-action/eu-emissions-trading-system-eu-ets/social-climate-fund_en for details on the European Climate Law, ETS 2 and the Social Climate Fund.

15 See Chapters 9 and 16 for discussions of the current official and proposed social cost of carbon values.

16 For interesting assessments of the MSR, please see Flachsland et al. (2020) and Perino et al. (2022).

17 Box 18.2 offers an example of an important confounder researchers needed to adjust for as they analyzed the effect of the EU ETS. Since the EU cap-and-trade system was launched in 2005, enormous solar PV innovation has occurred, and prices have declined by over 90%. This confounding variable – huge declines in world solar PV prices – perhaps partially explains why the estimated effect of the EU ETS was not larger.

Further Reading and References

Bayer, P. and M. Aklin. 2020. "The European Union Emissions Trading System Reduced CO_2 Emissions Despite Low Prices." *Proceedings of the National Academy of Sciences* 117 (16): 8804–8812.

Borras, S., R. Hall, I. Scoones, B. White and W. Wolford. 2011. "Towards a Better Understanding of Global Land Grabbing: An Editorial Introduction." *The Journal of Peasant Studies* 38 (2): 209–216.

California Air Resources Board (CARB). Undated. *Climate Change*. Retrieved from https://ww2.arb.ca.gov/ our-work/topics/climate-change#:~:text=The%20state%20achieved%20its%202020,than%20m andated%20by%20AB%2032

Cludius, J., V. Duscha, N. Friedrichsen and K. Schumacher. 2019. "Cost-Efficiency of the EU Emissions Trading System." *Economics of Energy & Environmental Policy* 8 (1): 145–162.

Colmer, J., R. Martin, M. Muuls and U. Wagner. 2020. "Does Pricing Carbon Mitigate Climate Change? Firm-Level Evidence from the European Union Emissions Trading Scheme." Discussion Paper No. 232, Discussion Paper Series – CRC TR 224, Collaborative Research Center Transregio 224.

Ecosystem Marketplace. 2023. *Ecosystem Markets Insight Report. Paying for Quality: The State of Voluntary Carbon Markets 2023*. Forest Trends Association.

European Commission. 2021. *Report from the Commission to the European Parliament on the Functioning of the European Carbon Market in 2021*. COM(2022) 516 final. Available in all the official EU languages at https://eur-lex.europa.eu/legal-content/EN/TXT/?uri=COM%3A2022%3A516%3AFIN

Flachsland, C., M. Pahlec, D. Burtraw, O. Edenhofer, M. Elkerbout, C. Fischer, O. Tietjen and L. Zetterberg. 2020. "How to Avoid History Repeating Itself: The Case for an EU Emissions Trading System (EU ETS) Price Floor Revisited." *Climate Policy* 20 (1): 133–142.

Joltreau, E. and K. Sommerfeld. 2019. "Why Does Emissions Trading under the EU Emissions Trading System (ETS) Not Affect Firms' Competitiveness? Empirical Findings from the Literature." *Climate Policy* 19 (4): 453–471.

Laing, T., M. Sato, M. Grubb and C. Comberti. 2014. "The Effects and Side-Effects of the EU Emissions Trading Scheme." *WIREs Climate Change* 5 (4): 509–519.

Perino, G., M. Willnery, S. Queminz and M. Pahle. 2022. "The European Union Emissions Trading System Market Stability Reserve: Does It Stabilize or Destabilize the Market?" *Review of Environmental Economics and Policy* 16 (2): 338–345.

Ritchie, H., P. Rosado and M. Roser. 2023. "Solar Photovoltaic Module Price." Part of *Energy Data*, adapted from International Renewable Energy Agency, Nemet, Farmer and Lafond. Retrieved from https://ourworldindata.org/grapher/solar-pv-prices May 2024.

Teixidó, J., S.F. Verde and F. Nicolli. 2019. "The Impact of the EU Emissions Trading System on Low-Carbon Technological Change: The Empirical Evidence." *Ecological Economics* 164: 106347.

United Nations (UN). 2018. *Achievements of the Clean Development Mechanism: Harnessing Incentive for Climate Action 2001-2018*. United Nations: New York.

World Bank. Undated. *Carbon Pricing Dashboard*. Retrieved from https://carbonpricingdashboard.worldbank.org/ October 2023.

19 Recap, Conclusions and Key Gaps

What You Will Learn in this Chapter

- The key linkages between topics in the book
- The main takeaways and lessons from the volume
- About key gaps in the book and important outstanding issues
- What are the most important unresolved intellectual questions related to addressing environmental moral hazard?

19.1 Recap and Conclusions

This chapter lays out and attempts to weave together some of the intellectual and experience threads developed in this book. I will also draw tentative general conclusions, point toward some of the many topics we were unable to address and highlight key gaps in understanding. As the first goal of this final chapter is to lay out the issues, the recap subsection will be significantly longer than in previous chapters.

As a reminder, in three sections, this book covered four overarching topics:

1. The most critical environmental challenges in the world today.
2. The key economic fundamentals required to analyze those issues.
3. The challenge of managing and protecting natural resources.
4. Instruments that have been used around the world to successfully manage and protect natural resources.

The book began with a discussion of the most important environmental challenges humans, who find themselves running Planet Earth, are facing. The planetary boundaries literature offered us some important, empirical guidance related to where human focus and effort are especially needed. Authors such as Johan Rockström, Will Steffen, Katherine Richardson and Paul Crutzen helped us understand that humans are blowing by some very important and highly interconnected planetary boundaries, including limits related to biogeochemical flows (nitrogen and phosphorus), biosphere integrity (especially genetic diversity due to excess extinctions), land use change and climate change.

DOI: 10.4324/9781003308225-22

With such a daunting set of challenges, we need a variety of flexible tools, diverse perspectives and creative insights to make progress. This book delves into key aspects of economics' significant contributions toward addressing environmental problems. Viewing nature from the human perspective – as "resources" that generate "ecosystem services" – is a part of that contribution. It is also a simplification that can blind us to more complex and nuanced ways of thinking about the environment, but it has the advantage of centering the human behaviors writ large that are driving environmental problems around the world. Other disciplines make critical contributions by looking at the inherently multidisciplinary challenge of saving the planet from other perspectives.

Environmental issues and necessary responses vary around the world. Looking across the globe, we see enormous variance in local-level challenges, progress and effects of global-level problems. Some of these differences come from the human side. For example, in richer countries a critical challenge is reducing and ultimately eliminating fossil fuels, decarbonizing and switching to renewable energy. Ramping up renewables to avoid powering future economic growth with coal and oil is also critical in the low-income parts of the Global South, but energy use and carbon emissions in those regions are typically very low and dominated by highly polluting biomass fuels. With more than 4 million people dying prematurely from indoor air pollution every year, increasing the use of a cleaner-burning bridge fuel, such as natural gas, makes eminent sense. A key is to avoid the fossil fuel-based technology lock-in and associated stranded assets (i.e. capital that is no longer usable) that are currently a challenge for the Global North.

Economics offers analytical tools that are potentially very important for addressing our most pressing environmental problems. The use of economic models helps us isolate the key human behaviors that lead to poor natural resource management. They also can contribute to the development of tools to improve management. In this book, we have relied mainly on a very simple, but seemingly powerful, behavioral model in which people try to do the best they can for themselves, their firms and their families. Examples of models where humans are framed as being largely self-serving include the supply-demand model, models of open access and externalities, the prisoner's dilemma model and even models of carbon taxes.

Self-serving behavioral models are very useful because they capture critical elements of the human character. They also have limits, as we saw from experimental tests (what were your red card-black card and/or public goods game results?). People on average have both altruistic and self-centered elements, which should make us wonder how socially beneficial tendencies can be amplified. As Professor Elinor Ostrom put it so well in her Nobel Prize address: the goal of policy should be to "facilitate the development of institutions that bring out the best in humans" (Ostrom, 2010, p. 435).

Changing norms and developing new social paradigms that dampen self-centered, exploitative incentives and move the human species away from environmental moral hazard-type behaviors are very interesting areas of ongoing and future work. Changing norms is at the heart of arguments to reduce consumption, particularly in high-income countries where consumption is greatest. Personal consumption is very important for market economies and is the destination for the bulk of production across the world. It is also the source of most of our incomes.

Over-consumption as a driver of natural resource degradation is difficult to define in economics, partly because there are so many factors that mitigate damages. In addition to affluence and population, which are at the heart of consumption, other factors include economic structure,[1] technology and siting decisions. Consumption is central to the economics of sustainability, because using something up now means fewer resources for investment. Low investment means a poor future, whether in terms of fewer private goods and services or less common pool ecosystem services. Low investment reduces total human welfare. This perspective is certainly well-defined in economics.

There are several hypotheses related to over-consumption that could use much more empirical testing. Empirical analysis is critical, because people have divergent perspectives on the same issues. Hypotheses that perhaps are just viewpoints can seem like reality when the situation is actually much more nuanced. The hypothesis that high levels of consumption do not make people happier is one example of an idea that is very worthy of additional careful empirical testing. The human species has certainly made enormous progress since industrialization started a mere 250 years ago. These advances cover everything from human lifespan and food security to iPhone apps, and many of those improvements are related to consumption. Have the marginal benefits of consumption declined so much for enough people on the planet that, on average, additional consumption does not make people happier? In the future, will consumption even increase human welfare? We need to find out.

Environmentally focused sustainability measures indicate that the best-performing countries are in the high-income world and some of the world's lowest-income countries are at the bottom. There has been significant environmental progress over time in much of the high-average-income Global North related to local air pollution, water quality, forest cover and habitat protection, among other issues. Some say that this progress is due to importing provisioning ecosystem services, including pollution-intensive goods, from the Global South. This is also a hypothesis that needs more investigation, particularly as economies in the Global North have become dominated by services. Existing evidence appears to be quite mixed.

The work by Hannah Ritchie at Our World in Data suggests that a consumption- rather than production-based CO_2 inventory would make environmental stars, such as Denmark, France and Sweden, look less stellar. Such findings, combined with questions about the greenness of the service sector, and calculations made by Sir Partha Dasgupta in his important Review, make me want to take another look at consumption as a driver of natural resource degradation.

The challenge of protecting natural resources is daunting, not only in terms of the environmental challenges we face, but also because of how humans interact with natural assets. Natural resources generally simultaneously produce many ecosystem services, benefitting a diverse range of humans. Unfortunately, the most valuable market services, including provisioning and sink ecosystem services, conflict with common pool ecosystem services that degrade and are difficult to exclude.

All the most pressing environmental problems highlighted by the planetary boundaries literature relate to common pool ecosystem services. These services deplete, but it is difficult to exclude people from depleting them. This combination of characteristics is a problem, because there are billions of us, all making decisions every day about how much to deplete common pool ecosystem services. There are also major incentives for self-serving people to focus on market-valuable provisioning services and the ability of nature to absorb our pollution, to the detriment of common pool services. Such behaviors, combined with ample room to conceal our motivations and actions, are what we mean by environmental moral hazard.

That common pool services are at best weakly excludable also means that markets will not form around them. In contrast to oil, which is a fully excludable private ecosystem service, genetic diversity has a tough time engaging businesses, because it is difficult to make money off it. Without profit motives, businesses are unlikely – or at least less likely – to engage and there will be limited or no markets in common pool ecosystem services. No markets means no market prices.

The most important problem with no market prices is there is no real-time information about scarcity. As we learned in Chapter 3, our modeling suggests that if markets are competitive, prices reflect the marginal costs and benefits of goods and services, which are the two elements of scarcity. Market prices therefore provide critical information about scarcities. Without prices, people can mistakenly infer that the real price is zero, implying that common pool ecosystem services are either undesirable, infinitely abundant or both. In other words, without prices, common pool ecosystem services can be viewed as valueless.

There is also the matter of *estimating* value, which is the metric of scarcity, without market prices. As we saw in Chapter 4, economists and others have developed a variety of methods for inferring ecosystem service values when there are no market prices. Tools such as hedonic pricing, avoided cost, travel cost and contingent valuation can help us infer values. They are also a lot more work to compute than merely looking at market prices for private provisioning services or the cost of pollution control equipment, which link to the ecosystem services most likely to compete with common pool services. If our self-serving behavioral model holds water, there will likely always be key asymmetries in value metrics and incentives for moral hazard.

The existence of moral hazard incentives does not tell us how important they are. How can we model incentives and perhaps even measure their impacts? In economics, we have three main models of environmental moral hazard. Perhaps the oldest and maybe the most important method takes a step back from ecosystem services and begins with markets for goods and services. The externality approach views damage to natural resources and common pool ecosystem services as spillover costs from perhaps otherwise well-functioning markets. These market failures imply that quantities of marketed goods and services would be too high and prices too low. In other words, external costs are not fully considered by markets in determining equilibrium prices and quantities.

One way to fix such a problem is to inject a price-based instrument into those markets, adding to the cost side and allowing the market to adjust to those increased costs. This approach to environmental policy is often called *getting the prices right* and is a powerful concept that motivates market-based instruments, such as carbon taxes and cap-and-trade. The externalities model is very useful because it leads to clean policy solutions and we can compute the social value[2] of internalizing externalities. Thinking about environmental moral hazard as an externalities problem therefore tells us what we lose by ignoring externalities and offers important guidance on what we need to do to address them.

A second way to frame environmental moral hazard is as a problem of open access to provisioning or waste disposal ecosystem services. This framework is a bit broader than the externalities approach because it is not confined to looking at markets. Market externalities can in fact be viewed as a particular form of open access to ecosystem services that occurs within markets.

The key difference in modeling environmental moral hazard as a property right rather than as an externalities problem is the implied solution. Whereas the existence of externalities implies a tweaking of prices to include all costs, open access points toward firming up property rights to improve incentives.

Though it does not have much of a foothold in the world of pollution control, the property rights view of natural resources has been enormously influential, especially with regard to renewable natural resources, such as forests, fish and wildlife. Reducing environmental moral hazard due to open access is in fact the basis of several policies we discussed in Chapter 13, which focused on privatization and community management of natural resources. Such approaches can be very successful because they help line up private incentives with social goals, potentially reducing environmental moral hazard in the process.

The most flexible model of environmental moral hazard behavior is to think about the challenge as a collective action problem. We all have experienced collective action, and we know it can be useful but also frustrating. On the positive side, getting diverse stakeholders together can increase resources, create synergies and yield better outcomes. On the other hand, we have all experienced the free riding that can accompany collective action. This possibility of free riding behavior in collective action situations is formalized in the prisoner's dilemma model. Importantly, the prisoner's dilemma model relies on the self-focused incentives we have employed throughout this book. The model then predicts that when private interests conflict with social goals, socially suboptimal results will occur.

Until Elinor Ostrom and others helped us consider collective action more deeply, it was almost assumed that meaningful collective action was basically impossible. Ostrom conducted both theoretical and empirical analysis and found that successful local-level collective action was not only possible but was occurring around the world. She analyzed the conditions for success and found that eight design principles contributed to successful collective action. Lab and field experiments (perhaps including your experiments!) have since confirmed the possibility for local-level collective action to contribute, perhaps imperfectly, to environmental protection.

We center the word "possibility" here because we know that collective action equilibria can be unstable and highly dependent on the past. Cooperation can start out strong, continue in that way, subsequently fall apart or never get off the ground. Though Ostrom identified critical design features, local-level collective action success is rarely assured.

Some institutions are making environmental collective action work. The Community Forestry Programme in Nepal appears to have contributed to a huge increase in forest cover since it was formalized in 1993. Artisanal fisheries collective action also seems to be playing important roles in protecting fish stocks and buoying the incomes of fishers. Communities also have often played productive roles in wildlife management, including in Zimbabwe and Uganda. The scope of community-focused environmental collective action is if anything increasing, particularly in the Global South.

Though local-level collective action is extremely important around the world, we cannot lose sight of the reality that governments are our main collective action institutions. A lot of the work of defending common pool ecosystem services is related to governance at local, regional, state/province and national levels. Governments come in several forms, have different levels of effectiveness, provide a variety of types and levels of service to their people and are funded by many mechanisms.

They also intervene in their economies in many ways, including by deliberately affecting prices, often using subsidies. Subsidies can be implicit, where preferential government pricing, regulations, import restrictions or other support boosts particular economic sectors. They can also be explicit, using public budgetary resources to support economic sectors or make goods and services cheaper for buyers. Most subsidies are implicit subsidies. Irrigation water subsidies based on pricing below market value, or even less than delivery cost, are one example of an implicit subsidy. Irrigation is an important service from ground and surface water, but it is not the only one, and irrigation can conflict with other ecosystem services, such as habitat services and recreation; subsidies make irrigation as a use for water more attractive than it would otherwise be, increasing that service and hampering others.

Energy subsidies by fossil fuel-rich countries, often to bring down the cost of gasoline and heating for consumers, are a pernicious type of subsidy. The logic is very easy to understand and compelling. A country is blessed with oil, gas or coal reserves. Why should a government that is working in the people's interest export it all to other countries? Shouldn't the people benefit, and why should they have to compete with others just because the international market sets such a high price? The people should pay less for *their* resources!

This logic makes sense, except it ignores opportunity costs. These costs have several dimensions. First, there is what could have been otherwise done with the money. Maybe that means forgone investments in the fossil fuel sector or fewer other investments. We cannot forget that lower prices mean higher consumption of fossil fuels, fewer energy-saving investments and a general tilting of economies in energy-intensive directions. These shifts will almost certainly increase air pollution, such as greenhouse gases, from burning fossil fuels, reducing critical ecosystem services such as human health. According to the World Health Organization, about 7 million premature deaths per year are caused by air pollution and fossil fuel subsidies are a contributor to that total. In some cases, targeted fossil fuel subsidies (e.g., for LPG cooking cylinders in areas that are highly dependent on biomass fuels) can be appropriate when they significantly reduce other social costs, but the Soviet Union and its allied states' experience should make us particularly wary of fossil fuel subsidies.

Subsidies generate some gnarly questions and are often difficult to roll back. Irrigation water subsidies almost certainly increase water usage and waste, but they also help ensure the food supply from fertile drylands. Are they worth the opportunity costs? People make durable investments in vehicles, manufacturing equipment, electricity generation capital, agricultural land and equipment that lasts many decades. They are therefore understandably upset when governments pull the plug on subsidies after they made those investments. Such was the situation when in May 2023 the Government of Nigeria eliminated petrol subsidies that had been in place since the 1970s, because they were too costly. Not surprisingly, there were protests at this policy shift and rapid increases in prices throughout the economy as transportation costs soared.

The governments, who may be implementing energy subsidies, are also the key institutions responsible for pollution control. Though Nobel prizewinner Ronald Coase argued that perhaps it should be otherwise, pollution problems are typically not addressed wholly within the private sector. Key reasons are related to problems of transaction costs and information, which explain why the so-called Coase Theorem is typically unlikely to hold.

Modeling pollution within our standard framework is not a problem, but solid information about the marginal costs and benefits of pollution (abatement) is critical for efficiently regulating pollution. Both metrics are, unfortunately, difficult to come by, and so pollution goals are typically set independently of the tools used to achieve those goals. This approach is more limited than, for example, the product-focused analysis of externalities we discussed in Chapter 9. In that example, if we indeed knew the marginal damages of pollution when output increased, we could choose a tax rate that would internalize the externality – in terms of overproduction. Of course, there are typically many ways to reduce pollution besides reducing output. We therefore needed to directly analyze pollution (abatement).

So, what can economics contribute, if not tools for achieving efficiency? It turns out, rather a lot. Abatement costs likely differ dramatically across countries, economic sectors and polluting facilities. If it were possible to allocate abatement effort to sources that can reduce most cheaply, environmental goals could be achieved more cheaply and/or we could get more abatement for our money. Focusing pollution policy on cost-effectiveness also opens up the possibility of overachieving goals.

Cost-effective instruments allow some flexibility in abatement, perhaps even including the level of abatement. In this book we considered pollution taxes and pollution markets, which are instruments that offer the potential to incentivize cost-effective pollution abatement. Taxes implicitly allocate rights to governments and polluters then purchase those rights by paying pollution taxes. Pollution markets, including cap-and-trade, create secure (though perhaps not costless) rights to pollution. Pollution rights then become assets that can be used, bought or sold.

Pollution taxes have been used to control a variety of pollutants, including CO_2. The object of taxation is sometimes pollution itself, but more often is one or two steps removed from actual pollution. For example, carbon taxes price the carbon content in fuels and other GHGs rather than the emissions themselves. This approach makes sense, especially for fossil fuel-based CO_2 emissions, because the carbon content in fuels is well-known; carbon content therefore offers a solid stand-in for CO_2 emissions.

A bit less than 25% of global GHG emissions are priced using either green pricing or green markets and 1/4 of those priced emissions are covered by carbon tax systems. Though not all evidence points in the same direction, a number of studies suggest that green pricing can have important incentive effects, particularly when first introduced – even if rates are lower than later – suggesting that it incentivizes low-cost emissions reductions.

The possibility of green tax reform is a tantalizing and to some a logical one, but the responsiveness of people to green pricing can make pollution taxes an unstable source of public finance. In no country in the world do green taxes make up more than 4% of GDP and the average OECD country gets only about 6% of their government revenues from environmental taxes. Unfortunately, it does not seem that environmental taxes are going to fund the world's governments anytime soon.

Green markets are also very important market-based instruments for environmental management. We focused on green markets for pollution in the US, which were first developed under the Clean Air Act, and for carbon trading. Much of the road testing of pollution offsets, banking, netting and cap-and-trade occurred at the US federal level, and much of that experience from the 1970s to the 2000s was very positive. The major US cap-and-trade programs have been extensively studied and the evidence suggests that those systems helped the US overachieve environmental goals and saved money in the process. Over time, for a number of reasons, non-carbon green markets have either become US state- or regional-level instruments or been phased out.

Voluntary carbon markets offset about 250 million metric tons CO_2e in 2022, which is much less than compliance carbon markets (9 billion metric tons in 2023). CO_2 is an especially good candidate for green market mechanisms and cap-and-trade regulation in particular, because it is truly a uniformly mixed pollutant. Locations of emissions and damages are therefore unrelated, implying that toxic CO_2 hot spots are unlikely to build up. Cap-and-trade systems are currently the main instruments for controlling GHGs in the European Union, the US State of California, the Republic of Korea and New Zealand, among others.

The EU ETS is an especially important example of a carbon cap-and-trade system. It is the largest and oldest such system in the world and is currently in its fourth phase. Innovations, such as the Market Stabilization Reserve, have helped stabilize carbon prices and contributed to market stability and predictability. Evidence suggests that the EU ETS, which covers about 40% of EU emissions, has reduced emissions by 4%-12% compared with a potential counterfactual where no cap-and-trade existed.

19.2 Key Gaps

This book has covered a number of critical issues related to humans' economic interactions with the natural world. Unfortunately, we could not discuss all issues and policy responses. The book had only limited mention of the economics of fisheries, which is a rich and complex topic due to movement, sometimes complex life cycles and even social relationships. Policy instruments focusing on more carefully defining property rights and allowing rights to be traded, have been especially useful for trying to stem declines in commercial fish stocks. Many of the world's capture fisheries remain under very severe threats.

We also could not cover water provision and pricing outside of water for irrigation. Understanding of urban water pricing has increased substantially during the past few decades. Using price and non-price instruments to reduce water use, especially in places such as Cape Town, South Africa, which has seen significant water shortages in recent years, has become particularly important.

We focused our discussion of market-based instruments on pollution control. As I intimated in some chapters, though, taxes, offsets and cap-and-trade are used in a variety of non-pollution situations, which we were not able to discuss. For example, in addition to numerous other requirements, housing developers may be required to pay fees for the right to build housing developments. These fees are used to charge housing markets for critical infrastructure, such as electricity, roads, water and sewerage service. They can also help fund incremental school needs, libraries and other common infrastructure.

Payments for ecosystem services (PES) is another major policy arena where pricing attempts to affect behavior, but this time by creating financial revenue streams from non-marketed ecosystem services, especially from forest resources. Costa Rica has long had a system of paying private landowners to keep land in forests. In 1987, approximately 40% of Costa Rica's territory was covered with natural forests, but 30 years later 60% of Costa Rica was forested – a 50% increase. I don't mean to suggest that a causal relationship has been documented, but there certainly is a strong correlation between implementation of the PES system and more forest cover. Costa Rica was also the first Latin American country to receive compensation from the World Bank-administered Forest Carbon Partnership facility for reducing GHG emissions from deforestation and forest degradation. The first payment of $16.4 million was made under the UN FCCC program called the enhanced Reduced Emissions from Deforestation and Forest Degradation (REDD+) (World Bank, 2022). This PES program offers financial compensation to countries in the Global South for reducing carbon emissions from loss of forest biomass.

Cap-and-trade in the form of tradable catch rights has been widely applied to management of capture fisheries. Such individual transferable quote (ITQ) systems attempt to give secure catch rights to fishers, and also allow those rights to be rented or sold. Similarly, offsets are sometimes used to manage wetland preservation requirements. Such programs allow destruction of wetlands that have especially high commercial value in exchange for permanently preserving other sites with greater area or higher ecosystem values.

Much of Section III of the book focused on instruments that attempted to alter self-serving incentives for environmental damage. They were therefore largely centered on altering prices, improving property rights or both. These measures are extremely useful, but non-price instruments that try to dial into people's altruistic sides can also perhaps play roles in changing long-run norms and supporting necessary "paradigm shifts."

Providing information about peers' environmentally beneficial behaviors has sometimes been found to reduce energy consumption. Public information about stationary source emissions – especially if companies are publicly traded in stock markets – can reduce toxic emissions. Programs that bring environmental auditors to polluting firms, grade them on their pollution prevention and waste minimization behaviors, and make that information public – without having any financial penalties at all – have also been shown to improve performance.

Excessive pollution compared to what is possible can be deliberate, but can also be due to ignorance, lack of investigation and inertia. We know from a variety of sources, including McKinsey & Company (2009), that GHG reduction costs vary enormously in all directions. In some situations, they can even be negative, meaning people benefit financially at the same time as common pool ecosystem services are improved. Why don't we always take those steps? Clearly, human behavior vis-à-vis ecosystem services, self-interest, attention and habit formation are interrelated in complicated ways.

Research on instruments to help people do right by the environment is an extremely active area of investigation that we were not able to cover in the book. Nudges that simply move opportunities for low-cost or no-cost environmentally beneficial behaviors up in people's consciousness can help them understand what is possible. Sometimes small financial subsidies or insignificant fees can be nudges. For example, local jurisdictions that have paper bag fees of even 5 US cents per bag can see major reductions in demand. Such effects are probably in part simply due to reminding users to bring reusable bags to stores and helping to create habits.

The maxim "reduce, re-use, recycle" has come to describe important dimensions of our environmental challenge. These objectives are strongly related to the economics of solid waste disposal and management, which are large topics we were not able to touch. A variety of important innovations especially target reuse and recycling. In a number of places around the world, product sharing programs have been implemented to allow joint use of vehicles, tools and other products. Ride sharing apps are one important example, but institutions other than markets have also supported product sharing. Community groups, religious organizations and libraries, which have long offered book sharing, have all created programs to support the reuse of products across individuals and organizations.

Many governments – and other collective action institutions – have programs to support and promote the recycling of materials. In some situations, such as the recycling of lead-acid batteries or aluminum, recycling may be privately profitable and those who recycle get paid for those materials. In many cases, though, recycling can be costly, even as it reduces disposal costs by diverting materials from costly landfills. Recycling programs take a variety of forms and are often paired with other waste management policies, such as volumetric pricing of garbage disposal. Lower-cost systems require recyclers to drop off materials at central sites, but higher-cost approaches collect recyclable materials from homes, businesses and other institutions. Some systems require recyclers to carefully separate different types of materials and others, such as in my hometown of Portland, Oregon in the US, allow everything to be dumped together to be sorted at recycling centers.

In many jurisdictions around the world, recycling deposits are required on glass bottles and aluminum cans used for beverages. These deposits, which can be significant, help ensure that bottles and cans are returned to collection centers for reuse or recycling. Deposits, and the collection of recyclable materials more generally, also serve as important sources of income for people around the world.

With the help of technology, the goal to reduce and reuse materials and products has also become more important. People who are interested to reuse products and reduce waste have come together in groups to pass along a variety of goods, including partially used consumable products, to others. Online mechanisms, such as Facebook, Craigslist and Nextdoor, among others, appear to be very important for facilitating the reuse of products from furniture to food products and supporting reductions in waste. These institutions can create positive social linkages between people, support conservation norms and may help people maintain lifestyles while reducing their environmental footprints.

In sum, while we were able to address many subjects in this book, there are also many important topics we were unable to cover. Each one is a world in itself, with books and articles written about them from a variety of disciplinary and multidisciplinary perspectives; as I have tried to make clear throughout this book, understanding the human relationship with nature is an inherently multidisciplinary undertaking, with many disciplines offering important insights.

Key insights from economics include recognizing the likelihood of environmental moral hazard and the careful thinking through of potential ways to address such antisocial behaviors. Without strong, self-benefiting incentives, are humans altruistic enough to save the planet in an age of climate change and seriously pushed planetary boundaries? We have highlighted important success stories in which people cooperated and allowed their governments to impose sacrifices, such as carbon pricing, on them in the interest of reducing environmental footprints.

There is also the US Inflation Reduction Act, which relied heavily on subsidies to steer the US in a more climate-friendly direction. Many countries in the Global South have also made their Paris Agreement Nationally Determined Contributions (NDCs) contingent on financing from Annex 1 countries. Are such external financing approaches the main path to net-zero emissions or does emissions pricing have a serious future?

Perhaps the key future challenge from an economic standpoint is how to get most of the 8 billion people and all their institutions to do their best to reduce environmental pressures in critical areas. The verdict is still to come on whether instruments can be crafted that sufficiently align social and private incentives and/or amplify the inherent altruistic tendencies of humans.

At many points in this book, we have relied on analysis and data assembled by the non-profit Our World in Data, which is affiliated with Oxford University. I happened to be listening to a podcast hosted by the *New York Times* opinion writer Ezra Klein. Our World in Data deputy editor and lead researcher Dr. Hannah Ritchie was on the program. She had just published a book titled *Not the End of the World: How We Can Be the First Generation to Build a Sustainable Planet*.

Klein and Ritchie engaged in a dialogue about whether people would really make sacrifices about things they care about, such as the food they eat, to save the planet. Despite the optimistic title of her book, Ritchie concluded that people could not be counted on to change habits, such as eating less beef, which has significant land and carbon footprints, unless they were provided truly equivalent alternatives. The good news, she noted, is that people don't care where their electricity comes from. They just want their plugs to work. Fortunately, over time technical change has offered us a number of feasible carbon-free energy sources that are potentially just as good as fossil fuels. She concludes that greening the world's energy system may just be possible (Klein and Ritchie, 2024).

Issues for Discussion

1. In this chapter, we discussed the key topics presented in this book. Can you name two issues important to you that were in the book but omitted from this chapter? What makes them important for you?

2. Several topical gaps were covered in this chapter. Each one is worthy of full consideration and has been subject to important investigation by economists and others. Can you name two gaps that I missed? Why do you think these issues should be considered "gaps?"

3. Do you think Hannah Ritchie is correct that people will not make significant sacrifices to reduce our most important demands on the planet? Are human innovation and green technologies really the solution to the climate crisis?

Notes

1   Though recent research questions whether services are really greener than goods.
2   In terms of consumer surplus, producer surplus and government revenues.

Further Reading and References

Klein, E. and H. Ritchie. 2024. "Is Green Growth Possible?" *The Ezra Klein Show*, April 30. Retrieved from https://www.nytimes.com/2024/04/30/opinion/ezra-klein-podcast-hannah-ritchie.html
McKinsey & Company. 2009. "Pathways to a Low-Carbon Economy: Version 2 of the Global Greenhouse Gas Abatement Cost Curve." Retrieved from https://www.mckinsey.com/capabilities/sustainability/our-insights/pathways-to-a-low-carbon-economy December 10, 2022.
Ostrom, E. 2010. "Beyond Markets and States: Polycentric Governance of Complex Economic Systems." *American Economic Review* 100 (3): 408-444.
World Bank. 2022. "Costa Rica's Forest Conservation Pays Off." November 22, 2022. Retrieved from https://www.worldbank.org/en/news/feature/2022/11/16/costa-rica-s-forest-conservation-pays-off May 2024.

INDEX

Printed in the United States
by Baker & Taylor Publisher Services

Printed in the United States
by Baker & Taylor Publisher Services